W9-CNF-620

The IMA Volumes in Mathematics and Its Applications

Volume 27

Series Editors
Avner Friedman Willard Miller, Jr.

Institute for Mathematics and its Applications
IMA

The **Institute for Mathematics and its Applications** was established by a grant from the National Science Foundation to the University of Minnesota in 1982. The IMA seeks to encourage the development and study of fresh mathematical concepts and questions of concern to the other sciences by bringing together mathematicians and scientists from diverse fields in an atmosphere that will stimulate discussion and collaboration.

The IMA Volumes are intended to involve the broader scientific community in this process.

Avner Friedman, Director
Willard Miller, Jr., Associate Director

* * * * * * * * * *

IMA PROGRAMS

1982-1983 **Statistical and Continuum Approaches to Phase Transition**
1983-1984 **Mathematical Models for the Economics of Decentralized Resource Allocation**
1984-1985 **Continuum Physics and Partial Differential Equations**
1985-1986 **Stochastic Differential Equations and Their Applications**
1986-1987 **Scientific Computation**
1987-1988 **Applied Combinatorics**
1988-1989 **Nonlinear Waves**
1989-1990 **Dynamical Systems and Their Applications**
1990-1991 **Phase Transitions and Free Boundaries**

* * * * * * * * * *

SPRINGER LECTURE NOTES FROM THE IMA:

The Mathematics and Physics of Disordered Media
Editors: Barry Hughes and Barry Ninham
(Lecture Notes in Math., Volume 1035, 1983)

Orienting Polymers
Editor: J.L. Ericksen
(Lecture Notes in Math., Volume 1063, 1984)

New Perspectives in Thermodynamics
Editor: James Serrin
(Springer-Verlag, 1986)

Models of Economic Dynamics
Editor: Hugo Sonnenschein
(Lecture Notes in Econ., Volume 264, 1986)

Barbara Lee Keyfitz Michael Shearer
Editors

Nonlinear Evolution Equations That Change Type

With 96 Figures

Springer-Verlag
New York Berlin Heidelberg
London Paris Tokyo Hong Kong

Barbara Lee Keyfitz
Department of Mathematics
University of Houston
Houston, Texas 77204
USA

Michael Shearer
Department of Mathematics
North Carolina State University
Raleigh, North Carolina 27695
USA

Mathematical Subject Classification Codes: Primary: 35M05, 76A10, 35L65, 35L67; Secondary: 35K65, 65N99, 73E05, 76A05, 76H05, 76505, 82A25.

Library of Congress Cataloging-in-Publication Data
Nonlinear evolution equations that change type / [edited by] Barbara Lee Keyfitz, Michael Shearer.
p. cm. — (The IMA volumes in mathematics and its applications ; v. 27)
"Based on the proceedings of a workshop which was an integral part of the 1988-89 IMA program on nonlinear waves"—Foreword.
1. Evolution equations, Nonlinear. I. Keyfitz, Barbara Lee. II. Shearer, Michael. III. Series.
QA377.N664 1990
515'.353—dc20 90-9970

Printed on acid-free paper.

ISBN 0-387-97353-2 1990 $0.00 + 0.20

Camera-ready copy prepared by the IMA.
Printed and bound by Edwards Brothers, Inc., Ann Arbor, Michigan.
Printed in the United States of America.

9 8 7 6 5 4 3 2 1

ISBN 0-387-97353-2 Springer-Verlag New York Berlin Heidelberg
ISBN 3-540-97353-2 Springer-Verlag Berlin Heidelberg New York

The IMA Volumes
in Mathematics and its Applications

Current Volumes:

Volume 1: Homogenization and Effective Moduli of Materials and Media
Editors: Jerry Ericksen, David Kinderlehrer, Robert Kohn, J.-L. Lions

Volume 2: Oscillation Theory, Computation, and Methods of Compensated Compactness
Editors: Constantine Dafermos, Jerry Ericksen,
David Kinderlehrer, Marshall Slemrod

Volume 3: Metastability and Incompletely Posed Problems
Editors: Stuart Antman, Jerry Ericksen, David Kinderlehrer, Ingo Muller

Volume 4: Dynamical Problems in Continuum Physics
Editors: Jerry Bona, Constantine Dafermos, Jerry Ericksen, David Kinderlehrer

Volume 5: Theory and Applications of Liquid Crystals
Editors: Jerry Ericksen and David Kinderlehrer

Volume 6: Amorphous Polymers and Non-Newtonian Fluids
Editors: Constantine Dafermos, Jerry Ericksen, David Kinderlehrer

Volume 7: Random Media
Editor: George Papanicolaou

Volume 8: Percolation Theory and Ergodic Theory of Infinite Particle Systems
Editor: Harry Kesten

Volume 9: Hydrodynamic Behavior and Interacting Particle Systems
Editor: George Papanicolaou

Volume 10: Stochastic Differential Systems, Stochastic Control Theory and Applications
Editors: Wendell Fleming and Pierre-Louis Lions

Volume 11: Numerical Simulation in Oil Recovery
Editor: Mary Fanett Wheeler

Volume 12: Computational Fluid Dynamics and Reacting Gas Flows
Editors: Bjorn Engquist, M. Luskin, Andrew Majda

Volume 13: Numerical Algorithms for Parallel Computer Architectures
Editor: Martin H. Schultz

Volume 14: Mathematical Aspects of Scientific Software
Editor: J.R. Rice

Volume 15: Mathematical Frontiers in Computational Chemical Physics
Editor: D. Truhlar

Volume 16: Mathematics in Industrial Problems
by Avner Friedman

Volume 17: Applications of Combinatorics and Graph Theory to the Biological and Social Sciences
Editor: Fred Roberts

Volume 18: q-Series and Partitions
Editor: Dennis Stanton

Volume 19: Invariant Theory and Tableaux
Editor: Dennis Stanton

Volume 20: Coding Theory and Design Theory Part I: Coding Theory
Editor: Dijen Ray-Chaudhuri

Volume 21: Coding Theory and Design Theory Part II: Design Theory
Editor: Dijen Ray-Chaudhuri

Volume 22: Signal Processing: Part I - Signal Processing Theory
Editors: L. Auslander, F.A. Grünbaum, W. Helton, T. Kailath, P. Khargonekar and S. Mitter

Volume 23: Signal Processing: Part II - Control Theory and Applications of Signal Processing
Editors: L. Auslander, F.A. Grünbaum, W. Helton, T. Kailath, P. Khargonekar and S. Mitter

Volume 24: Mathematics in Industrial Problems, Part 2
by Avner Friedman

Volume 25: Solitons in Physics, Mathematics, and Nonlinear Optics
Editors: Peter J. Olver and David H. Sattinger

Volume 26: Two Phase Flows and Waves
Editors: Daniel D. Joseph and David G. Schaeffer

Volume 27: Nonlinear Evolution Equations that Change Type
Editors: Barbara Lee Keyfitz and Michael Shearer

Forthcoming Volumes:

1988-1989: *Nonlinear Waves*

Computer Aided Proofs in Analysis

Multidimensional Hyperbolic Problems and Computations (2 Volumes)

Microlocal Analysis and Nonlinear Waves

Summer Program 1989: *Robustness, Diagnostics, Computing and Graphics in Statistics*

Robustness, Diagnostics in Statistics (2 Volumes)

Computing and Graphics in Statistics

1989-1990: *Dynamical Systems and Their Applications*

An Introduction to Dynamical Systems

Patterns and Dynamics in Reactive Media

Dynamical Issues in Combustion Theory

FOREWORD

This IMA Volume in Mathematics and its Applications

NONLINEAR EVOLUTION EQUATIONS THAT CHANGE TYPE

is based on the proceedings of a workshop which was an integral part of the 1988-89 IMA program on NONLINEAR WAVES. The workshop focussed on problems of ill-posedness and change of type which arise in modeling flows in porous materials, viscoelastic fluids and solids and phase changes. We thank the Coordinating Committee: James Glimm, Daniel Joseph, Barbara Lee Keyfitz, Andrew Majda, Alan Newell, Peter Olver, David Sattinger and David Schaeffer for planning and implementing an exciting and stimulating year-long program. We especially thank the workshop organizers, Barbara Lee Keyfitz and Michael Shearer, for their efforts in bringing together many of the major figures in those research fields in which theories for nonlinear evolution equations that change type are being developed.

Avner Friedman

Willard Miller, Jr.

PREFACE

During the winter and spring quarters of the 1988/89 IMA Program on Nonlinear Waves, the issue of change of type in nonlinear partial differential equations appeared frequently. Discussion began with the January 1989 workshop on Two-Phase Waves in Fluidized Beds, Sedimentation and Granular Flow; some of the papers in the proceedings of that workshop present strategies designed to avoid the appearance of change of type in models for multiphase fluid flow. As the papers in this volume indicate, physical processes whose simplest models may involve change of type occur also in other dynamic contexts, such as in the simulation of oil reservoirs, involving multiphase flow in a porous medium, and in granular flow.

There is also considerable recent mathematical work on simple model problems involving systems of conservation laws in space and time that change type. Some of this work addresses the theoretical issues, in particular the loss of linearized well-posedness of initial value problems; but there are interesting numerical problems also. Much of the mathematical work was not previously known to applied mathematicians or fluid dynamicists looking at models for specific flows. In addition, recent work on both steady and unsteady models of viscoelasticity has indicated the importance of composite systems in the study of steady visco-elastic flows, and has exhibited change of type in these steady models; unsteady change of type (change of type in the evolution equations) has even been conjectured to describe some instabilities in viscoelastic flows. The general theme of the March 1989 Workshop on Evolution Equations that Change Type was the relationship between the analytical and numerical issues posed by equations that change type, and the applications modelled by these equations.

The papers in these proceedings by Coleman and by Cook, Schleiniger and Weinacht discuss the current status of modelling of viscoelastic fluids, including change of type for both steady and unsteady flows, while the paper of Crochet and Delvaux details how numerical computations can be performed on steady viscoelastic flows that change type. This includes adapting the concept of an upwind scheme from transonic flow calculations. Renardy's paper, and that of Malkus, Nohel and Plohr, obtain analytical results which help to compare different models of viscoelastic fluids. An explanation of how multiphase flow in porous media leads to conservation laws that change type can be found in Lars Holden's paper. There are dynamic models for phase transitions which exhibit change of type, and the propagation of phase boundaries in equations arising this way is analysed by Mischaikow and by Sprekels. Models of granular flow give rise to linearly ill-posed equations; the paper of Schaeffer and Shearer contains an analytical treatment of change of type in yield-vertex models of plasticity.

Mathematical background may be found in papers of Keyfitz and of Warnecke, which include comparison of classical steady transonic with unsteady models. Mathematical properties of model equations which exhibit change of type, and construction of solutions, are discussed by Holden, Holden and Risebro, by Hsiao and by Azevedo and Marchesin. Theoretical issues of well-posedness, weak formulations, and admissibility of shock waves arise naturally if one tries to relate the linear ill-posedness of the Cauchy problem to nonlinear considerations, or to formulate correct boundary conditions for equations of mixed type. An approach to this analysis through examples which are not strictly hyperbolic is given in the papers of LeFloch, of Liu and Xin, and of Shearer and Schecter. The example considered by Kranzer and Keyfitz is strictly hyperbolic, but is evidently related to nonstrictly hyperbolic problems. Well-posedness for a nonlinear model which is linearly ill-posed is described by Slemrod.

One of the conclusions which emerged from the workshop was that at least some dynamic instabilities in viscoelastic flows can be explained by a simpler mechanism than change of type, namely a bifurcation of attractors. However, change of type of the transonic kind, in steady flows, remains of interest in viscoelasticity.

Among promising mathematical approaches which were displayed at the workshop, Riemann problems played a prominent role in many of the talks, with new phenomena, loss of uniqueness of solutions, and constructive solutions being discussed in detail. One classic result on equations of mixed type, Friedrichs' 1958 theory of symmetric positive systems, emerged as a potential tool to discuss well-posedness of boundary-value problems.

New uses for the qualitative theory of planar dynamical systems appear in the work of Liu and Xin, of Malkus, Nohel, and Plohr, of Azevedo and Marchesin, of Shearer and Schecter and of Keyfitz; higher-dimensional vectorfields appear in the papers of Kranzer and Keyfitz and of Mischaikow.

As organizers of the workshop and editors of the proceedings, we extend a special word of thanks to Dan Joseph, whose papers on loss of hyperbolicity in viscoelastic models provided an important link between specialists in viscoelasticity and participants working in other areas related to equations that change type. In addition to introducing the participants to each other and organizing a lab tour, Dan presented a summary of Fraenkel's work on change of type in steady flow.

We are also pleased to thank Avner Friedman and Willard Miller, Jr. and the IMA staff for their smooth organization of the details of the workshop and the visits of the participants. Finally, we thank all the participants in this volume, who submitted their papers so promptly, and we thank the editorial staff of Patricia V. Brick, Stephan Skogerboe, Kaye Smith and Marise Ann Widmer who completed the manuscript preparation.

Barbara Lee Keyfitz

Michael Shearer

CONTENTS

Foreword ix

Preface xi

Multiple viscous profile Riemann solutions in mixed elliptic-hyperbolic models for flow in porous media 1
A.V. Azevedo and D. Marchesin

On the loss of regularity of shearing flows of viscoelastic fluids 18
Bernard D. Coleman

Composite type, change of type, and degeneracy in first order systems with applications to viscoelastic flows 32
L. Pamela Cook, G. Schleiniger and R.J. Weinacht

Numerical simulation of inertial viscoelastic flow with change of type 47
M.J. Crochet and V. Delvaux

Some qualitative properties of 2×2 systems of conservation laws of mixed type 67
H. Holden, L. Holden and N.H. Risebro

On the strict hyperbolicity of the Buckley-Leverett equations for three-phase flow 79
Lars Holden

Admissibility criteria and admissible weak solutions of Riemann problems for conservation laws of mixed type: a summary 85
L. Hsiao

Shocks near the sonic line: a comparison between steady and unsteady models for change of type 89
Barbara Lee Keyfitz

A strictly hyperbolic system of conservation laws admitting singular shocks 107
Herbert C. Kranzer and Barbara Lee Keyfitz

An existence and uniqueness result for two nonstrictly hyperbolic systems 126
Philippe Le Floch

Overcompressive shock waves 139
Tai-Ping Liu and Zhouping Xin

Quadratic dynamical systems describing shear flow of non-Newtonian fluids 146
D.S. Malkus, J.A. Nohel and B.J. Plohr

Dynamic phase transitions: a connection
matrix approach 164
Konstantin Mischaikow

A well-posed boundary value problem
for supercritical flow of viscoelastic
fluids of Maxwell type 181
Michael Renardy

Loss of hyperbolicity in yield vertex plasticity models
under nonproportional loading 192
David G. Schaeffer and Michael Shearer

Undercompressive shocks in systems of
conservation laws 218
Michael Shearer and Stephen Schecter

Measure valued solutions to a backward-forward
heat equation: a conference report 232
M. Slemrod

One-dimensional thermomechanical phase transitions
with non-convex potentials of Ginzburg-Landau type 243
Jürgen Sprekels

Admissibility of solutions to the Riemann problem
for systems of mixed type
-transonic small disturbance theory- 258
Gerald Warnecke

MULTIPLE VISCOUS PROFILE RIEMANN SOLUTIONS IN MIXED ELLIPTIC-HYPERBOLIC MODELS FOR FLOW IN POROUS MEDIA*

A. V. AZEVEDO[1,2] AND D. MARCHESIN[2,3]

Abstract. We consider the Riemann problem for a system of two conservation laws of mixed type. We show by constructing two distinct solutions for a non trivial class of Riemann problem data that the viscous profile entropy condition is insufficient to guarantee uniqueness of the solution. This model possesses transitional shocks - or saddle to saddle connections of the associated dynamical system - of a kind not yet observed in conservation laws with quadratic polynomial flux functions.

Introduction. Weak solutions of systems of conservation laws are not uniquely determined from the initial data, unless conditions drawn from the physics - or entropy criteria - are used to supplement the partial differential equations and select physically meaningful solutions. The entropy criterion which seems to generalize most useful criteria for hyperbolic systems of conservation laws (such as Lax's [15] and Oleinik's [17]) requires the discontinuities to be the small viscosity limit of traveling wave solutions of the partial differential equations, retaining parabolic terms that describe smaller effects usually neglected in the conservation law formulation.

In this work we show a clear example of the limitations of the viscosity profile condition as an entropy criterion for mixed elliptic hyperbolic systems of two conservation laws. This is done most conveniently in the class of scale invariant solutions corresponding to a particularly simple class of initial data, called the Riemann problem. We show that for an open set of Riemann problem Cauchy data, there exist two distinct solutions, which are equally acceptable. Therefore one cannot expect that the Riemann solution for this model has a global continuous dependence on the initial data.

This is not the first example where the viscous profile condition fails to provide an adequate uniqueness criterion in mixed problems. This is the case of gas dynamics with phase transitions ([22], [23]) or similar systems of two equations possessing an infinite strip where the characteristic speeds are complex. However, this elliptic strip disappears completely and the problem becomes strictly hyperbolic when the flux functions are modified by making the thermodynamic equation of state convex. This behavior is competely different from three-phase flow which gives rise to the model we study. There is a compact elliptic region which can at best shrink to a non removable hyperbolic singularity when the permeabilities describing the flow are modified [25], [16]. At such an umbilic point the characteristic speeds coincide; everywhere else in a neighborhood of this point the problem is strictly hyperbolic.

---

*This research was supported in Brazil by CNPq, FAPERJ and in the U.S.A. by IMA with funds provided by NSF

[1] Univ. de Brasília, Brazil.

[2] Pontifícia Univ. Católica do Rio de Janeiro, Brazil.

[3] Instituto de Matemática Pura e Aplicada, E.D. Castorina 110, Rio de Janeiro, 22460, Brazil.

For the strictly hyperbolic gas dynamics case as well as for the case of models with an isolated umbilic point related to three-phase flow, the viscous profile entropy condition is adequate [5], [20], [14], [10], [24], [13]. Therefore the failure of the viscous profile entropy condition in the presence of elliptic region may conceivably be due to inadequate modeling giving rise to systems of conservation laws which turn out to have mixed elliptic-hyperbolic behavior. These difficulties may be related to the lack of convergence of Glimm's method for certain mixed problems [9], [19].

Nonuniqueness in the context of nonlinear parabolic equations is reported in [26].

We study a system of two conservation laws with quadratic polynomial flux functions such that the characteristic speeds are complex in a bounded part of the set of all possible states. We believe that this system models the local behavior of the solution of Stone's model [7], which is a description of immiscible three-phase flow commonly used in petroleum reservoir engineering. In the present work, shocks are considered admissible if they possess viscous profiles generated by a particularly simple parabolic term, where the viscosity matrix is a multiple of the identity. Nonunique solutions of this Riemann problem disregarding the viscous profile admissibility condition were obtained in [12].

We show by constructing two distinct solutions for a non trivial class of Riemann problem data that in this model the viscous profile entropy condition is insufficient to guarantee uniqueness of the solution. Conclusions that can be drawn are that either better entropy conditions are required in the presence of elliptic regions, or that fundamentally more accurate models must be developed to describe phenomena such as three-phase flow.

We believe that the special properties observed in this model are related to the existence of a kind of transitional shock not yet observed in conservation laws with quadratic polynomial flux functions. Transitional shocks are represented by orbits connecting two saddles in the associated dynamical system. Classes of quadratic polynomial dynamical systems possessing saddles connected by straight line orbits have been observed by Gomes [11], Shearer [24], Isaacson, Marchesin and Plohr [13]. However, the transitional shocks found in the model considered here are pairs of saddles connected only by an orbit which is not a straight line.

The behavior of solutions represented by states in the elliptic region is not clear yet. Numerical experiments reported in [3] indicate that states inside the elliptic region tend to leave it. We obtained an analytical proof of this fact using the viscosity entropy criterion. This result was obtained simultaneously and independently by H. Holden, L. Holden and N. Risebro.

The plan of this work is the following. The model is briefly described in §1. Entropy criteria including the viscous profile condition are briefly reviewed in §2. A pair of distinct solutions for the Riemann problem is constructed in §3. The Appendix contains some technical results used in the previous sections. Results related to the instability of states in the elliptic region are also contained in the Appendix.

1. The Model. Riemann problems originating in the theory of three phase flow in porous media are systems of conservation laws of the form

$$U_t + F(U)_x = 0, \tag{1.1}$$

with initial data

$$U(x, t=0) = U_0(x) = \begin{cases} U_l, & \text{if } x < 0 \\ U_r, & \text{if } x > 0. \end{cases} \tag{1.2}$$

where

$$U(x,t) = \begin{bmatrix} u(x,t) \\ v(x,t) \end{bmatrix}$$

is a solution of (1.1) for $t > 0$ and the flux function F is a prescribed C^2 function from $\mathbf{R}^2$ into $\mathbf{R}^2$

$$F(u,v) = \begin{bmatrix} f(u,v) \\ g(u,v) \end{bmatrix}. \tag{1.3}$$

If the eigenvalues $\lambda_1(U)$ and $\lambda_2(U)$ of $dF(U)$ are equal at (u_0, v_0) and distinct in a neighborhood of (u_0, v_0), we say that (u_0, v_0) is an umbilic point.

In [20], it was shown that if F is a quadratic polynomial mapping such that (1.1) is hyperbolic except at the umbilic point $U = 0$, the system (1.1) has the normal form

$$F(U) = \begin{bmatrix} \frac{1}{2}(au^2 + 2buv + v^2) \\ \frac{1}{2}(bu^2 + 2uv) \end{bmatrix},$$

which is the gradient of a cubic polynomial.

General linear perturbations of (1.1),

$$F(U) = \begin{bmatrix} \frac{1}{2}(au^2 + 2buv + v^2) + cu + dv \\ \frac{1}{2}(bu^2 + 2uv) + eu + fv \end{bmatrix},$$

split the umbilic point into a bounded region where the eigenvalues are complex conjugate [12], [18], [21]. This region is called an elliptic region and (1.1) becomes a conservation law of mixed type. For the elliptic region E to exist, it is important that d and e be distinct, to ensure that F is not a gradient. In our example, we consider $a = -1, b = 0, c = f = 0$ and $d = -e = \rho$, where the inequality $\rho > 0$ is satisfied, so that E is a circle of radius ρ. This case is a perturbation of symmetric case I ([20]). Therefore, F is given by

$$F(U) = \begin{bmatrix} \frac{1}{2}(-u^2 + v^2) + \rho v \\ uv - \rho u \end{bmatrix}. \tag{1.4}$$

We remark that adding a constant to F is irrelevant.

This model was proposed and studied in detail by Holden [12].

The eigenvalues of $dF(U)$ are

$$\lambda_1(U) = -\sqrt{u^2 + v^2 - \rho^2}\,, \quad \lambda_2(U) = \sqrt{u^2 + v^2 - \rho^2}\,, \tag{1.5}$$

with corresponding right eigenvectors associated to $\lambda_i (i = 1, 2)$:

$$r_i(U) = \begin{bmatrix} v + \rho \\ u + \lambda_i(U) \end{bmatrix}, \quad \text{if} \quad v \neq -\rho.$$

Since the solutions of (1.1) satisfy $U(ax, at) = U(x, t)$, for $a > 0$, we look for solutions of the form

$$U(x, t) = U(\xi), \quad \xi = \frac{x}{t}.$$

Such smooth solutions satisfy

$$-\frac{x}{t}U' + dF(U)U' = 0,$$

where the prime denotes the differentiation with respect to ξ. Therefore, continuous solutions of (1.1) are constructed locally using the integral curves of the differential equation

$$\begin{aligned} U' &= r_i(U), \\ \lambda_i(U) &= \frac{x}{t}, \quad i = 1, 2. \end{aligned} \tag{1.6}$$

A rarefaction curve $R_i(U_l)$ from U_l associated with a family i $(R_i(U_l), (i = 1, 2))$ is an integral curve of (1.6) starting at U_l along which $\lambda_i(U)$ is nondecreasing. It represents an i-rarefaction fan in physical space, defined by inverting the relation $\lambda_i(U) = \xi = \frac{x}{t}$. The eigenvalue $\lambda_i(U)$ is called the speed of the rarefaction at U, or characteristic speed.

A discontinuous solution of (1.1)

$$U(x, t) = \begin{cases} U_l, & x < st \\ U_r, & x > st \end{cases}$$

which propagates with speed $s = s(U_l, U_r)$ and separates two constant states U_l and U_r, is called a shock (U_l, s, U_r). For such a discontinuous solution s, U_l and U_r satisfy the Rankine-Hugoniot relation, which can be derived from the weak formulation of solutions of (1.1):

$$\mathcal{H}(U_r, s, U_l) = F(U_r) - F(U_l) - s(U_r - U_l) = 0 \tag{1.7}$$

For a fixed U_l, the set of U_r satisfying (1.7) is a one parameter family. A branch along which s decreases is called a shock curve; it is a parametrization in state space of physical shock waves. We obtain the Hugoniot curve eliminating s in (1.7); it is the solution of

$$H(U_l) = \{U : [f(u, v) - f(u_l, v_l)](v - v_l) - [g(u, v) - g(u_l, v_l)](u - u_l) = 0\}. \tag{1.8}$$

The Hugoniot curve undergoes topological change of shape at certain lines called secondary bifurcation lines [13], [18], given by B_1, B_2, B_3 where $v = \rho, v = \sqrt{3}u - 2\rho, v = -\sqrt{3}u - 2\rho$, respectively.

In many systems satisfying certain hypotheses [15], for a shock to be physically realizable, its speed as well as the characteristic speeds at the left and right of the discontinuity must satisfy certain inequalities, called *Lax's entropy conditions.* This gives rise to the following nomenclature for discontinuities

$$
\begin{aligned}
1 - shock: \quad & (S_1) \quad \lambda_1(U_r) < s < \lambda_1(U_l), \quad s < \lambda_2(U_r) \\
2 - shock: \quad & (S_2) \quad \lambda_2(U_r) < s < \lambda_2(U_l), \quad s > \lambda_1(U_l).
\end{aligned}
$$

Shocks for which either the left or right state is inside the elliptic region, do not fall directly in the above framework. There is no shock both states of which are inside the elliptic region (Prop. A.9).

An elementary wave associated with a family $i (i = 1, 2)$ is a rarefaction or a shock associated with a family i. If $i = 1$ $(i = 2)$ the elementary wave is called slow (fast).

Intermediate states in a Riemann solution represent a region with constant state in physical space. A solution of a Riemann problem (1.1) and (1.2) comprises wave groups each containing a sequence of adjacent elementary waves of the same family. In certain cases, transitional shocks or rarefactions not associated with any particular family may also exist. Constant states separate different wave groups. The speed increases from U_l to U_r along the solution of a Riemann problem.

2. Viscosity Admissibility Criterion. Weak solutions of (1.1) and (1.2) are required for an existence theory but they are not uniquely determined by the initial data. Other conditions, called entropy criteria, are necessary to obtain uniqueness. A typical criterion is to consider (1.1) as an approximation to an equation of the form

$$U_t + F(U)_x = \epsilon[D(U)U_x]_x, \qquad \epsilon > 0, \tag{2.1}$$

where $D(U)$ is the 2×2 viscosity matrix which models certain physical effects that are neglected in the conservation law (1.1), in the limit as $\epsilon \to 0^+$. We consider admissible the shocks (U_l, s, U_r) that are limits of traveling waves of (2.1) as $\epsilon \to 0^+$, that is, limits of solutions of the form

$$U = U(\zeta), \qquad \zeta = (x - st)/\epsilon \tag{2.2}$$

of (2.1), which tend to U_l and U_r as ζ approaches $-\infty$ and $+\infty$, respectively [5].

We assume that $D(U)$ has eigenvalues with positive real part. If $dF(U)$ has real distinct eigenvalues, this assumption guarantees that short wavelength perturbations of constant solutions decay exponentially in time. In this paper, we consider $D(U)$ as the identity, so that (2.1) can be written in the form

$$U_t + F(U)_x = \epsilon U_{xx}. \tag{2.3}$$

Substituting (2.2) into (2.3) and integrating the result, we have

$$\dot{U} = -s[U - U_l] + F(U) - F(U_l) \tag{2.4}$$

where the dot denotes differentiation with respect to ζ.

We remark that as ζ tends to infinity the right hand side of (2.4) tends to zero, satisfying the Rankine-Hugoniot relation (1.7). We refer to (2.4) (for fixed U_l) as the field $\mathfrak{X}_s(U, U_l)$. Thus, we write

$$\mathfrak{X}_s(U, U_l) = \begin{bmatrix} \frac{1}{2}(-u^2 + v^2) + \rho v - \frac{1}{2}(-u_l^2 + v_l^2) - \rho v_l - s(u - u_l) \\ uv - \rho u - u_l v_l + \rho u_l - s(v - v_l) \end{bmatrix}$$

In [8] Gel'fand and in [6] Courant and Friedrichs show that studying discontinuous solutions of (1.1) and (1.2) as limits of solutions of (2.3) is equivalent to studying the existence of an orbit γ of $\mathfrak{X}_s(U, U_l)$, such that

$$U_l = \alpha(\gamma) \qquad \text{and} \qquad U_r = \omega(\gamma) \tag{2.5}$$

where $\alpha(\gamma)$ and $\omega(\gamma)$ are the α-limit and ω-limit sets of γ, respectively.

The viscosity entropy criterion consists in considering a shock (U_l, s, U_r) as admissible if there exists an orbit of $\mathfrak{X}_s(U, U_l)$ satisfying (2.5); note that admissible Lax 1-shocks are repeller-saddle connections while Lax 2-shocks are saddle-attractor connections.

We remark that for F given by (1.4) the eigenvalues $\mu_i = \mu_i(U), (i = 1, 2)$ of $d_{(U)}\mathfrak{X}_s$ are given by

$$\mu_i = \lambda_i - s. \tag{2.6}$$

Note that if $U_j (j = 1, 2)$ lie in the elliptic region E, then $\Re\mu_i = -s$ and U_j is an attractor if $s > 0$ or a repeller if $s < 0$.

If U_l is a repeller inside E, U_r is a saddle outside E and there exists an orbit of (2.4) connecting U_l to U_r we call (U_l, s, U_r) a *1-complex shock.* Similarly, if U_l is a saddle outside E, U_r is an attractor inside E, and there is an orbit connecting U_l to U_r we call (U_l, s, U_r) a *2-complex shock.* We will continue to denote a 1-complex shock (2-complex shock) by $S_1(S_2)$, respectively.

It is known that there exist shocks obeying Lax's entropy condition which are inadmissible because there is no orbit connecting the singularities U_l and U_r; there are also saddle-saddle connections representing shocks with viscous profile which do not obey Lax's inequalities [13]. These are called *transitional* shock waves. They are discontinuous solutions (U_-, s, U_+) with U_- connected to U_+ by an orbit, such that

$$\lambda_1(U_-) < s < \lambda_2(U_-),$$
$$\lambda_1(U_+) < s < \lambda_2(U_+).$$

Obviously, such waves appear only outside E. We will denote transitional shock waves by X. Such waves are essential to ensure the existence of solutions of Riemann problems [20], [24], [10].

For quadratic polynomial gradient systems, Chicone [4] showed that transitional shocks are represented straight line orbits. When the system is not a gradient,

Frommer and Bautin [27] obtained an example with a saddle-saddle connection which is not a straight line; however, the two singularities are still connected by another straight line orbit (see Fig. A.1 in the Appendix). We will show in §4 an example of two saddles connected only by one orbit which is not a straight line segment. Non straight line saddle-saddle connections allow for more complicated Riemann solution structures than those already known [13].

In the remaining chapters, if two states U_- and U_+ are connected by a wave $W(W = S_1, S_2, R_1, R_2, X)$, we will use the notation

$$U_- \overset{W}{\to} U_+.$$

3. Nonuniqueness. We will show that there exists an open region of pairs (U_l, U_r) such that the Riemann problem with data (U_l, U_r) in this set admits two distinct solutions, both of which are admissible. We construct the Riemann solutions as sequences of elementary waves. The first solution consists of a 1-rarefaction, a pair of transitional shocks and a 2-complex shock; the second solution consists of a 1-rarefaction and a 2-complex shock. We will prove this result in two steps. In the first step (Prop. 3.1), we restrict U_l to a segment of a straight line and U_r to an open region Ω_2; in the second step (Prop. 3.2), we consider U_l in a region Ω_1 and U_r in another region Ω_2.

To prove Prop. 3.1, we construct an open segment of points U_l in B_2. We define certain points, shown in Fig. 3.1, as follows (this can be done explicitly, since the Hugoniot curves are conic sections):

i) $A = (a_1, a_2) \in B_2, B = (b_1, b_2) \in B_3$, such that $a_2 > \rho, B \in H(A)$ and $H(B)$ is tangent to the axis $u = 0$,

ii) $U_l = (u_l, v_l) \in B_2$, with $\rho < v_l < a_2$,

iii) $M_3 = (m_3, n_3) \in B_3$, such that $M_3 \in H(U_l)$.

Now we define $\Omega_2(U_l)$ as the region bounded by the axis $u = 0$ and the curve $H(M_3)$. Note that $\Omega_2(U_l)$ is contained in the triangle with vertices O, α, β, where $O = (0,0), \alpha = (0, \rho), \beta = (\frac{\rho}{\sqrt{3}}, \rho)$.

To construct the solution of the Riemann problem, we consider (see Fig. 3.2)

iv) $U_r = (u_r, v_r) \in \Omega_2(U_l)$,

v) $M_2 = (m_2, n_2) = H(U_r) \cap B_3$; note that $-\sqrt{3}\rho < m_3 < m_2 < b_2 < 0$,

vi) $M = (m, n) = H(U_r) \cap R_1(U_l)$,

vii) $M_1 = (0, -2\rho)$,

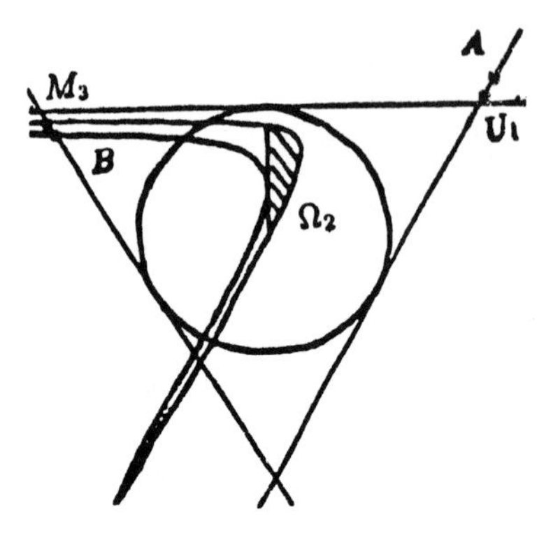

Fig. 3.1: Points used to construct Ω_1 and Ω_2; also, the region Ω_2.

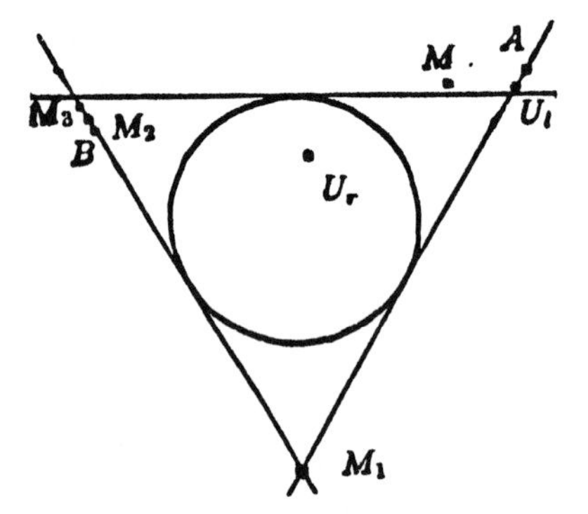

Fig. 3.2: Points used to construct the solution of the Riemann problem.

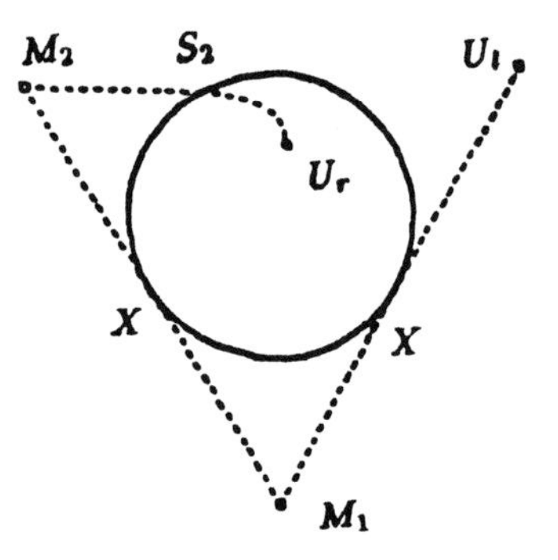

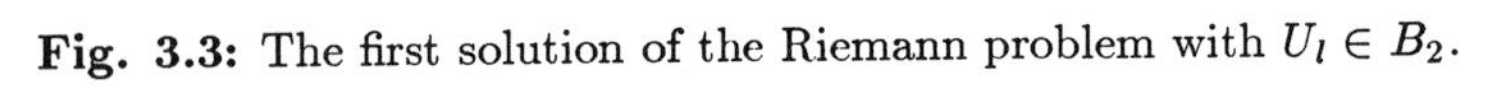

Fig. 3.3: The first solution of the Riemann problem with $U_l \in B_2$.

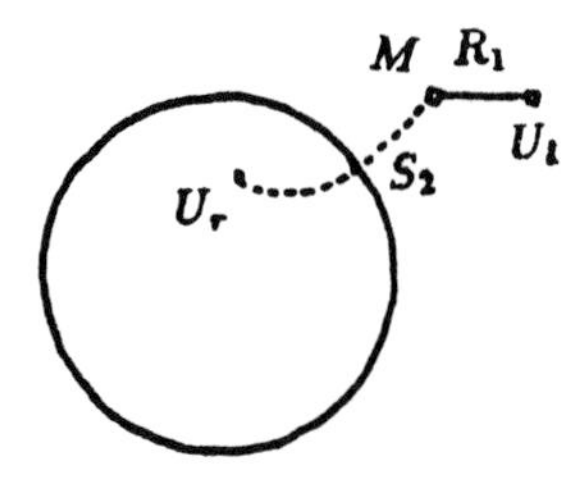

Fig. 3.4: The second solution of the Riemann problem with $U_l \in B_2$.

PROP. 3.1. *Let U_l and U_r be as above. Then the Riemann problem (1.1) and (1.2), with F given by (1.4), has at least two distinct solutions (I and II) which are admissible.*

Proof. I) First solution: $U_l \xrightarrow{X} M_1 \xrightarrow{X} M_2 \xrightarrow{S_2} U_r$ (see Fig. 3.3):

i) U_l can be connected to M_1 by a transitional shock (for v_l close to ρ; otherwise, M_1 is an attractor such that the discontinuity (U_l, s, M_1) is not a transitional shock), since from (1.5)

$$\lambda_1(U_l) = -\lambda_2(U_l) = -(2u_l - \sqrt{3}\rho),$$
$$\lambda_1(M_1) = -\lambda_2(M_1) = -\sqrt{3}\rho,$$

and the wave speed is

$$s = \frac{u_l(v_l - \rho)}{v_l + 2\rho} = u_l - \sqrt{3}\rho.$$

Clearly,

$$\lambda_1(U_l) < s < \lambda_2(U_l) \qquad \text{and} \qquad \lambda_1(M_1) < s < \lambda_2(M_1).$$

By Prop. A.2, there is an orbit connecting U_l and M_1. The eigenvectors $r_1(U_l)$ and $r_2(M_1)$ are parallel to B_2, and the orientation of this orbit is from U_l to M_1. Therefore, we have joined U_l to M_1 by a transitional shock with speed $s = u_l - \sqrt{3}\rho$.

ii) M_1 can be connected to M_2 by a transitional shock, where

$$\lambda_1(M_2) = -\lambda_2(M_2) = -\sqrt{((m_2)^2 + (n_2)^2 - \rho^2)},$$

and the wave speed is

$$s = \frac{m_2(n_2 - \rho)}{n_2 + 2\rho} = m_2 + \sqrt{3}\rho.$$

Clearly,

$$\lambda_1(M_1) < s < \lambda_2(M_1) \qquad \text{and} \qquad \lambda_1(M_2) < s < \lambda_2(M_2).$$

So we have joined M_1 to M_2 by a transitional shock with speed $s = m_2 + \sqrt{3}\rho$.

The proof that this shock is admissible is analogous to that in i). Also, the speed $s(M_1, M_2)$ is larger than $s(U_l, M_1)$ by Prop. A.4.

iii) M_2 can be connected to U_r by a 2-complex shock with speed

$$s = u_r + \frac{(u_r - m_2)(m_2 - \rho)}{v_r - m_2}$$

By Prop. A.6, this shock is admissible and by Prop. A.7 the speed $s(M_2, U_r)$ is larger than $s(M_1, M_2)$.

Therefore, we have presented a sequence of admissible waves

$$U_l \overset{X}{\rightarrow} M_1 \overset{X}{\rightarrow} M_2 \overset{S_2}{\rightarrow} U_r$$

with strictly increasing speed; this is the first Riemann solution.

II) Second solution: $U_l \overset{R_1}{\rightarrow} M \overset{S_2}{\rightarrow} U_r$ (see Fig. 3.4):

i) U_l can be connected to M by a slow rarefaction wave. Since $\lambda_1(U) < 0$ for all U outside the elliptic region, all speeds in this wave are negative. So we have

$$U_l \overset{R_1}{\rightarrow} M,$$

with negative characteristic speeds.

ii) M can be connected to U_r by a 2-complex shock with speed given by

$$s = \frac{u_r(v_r - \rho) - m(n - \rho)}{v_r - n},$$

which is greater than zero, since $v_r < \rho, n > \rho, m > 0, u_r > 0$ and $v_r < n$. By Prop. A.5, this shock is admissible.

Therefore

$$U_l \overset{R_1}{\rightarrow} M \overset{S_2}{\rightarrow} U_r$$

is another sequence of admissible waves with increasing speed.

Clearly, the solutions I and II are distinct. □

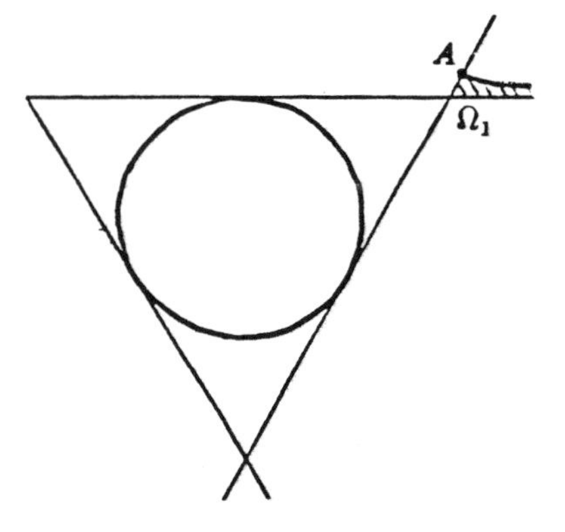

Fig. 3.5: The region Ω_1.

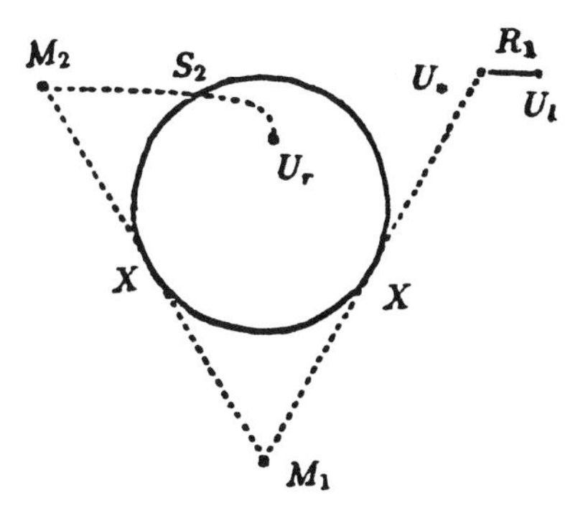

Fig. 3.6: The first solution of the Riemann problem with $U_l \in \Omega_1$.

To construct Ω_1 (the set of points U_l possessing solutions similar to I and II), we consider the point A given in i) and define Ω_1 as the region bounded by B_1, B_2 and the 1-integral curve through A (Fig. 3.5). To construct the second sequence, we consider U_l inside Ω_1 and define U_* as the intersection of $R_1(U_l)$ with B_2. Similarly, $\Omega_2(U_*)$ is defined as in Prop. 3.1.

PROP. 3.2. *Let $U_l \in \Omega_1$ and $U_r \in \Omega_2(U_*)$, then the Riemann problem (1.1), (1.2) has at least two distinct solutions which are admissible.*

Proof. Since the rarefaction $R_1(U_l)$ crosses B_2 at U_* lying underneath A [18], we can join U_l to U_* by a slow rarefaction; from U_* we follow the same construction as described in the proof of Prop. 3.1 (see Fig, 3.6 and Fig. 3.7). □

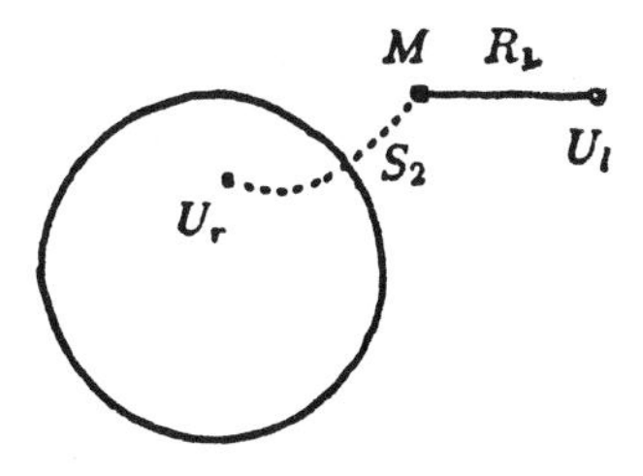

Fig. 3.7: The second solution of the Riemann problem, $U_l \in \Omega_1$.

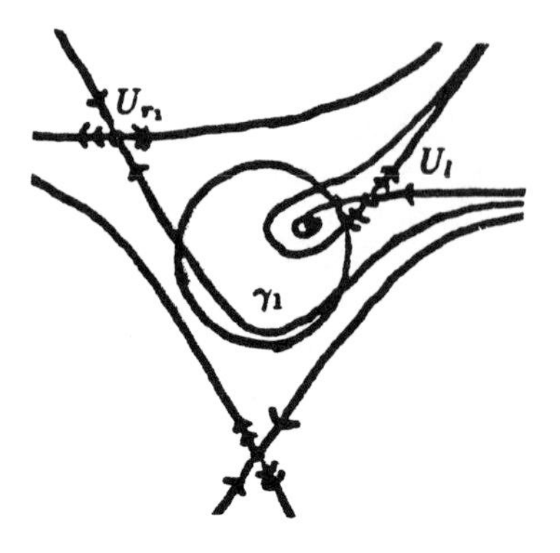

Fig. 4.1: The dynamical system with $\alpha(\gamma_1) = \infty$.

4. Saddle-Saddle Connections. In order to check that a discontinuity satisfying the Rankine-Hugoniot condition (1.8) is in fact admissible one needs to verify the existence of orbits connecting singularities of a vector field. In this context, we must look for saddle-saddle connections to verify whether a crossing shock is admissible. An important result is Chicone's theorem [4], which states that saddle-saddle connections of quadratic polynomial gradient systems lie on straight lines. Frommer and Bautin [27] showed an example of a quadratic system on the plane which is not a gradient, where there exist two distinct orbits connecting a pair of saddles. One orbit is a straight line, but the other is not. A generalization of this fact for non-gradient systems would be a very useful result, since by Prop. A.1 the only invariant lines are the secondary bifurcations, and therefore the transitional shocks would occur on these lines. However, by performing computer experiments using the Riemann Problem Solver Package of E. Isaacson, D. Marchesin and B. Plohr, we discovered pairs of saddles connected by only one orbit which is not a straight line segment.

If we consider U_l and U_{r1} as in the Fig. 4.1, we obtain the configuration shown there. The configuration shown in Fig. 4.2 is obtained considering a point U_{r2} on the Hugoniot locus, which is a perturbation of U_{r1}. We take $s < 0$ and obtain U_{r2} by increasing s in such way that it remains negative. We remark that U_l, U_{r1} and U_{r2} are saddles; since $s < 0$ and $divX_s(U, U_l) = -2s$, by Bendixson's criterion [1], there is no closed orbit. This is important: computer experiments displaying configurations such as those shown in these two figures would be unreliable if closed orbits were possible.

In the first case (Fig. 4.1), the stable manifold γ_1 of U_{r1}, which crosses the v axis, has a singularity at infinity as α-limit, while in the second case (Fig. 4.2), the manifold γ_2 has a repeller U_i as α-limit. Therefore, by continuity, there is a saddle-saddle connection between U_l and a point U_r between U_{r1} and U_{r2}. It is clear that there is no straight line orbit connecting these two saddles.

We will give more information about these connections in [2].

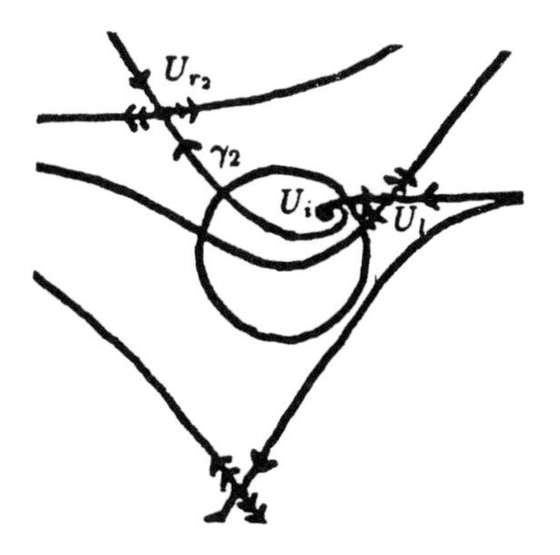

Fig. 4.2: The dynamical system with $\alpha(\gamma_2) = U_i$.

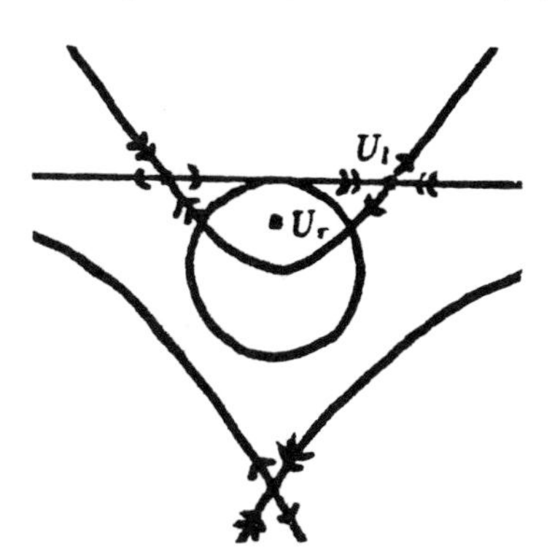

Fig. A.1: The dynamical system of $\mathfrak{X}_0(U,(0,a_2))$.

Appendix. In this Appendix, we present the propositions and proofs used in the previous sections.

The proof of the next proposition can be found in [13]; a direct proof is easy.

PROP. A.1. *The only invariant straight lines of $\mathfrak{X}_s(U,U_l)$ given by (2.4), are B_1, B_2 and B_3.*

PROP. A.2. *Let A and B be saddles of $\mathfrak{X}_s(U,U_l)$ lying on the same secondary bifurcation line. Then on this line there is an orbit joining A to B.*

Proof. The result follows by Bezout's theorem; A and B are the only singularities on the same secondary bifurcation, which is an invariant line of $\mathfrak{X}_s(U,U_l)$. □

PROP. A.3. *If U_- lies on $B_2, M_1 = (0,-2\rho)$, and $U_+ \in B_3$ is such that $s(U_-,M_1) = s(M_1,U_+)$ then $U_+ = H(U_-) \cap B_3$.*

Proof. Since $s_1 = s(U_-,M_1) = u_- - \sqrt{3}\rho$ and $s_2 = s(M_1,U_+) = u_+ + \sqrt{3}\rho$, we have $u_+ = u_- - 2\sqrt{3}\rho$. We obtain the conclusion using that
$U_+ = (u_- - 2\sqrt{3}\rho, -\sqrt{3}u_- + 4\rho)$ to verify that $U_+ \in H(U_-)$. □

PROP. A.4. *If we consider U_l, M_1, M_2 as in Prop. 3.1, then $s(U_l, M_1) < s(M_1, M_2)$.*

Proof. The proposition follows from Prop. A.3 because the speed decreases along the shock curve. □

PROP. A.5. *Given M and U_r as in the Prop. 3.1, then there exist an orbit γ with M as α-limit of γ and U_r as ω-limit of γ.*

Proof. To show that there exists an orbit of $\mathfrak{X}_s(U, U_l)$ connecting the pair M, U_r given in Prop. 3.1, we consider $A = (0, a_2)$ for $0 < a_2 < \rho$ and $B = (b_1, b_2) \in H(A)$ and recall that the shock speed s is given by

$$s = s(A, B) = \frac{b_1(b_2 - \rho)}{a_2 - b_2},$$

if $a_2 \neq b_2$ and $\mathfrak{X}_s(U, U_l)$ is given, for $s = 0$, by

$$\mathfrak{X}_0(U, U_l) = \begin{bmatrix} \frac{1}{2}(-u^2 + v^2 - a_2^2) + \rho(v - a_2) \\ u(v - \rho) \end{bmatrix}.$$

If $b_1 = \sqrt{3}\rho$ and $b_2 = \rho, s = 0$ and the dynamical system is shown in Fig. A.1 [27].

Now, we take B' as a small perturbation of the point B, such that $b_2' > \rho$ and $B \in H(A)$. Thus $s > 0$ and $\mathfrak{X}_s(U, U_l)$ is a perturbation of $\mathfrak{X}_0(U, U_l)$, where the perturbation is given by

$$-s \begin{bmatrix} u - a_1 \\ v - a_2 \end{bmatrix} = -s \begin{bmatrix} u \\ v - a_2 \end{bmatrix}$$

which is vector field pointing to U_r. Therefore U_r is an attracting focus.

In this case, since A is an attractor and $divX_s(U, U_l) = -2_s \neq 0$ there is no closed orbit, by Bendixson's criterion. Hence the unstable manifold of B tends to A. Therefore, there is a saddle-attractor connection.

Since saddle-attractor connections are stable, under a small perturbation of A we still have orbits connecting A and B. Therefore the proposition is proved for $A = U_r$ and $B = M$. □

PROP. A.6. *Given M_3 and U_r as in Prop. 3.1, then there exists an orbit γ with M as α-limit and U_r as ω-limit of γ.*

The proof is omitted because its idea is the same as that in Prop. A.5.

PROP. A.7. *Let M_1, M_2 and U_r be as in Prop. 3.1. Then*

$$s(M_2, U_r) > s(M_1, M_2)$$

Proof. We must show that

$$-\frac{n_2 - \rho}{\sqrt{3}} < u_r + \frac{(u_r - m_2)(n_2 - \rho)}{v_r - n_2}.$$

We note that $u_r > m_2, n_2 < \rho, v_r < n_2$ and $-2\rho < -\rho < \sqrt{3}u_r + v_r < (\sqrt{3}+1)\rho$. So, since $M_2 \in B_3$, we have

$$\sqrt{3}u_r + v_r > n_2 + \sqrt{3}m_2 = -2\rho,$$
$$\frac{(u_r - m_2)(n_2 - \rho)}{v_r - n_2} > -\frac{n_2 - \rho}{\sqrt{3}}$$
$$u_r + \frac{(u_r - m_2)(n_2 - \rho)}{v_r - n_2} > -\frac{n_2 - \rho}{\sqrt{3}}. \ \square$$

The next propositions present results related with instability of states inside the elliptic region E and with invariance of waves connecting states outside E. These results were obtained simultaneously and independently by H. Holden, L. Holden and N. Risebro. Before proving them, we remark that a state inside E can be connected to another state outside E only by a (complex) shock, since rarefactions curves do not enter E.

PROP. A.8. *Let U_l and U_r be two states outside the open elliptic region E of a generic system of two conservation laws. Then there is no admissible wave from U_l to U_r with an intermediate state in the elliptic region.*

Proof. For simplicity, we suppose that there exists only one intermediate state M in the Riemann solution.

By contradiction, we suppose that M is inside E and there exist two admissible shocks, one from U_l to M and another from M to U_r. So, as far as the dynamical system with U_l and M as singularities is concerned,M is an attractor and, by (2.6),

$$\Re(\mu(M)) < 0 \Rightarrow s(U_l, M) > 0.$$

On the other hand, for the dynamical system with M, U_r as singularities, M is a repeller and

$$s(M, U_r) < 0.$$

Therefore, the wave speed from left to right is decreasing, which is a contradiction. $\square$

PROP. A.9. *For quadratic polynomial conservation laws, with $D = I$ in (2.1), and homogeneous parts in cases $I, \ldots, IV$ [21], if U_l is inside the elliptic region, then $H(U_l) \cap E = U_l$.*

Proof. If $H(U_l) \cap E \neq U_l$, by contradiction, then there is state $U_r \in H(U_l)$ such that $U_r \in E$. Since F is quadratic, using the midpoint rule [13],

$$F(U_r) - F(U_l) - s(U_r - U_l) = dF(U_m)(U_r - U_l) - s(U_r - U_l) = 0,$$

where $U_m = (U_l + U_r)/2$ lies inside E (E is convex for cases $I, \ldots, IV$ [18]). This implies that $U_r - U_l$ is an eigenvector of $dF(U_m)$ where U_m lies inside E, with real eigenvalue s, which is a contradiction. $\square$

PROP. A.10. *Under the hypotheses of the previous proposition, if U_l and $U_r, U_l \neq U_r$, are states inside the elliptic region E, then if the Riemann solution exists, it has at least an intermediate state M which is outside E.*

Proof. By contradiction, we assume that there does not exist M outside E. Then M (which may be U_l) is inside E and we have a shock from U_l to M. This implies that $H(U_l) \cap E \neq U_l$ which contradicts Prop. A.9. □

REFERENCES

[1] A. A. ANDRONOV, A. A. VITT AND S. E. KHAIKIN, *Theory of Oscillations*, Addison-Wesley Pub. Co., Inc., Massachusetts (1966).

[2] A. V. AZEVEDO, *Doctoral Thesis*, PUC/RJ, Rio de Janeiro, Brazil (1990).

[3] J. BELL, J. R. TRANGENSTEIN AND G. SHUBIN, *Conservation Laws of Mixed Type Describing Three-Phase Flow in Porous Media*, SIAM Jour. Appl. Math. 46 (1986), pp. 1000–1017.

[4] C. CHICONE, *Quadratic Gradients on the Plane are Generically Morse-Smale*, Jour. Diff. Eq. 33 (1979), pp. 159–166.

[5] C.C. CONLEY, J.A. SMOLLER, *Viscosity Matrices for Two- Dimensional Nonlinear Hyperbolic Systems*, Comm. Pure Appl. Mat. XXIII (1970), pp. 867–884.

[6] R. COURANT AND K.O. FRIEDRICHS, *Supersonic Flow and Shock Waves*, Interscience Publishers, New York (1948).

[7] F.J. FAYERS AND J.D. MATTHEWS, *Evaluation of Normalized Stone's Methods for Estimating Three-Phase Relative Permeabilities*, Soc. Petrol. Engin. J. 24 (1984), pp. 225–232.

[8] I.M. GEL'FAND, *Theory of Quasilinear Equations*, English transl. in Amr. Mat. Soc. Trans., ser. 2 29 (1963), pp. 295–381.

[9] H. GILQUIN, *Glimm's Scheme and Conservation Laws of Mixed Type*, SIAM J. Sci. Stat. Comput. 10 (1989), pp. 133–153.

[10] M.E. GOMES, *Riemann Problems requiring a Viscous Profile Entropy Condition*, Adv. Appl. Math. 10 (1989), pp. 285–323.

[11] M.E. GOMES, *On Saddle Connections in Planar, Quadratic Dynamical Systems' with Applications to Conservation Laws*, Preprint (1989).

[12] H. HOLDEN, *On The Riemann Problem for a Prototype of a Mixed Type Conservation Law*, Comm. Pure Appl. Mat. XL (1987), pp. 229–264.

[13] E. ISAACSON, D. MARCHESIN AND B. PLOHR, *Transitional Waves for Conservation Laws*, CMS Technical Report #89-20, U. Wisconsin-Madison (1988), to appear in SIAM J. Math. Anal., 1990.

[14] E. ISAACSON AND J.B. TEMPLE, *The Riemann Problem Near a Hyperbolic Singularity II*, SIAM J. Appl. Math. 48 (1988), pp. 1287–1301.

[15] P. LAX, *Hyperbolic Systems of Conservation Laws II*, Comm. Pure Appl. Math. 19 (1957), pp. 537–566.

[16] D. MARCHESIN AND H. B. MEDEIROS, *A Note on the Stability of Eigenvalue Degeneracy in Nonlinear Conservation Laws of Multiphase Flow*, Current Progress in Hyperbolic Systems: Riemann Problems and Computations. (Bowdoin, 1988), Contemporary Mathematics 100, Amer. Math. Soc. (1989), pp. 215–224.

[17] O.A. OLEINIK, *On the Uniqueness of Generalized Solution of Cauchy Problem for Non Linear System of Equations Occurring in Mechanics*, Uspeki Mat. Nauk (Russian Math. Surveys) 12 (1957), pp. 169–176.

[18] C.F. PALMEIRA, *Line Fields Defined by Eigenspaces of Derivatives of Maps form the Plane to Itself*, Proceedings of the VI Conference International of Differential Geometry, Santiago de Compostela, Spain (1988).

[19] R. PEGO AND D. SERRE, *Instability in Glimm's Scheme for Two Systems of Mixed Type*, SIAM J. Numer. Anal. 25 (1988), pp. 965–988.

[20] M. SHEARER, D. SCHAEFFER, D. MARCHESIN AND P. PAES-LEME, *Solution of Riemann Problem for a Prototype* 2×2 *System of Non-Strictly Hyperbolic Conservation Laws*, Arch. Rat. Mech. Anal. 97 (1987), pp. 299–320.

[21] D. SCHAEFFER, M. SHEARER, *The Classification of* 2×2 *Systems of Non-Strictly Hyperbolic Conservation Laws, with Application to Oil Recovery*; with appendix by D. Marchesin, P. Paes-Leme, M. Shearer, D. Schaeffer, Comm. Pure Appl. Math. vol. XL (1987), pp. 141–178.

[22] M. SHEARER, *Admissibility Criteria for Shock Wave Solutions of a System of Conservation Laws of Mixed Type*, Proceeding of the Royal Society of Edinburgh 93A (1983), pp. 233–244.

[23] M. SHEARER, *Non-uniqueness of Admissible Solutions of Riemann Initial Value Problems for a System of Conservation Laws of Mixed Type*, Arch. Rat. Mech. Anal. 93 (1986), pp. 45–59.

[24] M. SHEARER, *The Riemann Problem for* 2×2 *Systems of Hyperbolic Conservation Laws with Case I Quadratic Nonlinearities*, J. Differential Equations 80 (1989), pp. 343–363.

[25] M. SHEARER, *Loss of Strict Hyperbolicity of the Buckley-Leverett Equations for Three-Phase Flow in a Porous Medium*, Numerical Simulation in Oil Recovery (ed. M.F. Wheeler), IMA vol. 11, Springer-Verlag (1988).

[26] N. VVEDENSKAYA, *An Example of Nonuniqueness of a Generalized Solution of a Quasilinear System of Equations*, Sov. Math. Dokl. 2 (1961), pp. 89–90.

[27] YE YAN-QIAN AND OTHERS, "*Theory of Limit Cycles*", Translations of Mathematical Monographs-AMS, Providence, Rhode Island (1984).

ON THE LOSS OF REGULARITY OF SHEARING FLOWS OF VISCOELASTIC FLUIDS

BERNARD D. COLEMAN*

Abstract. In this expository article a brief survey is given of a part of the theory of rectilinear shearing flows of simple fluids with fading memory. The topics discussed have relevance to the search for a theory of melt fracture of polymers at high rates of shear. The emphasis is laid on unbounded growth of shear-acceleration waves and breakdown of initially smooth solutions.

Key words. Viscoelasticity, non-Newtonian fluids, hyperbolic systems, melt fracture of polymers.

1. Preface. In a work published in 1968 [1], Morton Gurtin and I proposed that the phenomenon rheologists call "melt fracture" [2–10] or "elastic turbulence" may be an example of the formation of shock waves in viscoelastic fluids subjected to high rates of shear. Experimenters have not yet decided whether such is the case or whether the phenomenon is caused by failure of adherence to bounding surfaces. Moreover, the situation has been complicated by the realization that there are several phenomena that are occasionally grouped under the appellation, "melt fracture", but which, like the "spurt" phenomenon studied by Vinogradov and co-workers [11],[1] are associated with an apparent loss of monotonicity of the shear-stress function.

In the last two decades much progress has been made in the development of the theory of the formation and growth of singularities in viscoelastic fluids with monotone shear-stress functions. I attempt to summarize here parts of the old and new work on the subject that I believe will prove important for the attainment of an understanding of the origin of melt fracture.

2. General Theory of Simple Fluids in Shear. A motion is called a *rectilinear shearing flow* if there is a Cartesian coordinate system in which the components of the velocity field have the form

$$v^x = 0, \qquad v^y = v(x,t), \qquad v^z = 0. \tag{2.1}$$

Such a flow is isochoric.

Of frequent occurrence in the theory of rectilinear shearing flows in viscoelastic fluids is the real-valued function, λ^t, defined by

$$\lambda^t(s) = -\partial_x \int_{t-s}^{t} v(x,\xi)\,d\xi, \qquad (0 \le s < \infty); \tag{2.2}$$

λ^t is called the *history up to time* t (at x) *of the relative shear strain.*[2] Shortly after Noll [13] formulated his definition of a simple fluid, Coleman and Noll [14]

---

* Department of Mechanics and Materials Science, Rutgers University, Piscataway, New Jersey.

[1] See also Bagley, Cabott, and West [12].

[2] $\partial_x = \frac{\partial}{\partial x}$.

showed that for a general incompressible simple fluid it follows from the principle of material frame indifference that the components of the stress tensor $\boldsymbol{T}$ in a rectilinear shearing flow obey the relations

$$
\begin{cases} T^{xy}(t) = \mathfrak{t}(\lambda^t), \\ T^{yy}(t) - T^{xx}(t) = \mathfrak{n}_1(\lambda^t), \\ T^{yy}(t) - T^{zz}(t) = \mathfrak{n}_2(\lambda^t), \\ T^{xz} = T^{yz} = 0, \end{cases} \tag{2.3}
$$

in which $\mathfrak{t}$, $\mathfrak{n}_1$, and $\mathfrak{n}_2$ are real-valued functionals that depend on the material under consideration and satisfy the identities

$$
\mathfrak{t}(-\lambda^t) = -\mathfrak{t}(\lambda^t), \tag{2.4}
$$

$$
\mathfrak{n}_i(-\lambda^t) = \mathfrak{n}_i(\lambda^t), \qquad (i = 1, 2). \tag{2.5}
$$

If one assumes that the body force density $\boldsymbol{b}$ has a potential ψ, *i.e.*, that

$$
\boldsymbol{b} = -\text{grad}\psi, \tag{2.6}
$$

then an elementary analysis shows that it follows from (2.1)–(2.3) that the dynamical equations,

$$
\text{div}\boldsymbol{T} + \rho\boldsymbol{b} = \rho\dot{\boldsymbol{v}}, \tag{2.7}
$$

in which ρ is the mass density, are equivalent to the assertions that

$$
\partial_x T^{xy} + \alpha(t) = \rho\partial_t v, \tag{2.8}
$$

and

$$
T^{xx} - \rho\psi = \alpha(t)y + \beta(t) \tag{2.9}
$$

with α and β functions of t only. It is easily shown that α is the driving force (per unit volume) in the direction of flow. When the specific driving force α is specified, (2.8) is a functional-differential equation for the function v in (2.1) and (2.2).

Steady Rectilinear Shearing Flows

As we shall here be concerned with the dynamical stability of steady rectilinear shearing flows, a brief review of the properties of such solutions of (2.8) is needed.

If the flow is steady, then

$$
\partial_t v = 0, \tag{2.10}
$$

(2.2) reduces to

$$
\lambda^t(s) = -\kappa s \tag{2.11}
$$

with[3]

$$\kappa = \kappa(x) = v'(x), \tag{2.12}$$

and (2.8) becomes

$$\frac{d}{dx}\tau(\kappa) + \alpha = 0, \tag{2.13}$$

where τ is the *shear-stress function* which is familiar in the theory of steady viscometric flows[4] [15,16] and which is determined by the functional $\mathfrak{t}$ through the relation

$$\tau(\kappa) = \mathfrak{t}(\kappa\iota), \qquad \iota(s) \equiv s. \tag{2.14}$$

The number κ is called the *rate of shear.* It follows from (2.4) that the function τ, which takes real numbers into real numbers, is an *odd* function, *i.e.,*

$$\tau(-\kappa) = -\tau(\kappa), \tag{2.15}$$

and hence $\tau(0) = 0$. Thermodynamics requires that $\tau(\kappa)\kappa$ be never negative [16], and hence that $\tau'(0) \geq 0$. Let us here assume

$$\tau'(\kappa) > 0 \tag{2.16}$$

for all κ, which implies that τ is invertible. One writes τ^{-1} for the inverse of τ.

It follows from (2.10) and (2.13) that a steady rectilinear shearing flow can occur only when the specific driving force α is independent of t and

$$\tau(\kappa) = -\alpha x + \nu \tag{2.17}$$

with ν, like α, constant (in space and time). In view of (2.16), specification of α and ν determines the rate of shear as a function of x:

$$\kappa = \tau^{-1}(-\alpha x + \nu). \tag{2.18}$$

If the driving force α is zero, κ is independent of x, and, to within an added constant, v has the form

$$v = \kappa x. \tag{2.19}$$

A steady rectilinear flow obeying (2.19) is called a *simple shearing flow,* and is a motion that can take place in a fluid that lies between and adheres to two infinite

---

[3] An apostrophe indicates a derivative.
[4] For a survey of that theory see [17].

plates parallel to the (y, z)-plane with separation d, one of which, at $x = 0$, is at rest, and the other, at $x = d$, is moving with velocity V; for clearly, (2.19) with

$$\kappa = V/d \tag{2.20}$$

is compatible with such boundary conditions.

Steady channel flow is a steady rectilinear shearing flow between two infinite plates of separation d which are again parallel to the (y, z)-plane, but are both at rest. For such a motion the driving force α is not zero. If we place the (y, z)-plane halfway between the plates, then the assumption that the fluid adheres to the plates gives the boundary conditions

$$v(d/2) = v(-d/2) = 0. \tag{2.21}$$

In view of (2.15) and (2.16), these boundary conditions are compatible with (2.18) only if $\nu = 0$, and hence[5] $v'(x) = \kappa(x) = -\tau^{-1}(\alpha x)$, *i.e.*,

$$v(x) = \int_x^{\frac{1}{2}d} \tau^{-1}(\alpha x)\, dx. \tag{2.22}$$

3. Fading Memory and Instantaneous Elasticity. The postulate of fading memory introduced by Coleman and Noll [18,19][6] is an assumption of smoothness for constitutive functionals that relate certain variables, such as stress, to the history of others, such as strain. The postulate asserts that such functionals have continuous differentials with respect to a particular norm $\|\cdot\|$ on a space of histories. This norm has two important properties. (*i*) For a history such as λ^t, values $\lambda^t(s)$ at large s, *i.e.*, occurring in the "remote past", have less influence on $\|\lambda^t\|$ than values at small s. (*ii*) A material whose constitutive functionals are smooth with respect to this norm shows *instantaneous elasticity:* a jump in strain results in a jump in stress that depends smoothly on the jump in strain, albeit the response function governing such dependence is determined by the material's history prior to the jump.

As the present discussion is restricted to shearing motions governed by the functional-differential equation (2.8), *i.e.*,

$$\partial_x \mathfrak{t}(\lambda^t) + \alpha = \rho \partial_t v, \tag{3.1}$$

it is not necessary to present the theory of fading memory in full generality, and we may confine our attention to its implications for the functional $\mathfrak{t}$.

The elements λ^t of the domain of $\mathfrak{t}$ are histories of the relative shear strain. By (2.2), each such history is a function on $[0, \infty)$ obeying

$$\lambda^t(0) = 0, \tag{3.2}$$

---

[5] *Cf.* [15,17].

[6] A more axiomatic treatment is given in the work Coleman and Mizel [20]. The presentation given below draws heavily on that of Coleman and Gurtin [1].

and, therefore, the value $\mathfrak{t}(\lambda^t)$ of $\mathfrak{t}$ is determined by the restriction of λ^t to $(0, \infty)$. Thus, in a discussion of the functional $\mathfrak{t}$, we need not distinguish between a relative shearing history and its restriction to $(0, \infty)$.

For each (finite) $p \geq 1$ and each influence function k, *i.e.*, each positive, monotone-decreasing, measurable function k with $s^p k(s)$ integrable over $(0, \infty)$, let $\mathcal{L}_p(k)$, with norm $\|\cdot\|$, be the Banach space formed in the usual way from real-valued functions g on $(0, \infty)$ for which

$$\|g\|^p = \int_0^\infty g(s)^p k(s)\, ds \tag{3.3}$$

is finite.

In the present context, the postulate of fading memory asserts that there is a $p \geq 1$ and an influence function k such that $\mathfrak{t}$ has $\mathcal{L}_p(k)$ for its domain and is twice continuously differentiable in the following sense: For each g in $\mathcal{L}_p(k)$ there is, on $\mathcal{L}_p(k)$, a bounded linear form $\delta\mathfrak{t}(g|\cdot)$ and a bounded, symmetric, bilinear form $\delta^2\mathfrak{t}(g|\cdot,\cdot)$ such that

$$\mathfrak{t}(g + l) = \mathfrak{t}(g) + \delta\mathfrak{t}(g|l) + \frac{1}{2}\delta^2\mathfrak{t}(g|l, l) + o(\|l\|^2). \tag{3.4}$$

It is assumed that each of the functionals $\delta\mathfrak{t}(\cdot\,|\,\cdot)$ and $\delta^2\mathfrak{t}(\cdot\,|\,\cdot,\cdot)$ is jointly continuous in all its arguments.

Since $\delta\mathfrak{t}(g|\cdot)$ is both linear and continuous on $\mathcal{L}_p(k)$, there is a function K in $\mathcal{L}_q(k)$, with $q^{-1} + p^{-1} = 1$, such that for each l in $\mathcal{L}_p(k)$

$$\delta\mathfrak{t}(g|l) = \int_0^\infty K(s) l(s) k(s)\, ds. \tag{3.5}$$

Of course, K depends on g. One puts

$$G'(s; g) = K(s)k(s), \tag{3.6}$$

so that

$$\delta\mathfrak{t}(g|l) = \int_0^\infty G'(s; g) l(s)\, ds. \tag{3.7}$$

It is assumed that the mapping $G'(s;\cdot)$ is, for each s, a continuous functional on $\mathcal{L}_p(k)$ and that, for each g, $G'(\cdot\,; g)$ has a bounded derivative $G''(\cdot\,; g)$. It is further assumed that the mapping $g \mapsto G''(\cdot\,; g)k(\cdot)^{-1}$ is continuous from $\mathcal{L}_p(k)$ into $\mathcal{L}_q(k)$.

If we put

$$G(s; g) = -\int_s^\infty G'(\sigma; g)\, d\sigma = -\int_s^\infty K(\sigma; g)k(\sigma)\, d\sigma, \quad (0 < s < \infty), \tag{3.8}$$

then for each g,

$$\lim_{s \to \infty} G(s; g) = 0. \tag{3.9}$$

The function $G(\cdot\,;g)$ is called the *relaxation modulus* (for perturbations about the history g). As we shall see below, $G(0;g)$ is an "instantaneous modulus".

Let $\mathcal{R}$ be a configuration that occurs at some time, say $t = 0$, during the rectilinear motion of the fluid, and let $\gamma(x,t)$ and $u(x,t)$ be the shear strain and displacement at x at time t computed using $\mathcal{R}$ as the reference configuration:

$$\gamma(x,t) = \partial_x u(x,t), \qquad u(x,t) = \int_0^t v(x,\xi)\,d\xi. \tag{3.10}$$

By (2.2), the relative shear strain at time $t-s$ (taken relative to the configuration at time t) is

$$\lambda^t(s) = \gamma(x,t-s) - \gamma(x,t). \tag{3.11}$$

Suppose now that at a particular value of x the shear strain γ suffers a jump θ at time t, so that

$$\gamma(x,t^+) = \gamma(x,t^-) + \theta. \tag{3.12}$$

Then, the histories of the relative shear strain immediately before and immediately after the jump are given, for each $s > 0$, by

$$\lambda^{t^-}(s) = \gamma(x,t-s) - \gamma(x,t^-), \tag{3.13a}$$

$$\lambda^{t^+}(s) = \gamma(x,t-s) - \gamma(x,t^+) = \gamma(x,t-s) - \gamma(x,t^-) - \theta = \lambda^{t^-}(s) - \theta; \tag{3.13b}$$

i.e.,

$$\lambda^{t^+} = \lambda^{t^-} - \theta 1^\dagger \tag{3.14}$$

with $1^\dagger$ the constant function on $(0,\infty)$ with value 1. For the shear-stresses immediately before and after the jump in strain, we have

$$T^{xy}(t^-) = \mathfrak{t}(\lambda^{t^-}), \tag{3.15a}$$

$$T^{xy}(t^+) = \mathfrak{t}(\lambda^{t^+}) = \mathfrak{t}(\lambda^{t^-} - \theta 1^\dagger). \tag{3.15b}$$

Thus the jump in stress can be expressed as a function of the jump θ in strain and this function depends on the history λ^{t^-} as a parameter:

$$\begin{aligned} [\![T^{xy}(t)]\!] &:= T^{xy}(t^+) - T^{xy}(t^-) \\ &= \mathfrak{f}(\theta;\lambda^{t^-}) := \mathfrak{t}(\lambda^{t^-} - \theta 1^\dagger) - \mathfrak{t}(\lambda^{t^-}). \end{aligned} \tag{3.16}$$

The functions $\theta \mapsto \mathfrak{f}(\theta;\lambda^{t^-})$ and $\theta \mapsto \mathfrak{t}(\lambda^{t^-} - \theta 1^\dagger)$ are called *instantaneous response functions* and characterize the instantaneous elasticity of the material. For each relative history λ^t, the *instantaneous modulus* E_1 and *instantaneous second-order modulus* E_2 are defined by

(3.17a) $$E_1 = E_1(\lambda^t) = \frac{\partial}{\partial\theta}\mathfrak{f}(\theta;\lambda^t)\bigg|_{\theta=0},$$

(3.17b) $$E_2 = E_2(\lambda^t) = \frac{\partial^2}{\partial\theta^2}\mathfrak{f}(\theta;\lambda^t)\bigg|_{\theta=0}.$$

In view of (3.4) and (3.16), we have

(3.18a) $$E_1(\lambda^t) = \delta\mathfrak{t}(\lambda^{t^-}|1^\dagger),$$

(3.18b) $$E_2(\lambda^t) = \delta^2\mathfrak{t}(\lambda^{t^-}|1^\dagger, 1^\dagger),$$

and (3.7) and (3.8) yield

(3.19) $$E_1(\lambda^t) = G(0;\lambda^t).$$

It follows from (2.4) that E_1 is an even function of λ^t and E_2 an odd function:

(3.20) $$E_1(-\lambda^t) = E_1(\lambda^t), \qquad E_2(-\lambda^t) = -E_2(\lambda^t).$$

When g in (3.4) has the special form $g(s) = -\kappa s$, *i.e.*, when $g = -\kappa\iota = \lambda^t$, one writes simply $G(s;\kappa)$ for $G(s;g)$, $G'(s;\kappa)$ for $G'(s;g)$, and $E_i(\kappa)$ for $E_i(\lambda^t)$ in the above equations, *e.g.*,

(3.21a) $$E_1(\kappa) = \delta\mathfrak{t}(-\kappa\iota|\,1^\dagger) = E_1(-\kappa),$$

(3.21b) $$E_2(\kappa) = \delta^2\mathfrak{t}(-\kappa\iota|\,1^\dagger, 1^\dagger) = -E_2(-\kappa).$$

As E_2 is an odd function,

(3.22) $$E_2(0) = 0;$$

i.e., the instantaneous second-order modulus E_2 is zero for a fluid that has never been sheared. Let us here assume

(3.23) $$\frac{d}{d\kappa}E_2(\kappa)\bigg|_{\kappa=0} \neq 0.$$

Thermodynamics requires that the instantaneous first-order modulus E_1 be not negative for a fluid that has not been sheared, *i.e.*, $E_1(0) \geq 0$. For simplicity, let us assume that

(3.24) $$E_1(\lambda^t) > 0$$

for all λ^t in $\mathcal{L}_p(k)$.

4. Rectilinear Shear-Acceleration Waves. In our present subject the word *wave* refers to a propagating singular surface. A rectilinear shear wave is, at each instant, a planar surface perpendicular to the x-axis of the fixed Cartesian system in which (3.1) holds. If we write x_t for the value of x on this surface at time t, the velocity of the wave is

$$c = c(t) = \frac{d}{dt}x_t. \tag{4.1}$$

It is usual to assume that $v(x,t)$ and all its derivatives $\partial_t^i \partial_x^j v(x,t)$ are continuous functions of (x,t) for $(x,t) \neq (x_t,t)$ and that these quantities experience, at worst, jump discontinuities, $[\![\partial_t^i \partial_x^j v]\!]$, across the surface $x = x_t$.

The rectilinear shear waves for which $[\![v]\!] = 0$, but $[\![\partial_t v]\!] \neq 0$ and $[\![\partial_x v]\!] \neq 0$ are called *shear-acceleration waves.*

The velocity $c(t)$ of a shear-acceleration wave propagating in a fluid obeying the postulate of smoothness laid down in the previous section obeys the formula [1,21],

$$c(t)^2 = E_1(\lambda^t)/\rho, \tag{4.2}$$

in which $E_1(\cdot)$ is the functional in (3.17a), (3.18a), and (3.19), and λ^t is the history up to t (at x_t) of the relative shear strain; *i.e.*, $E_1(\lambda^t)$ is the instantaneous modulus at the wave. (The $\mathcal{L}_p(k)$-valued function $(x,t) \mapsto \lambda^t$ is continuous across such a wave.)[7]

The goal of the research reported in [1] was to derive an exact formula for the amplitude

$$a = a(t) = [\![\partial_t v]\!] \tag{4.3}$$

of a shear-acceleration wave assuming that the wave is propagating into a region undergoing a steady rectilinear shearing flow. Thus, taking $c(t)$ to be positive, Coleman and Gurtin supposed that for $x \geq x_t$, $t \geq 0$, and $-\infty < \sigma \leq t$

$$v^x(x,\sigma) = 0, \qquad v^y(x,\sigma) = v(x), \qquad v(x,\sigma) = 0, \tag{4.4}$$

and hence ahead of the wave, for all $\sigma \leq t$, the history λ^σ is given by $\lambda^\sigma(s) = -\kappa s$ with $\kappa = v'(x)$, and at the wave,

$$\lambda^t(s) = -\kappa_t s \qquad (0 \leq s < \infty) \tag{4.5}$$

with

$$\kappa_t = v'(x_t). \tag{4.6}$$

In such a case, (4.2) becomes, in the notation of equation (3.21a),

$$c(t)^2 = E_1(\kappa_t)/\rho. \tag{4.7}$$

---

[7] See [22], pp. 253, 254.

It was shown that the quantity,

$$b = b(t) = a(t)c(t)^{1/2}, \tag{4.8}$$

obeys a differential equation of Bernoulli type,

$$\frac{db}{dt} + \mu(t)b + \varphi(t)b^2 = 0, \tag{4.9}$$

with

$$\mu(t) = -\frac{G'(0;\kappa_t)}{2E_1(\kappa_t)}, \qquad \varphi(t) = \frac{E_2(\kappa_t)}{2E_1(\kappa_t)c(t)^{3/2}}. \tag{4.10}$$

Solution of this equation gave the following theorem [1, p. 178]: *The amplitude $a(t)$ of a shear-acceleration wave which since time $t = 0$ has been advancing into a region undergoing a steady rectilinear shearing flow* (4.4) *is given by the explicit formula*

$$a(t)c(t)^{1/2} = \frac{a(0)c(0)^{1/2}\exp(-\Psi(t))}{1 + a(0)c(0)^{1/2}\int_0^t \varphi(\sigma)\exp(-\Psi(\sigma))d\sigma}, \qquad \Psi(t) = \int_0^t \mu(\sigma)\,d\sigma \tag{4.11}$$

with μ and φ as in (4.10).

In the same paper it was observed that methods developed by Coleman, Greenberg, and Gurtin [23, §3] can be employed to show that if the motion of the fluid ahead of the wave is a rectilinear shearing flow that is *not* steady, *i.e.*, if v^y in (4.4) depends on time, then (4.11) still holds, but μ and φ are not given by (4.10).

The implications of equation (4.11) are particularly transparent in the case in which the motion ahead of the wave is a simple shearing flow so that (4.4) becomes

$$v^x(x,\sigma) = 0, \qquad v^y(x,\sigma) = V + \kappa_0 x, \qquad v^z(x,\sigma) = 0, \tag{4.12}$$

for all $x \geq x_t$, $t \geq 0$, and $-\infty < \sigma \leq t$. Here κ_0 is independent of x, and $\kappa_t = \kappa_0$ for $t \geq 0$. We then have

$$c = c(\kappa_0) = (E_1(\kappa_0)/\rho)^{1/2} = \text{const.}, \tag{4.13}$$

and (4.11) reduces to

$$a(t) = \frac{\Lambda}{\left(\frac{\Lambda}{a(0)} - 1\right)e^{\mu t} + 1} \tag{4.14a}$$

with

$$\mu = \mu(\kappa_0) = \frac{-G'(0;\kappa_0)}{2E_1(\kappa_0)} = \text{const.}, \tag{4.14b}$$

$$\Lambda = \Lambda(\kappa_0) = \frac{c(\kappa_0)G'(0;\kappa_0)}{E_2(\kappa_0)} = \text{const.}; \tag{4.14c}$$

$E_1(\kappa_0)$ is the initial value, and $G'(0;\kappa_0)$ the initial slope, of the stress relaxation function for shearing perturbations about κ_0. By (3.24), $E_1(\kappa_0; 0) > 0$; let us also assume $G'(0;\kappa_0) < 0$, so that $\mu(\kappa_0) > 0$.

When $E_2(\kappa_0) = 0$, as is the case when $\kappa_0 = 0$, (4.14a) reduces to

$$a(t) = a(0)e^{-\mu t}. \tag{4.15}$$

In particular, the amplitude of a shear-acceleration wave propagating into a region that has not been sheared previously decays to zero exponentially.

In view of (3.23), there will be a range of values of the rate of shearing for which

$$\kappa_0 \neq 0 \implies E_2(\kappa_0) \neq 0. \tag{4.16}$$

When $E_2(\kappa_0) \neq 0$, the number $|\Lambda(\kappa_0)|$ is finite and plays the role of a critical amplitude. Because we here take c to be positive and have $\mu(\kappa_0) > 0$, if $|a(0)| < |\Lambda(\kappa_0)|$ or if $a(0)E_2(\kappa_0) > 0$, then $a(t) \to 0$ monotonically as $t \to \infty$. If $a(0) = \Lambda(\kappa_0)$, then $a(t) = a(0)$. But, *if both* $|a(0)| > |\Lambda(\kappa_0)|$ *and* $a(0)E_2(\kappa_0) < 0$, *then* $|a(t)| \to \infty$ *monotonically and in a finite time* t_∞, *given by*

$$\begin{aligned} t_\infty &= \frac{-1}{\mu(\kappa_0)} \ln\left(1 - \frac{\Lambda(\kappa_0)}{a(0)}\right) \\ &= \frac{2G(0;\kappa_0)}{G'(0;\kappa_0)} \ln\left(1 - \frac{c(\kappa_0)G'(0;\kappa_0)}{E_2(\kappa_0)a(0)}\right). \end{aligned} \tag{4.17}$$

Thus, for a shear-acceleration wave propagating into a region undergoing the simple shearing motion (4.4), we have the following result: Although it is impossible for such a wave to grow in amplitude when $\kappa_0 = 0$, the wave can achieve infinite amplitude in finite time if $\kappa_0 \neq 0$, provided $[\![\partial_t v]\!]$ (or $-c[\![\partial_x v]\!]$) is of proper sign and exceeds in magnitude the critical amplitude $|\Lambda|$. One expects that for κ_0 near to zero, this critical amplitude, which is infinite for $\kappa_0 = 0$, will decrease as the magnitude of κ_0 is increased.

If the region ahead of the wave is undergoing a steady rectilinear shearing flow for which $v'(x)$ does not reduce to a constant κ_0 independent of x, the formula (4.11) does not reduce to (4.14) but can be analyzed[8] if $E_1(\kappa)$, $E_2(\kappa)$, and $G'(0;\kappa)$ are known as functions of κ and $\kappa = v'$ is specified as a function of x as it is in channel flow for which (2.21) and (2.22) hold and $\kappa = -\tau^{-1}(\alpha x)$. In channel flow, $|\kappa|$ is a maximum at the bounding surfaces $x = \pm\frac{1}{2}d$. From this and the observations made above for simple shearing flow, one expects that a shear-acceleration wave propagating into a fluid undergoing steady channel flow is more likely to attain infinite amplitude when it is near rather than far from the bounding surfaces. It is conjectured, but a proof is lacking, that the approach of $|[\![\partial_t v]\!]|$ to infinity as $t \to t_\infty$ signifies the appearance of a jump discontinuity in v, *i.e.*, the appearance of a shear-shock wave.

[8] As in [23, §3].

5. Formation of Singularities in Initially Smooth Motions. The results on the amplitude of acceleration waves suggest that when the functional $\mathfrak{t}$ obeys the postulate of fading memory adopted here, the solutions of equation (3.1) corresponding to smooth initial data with $|\partial_x v|$ or $|\partial_t v|$ large will be such that these quantities become infinite at some value of x in finite time. Before discussing results of this type, let us examine some special cases of fluids with fading memory.

Among the fluids obeying the postulate of fading memory as stated in Section 3 are those for which the functional $\mathfrak{t}$ has the form

$$\mathfrak{t}(\lambda^t) = \sum_{n \text{ odd}} \int_0^\infty \cdots \int_0^\infty K_n(s_1, \ldots, s_n)\lambda^t(s_1)\ldots\lambda^t(s_n)\, ds_1 \ldots ds_n \tag{5.1}$$

with each function K_n decaying rapidly to zero for large values of its arguments s_i, $i = 1, \ldots, n$, and with these functions tending to zero in an appropriate sense as $n \to \infty$.

The postulate is obeyed also by BKZ fluids [24], for which

$$\mathfrak{t}(\lambda^t) = \int_0^\infty H(\lambda^t(s), s)\, ds = \int_{-\infty}^t H\left(\gamma(\tau) - \gamma(t), t - \tau\right)\, d\tau, \tag{5.2}$$

provided H is sufficiently smooth and $|H(a, b)|$ decays sufficiently rapidly with increasing b at fixed a and does not grow too rapidly with increasing $|a|$ at fixed b.[9] Also obeying the postulate are the extensions of the BKZ theory proposed recently by Coleman and Zapas [25].

It should be observed, however, that constitutive relations of the form,

$$T(t) = \Psi(\gamma(t)) + \int_{-\infty}^t K(t - \sigma)\Phi\left(\gamma(\sigma)\right)\, d\sigma, \tag{5.3}$$

with γ a component of strain and T a component of stress, do not fall as special cases of the equation $(2.3)_1$, *i.e.*,

$$T^{xy}(t) = \mathfrak{t}(\lambda^t), \tag{5.4}$$

for the shear stress in a simple fluid. The large and growing literature on the qualitative theory of the dynamical equations of viscoelastic solids obeying relations of the form (5.3) is interesting in its own right and has on occasion suggested research applicable to fluids.[10]

Slemrod [28] was the first to obtain a theorem showing the non-existence of global smooth solutions to the dynamical equation (2.8) for rectilinear shearing flows of simple fluids under the postulate of fading memory. He considered the (albeit unlikely) special case of (5.1) in which

$$K_n(s_1, \ldots, s_n) = -\beta^n h_n e^{-\beta(s_1 + \cdots + s_n)}, \qquad (n = 1, 3, 5, \ldots) \tag{5.5}$$

[9] For BKZ fluids, the existence of $\delta^2\mathfrak{t}$ in (3.4) requires that p in (3.3) be ≥ 2.
[10] For surveys see [26] and [27].

and hence (5.4) reduces to

$$T^{xy}(x,t) = h\left(\int_0^\infty e^{-\beta s}\partial_x v(x,t-s)\,ds\right) \tag{5.6}$$

with

$$h(\xi) = \sum_{n \text{ odd}} h_n \xi^n, \qquad h(-\xi) = -h(\xi), \tag{5.7}$$

and he studied finite perturbations $\hat{v}$ of steady simple shearing flow. Thus he considered velocity fields v of the form,

$$v(x,t) = Vx/d + \hat{v}(x,t), \tag{5.8}$$

for which (5.6) becomes

$$T^{xy}(x,t) = h\left(\int_0^\infty e^{-\beta s}\partial_x \hat{v}(x,t-s)\,ds + \frac{V}{\beta d}\right). \tag{5.9}$$

It was assumed that $\hat{v}(x,t)$ is given as a smooth function for $-\infty < t \le 0$ and $0 \le x < d$. Slemrod's analysis was based on the clever observation that after the change of variables

$$\begin{cases} u(x,t) = \hat{v}(x,t) - \beta\displaystyle\int_0^\infty e^{-\beta s}\hat{v}(x,t-s)\,ds, \\ w(x,t) = \partial_x v(x,t), \end{cases} \tag{5.10}$$

the functional-differential equation (3.1) with $\alpha = 0$ becomes

$$\begin{cases} \partial_t w = \partial_x u \\ \partial_t u = \partial_x \hat{h}(w) - \beta u \end{cases} \tag{5.11}$$

with

$$\hat{h}(\zeta) = \frac{1}{\rho}\left[h\left(\zeta + \frac{V}{\beta d}\right) - h\left(\frac{V}{\beta d}\right)\right]. \tag{5.12}$$

After subjecting (5.11) to the prescribed data,

$$\begin{cases} u(d,t) = u(0,t) = 0, \\ u(x,0) = u_0(x), \\ w(x,0) = w_0(x), \end{cases} \tag{5.13}$$

and studying the evolution of the Riemann invariants along the characteristic curves of (5.11), he obtained results which may be summarized as follows: *If* $h'(0) > 0$, $h''(0) \neq 0$, *if* $\max_x |u_0(x)|$ and $\max_x |w_0(x)|$ are sufficiently small, *if* $|\partial_x u_0(x)|$ or $|\partial_x w_0(x)|$ are sufficiently large, and *if* appropriate sign conditions, depending on the sign of $\sigma''(0)$, hold for $\partial_x u_0$ and $\partial_x w_0$ where $|\partial_x u_0|$ and $|\partial_x w_0|$ attain their maxima, *then* (5.11), (5.13) has a classical solution with u and w in $C^1([0,d])$ for only a finite time; *i.e.*,

$$\max_x |\partial_x v(x,t)| + \max_x |\partial_x w(x,t)| \to \infty \tag{5.14}$$

in finite time, which yields a number t_∞ for which

$$\lim_{t \to t_\infty} \max_x [|\partial_x \hat{v}(x,t)| + |\partial_t \hat{v}(x,t)|] = \infty. \tag{5.15}$$

This result was extended to channel flows by Hattori [29] and further extended by Gripenberg [30]. (Related results for a fluid of the rate type are given in the recent monograph of Renardy, Hrusa, and Nohel.[11])

Many results on "blow up" of smooth solutions were subsequently obtained using constitutive equations appropriate for solids. In particular there is the numerical study of Markowich and Renardy [31], and the recent analyses of Dafermos [32], Ramaha [33], and Nohel and Renardy [34], all of which are described in Chapter II, Section 4 of the monograph of Renardy, Hrusa, and Nohel [26].

The method published by Dafermos [32] in 1986 is applicable to Cauchy problems for many types of viscoelastic fluids, including BKZ fluids, and, more generally, fluids obeying the postulate of fading memory. Without giving Dafermos' treatment the full description it deserves, this short discussion must now conclude with the observation that, when applied to rectilinear shearing flows of fluids with fading memory, the method developed by Dafermos permits one to show that *if*, for a perturbation $\hat{v}$ of a flow v in which E_1 and E_2 are both positive, one has (i) an appropriate norm of a relative-strain history (constructed from $\hat{v}$) everywhere small initially, (ii) $\sup_x(-\partial_x \hat{v})$ not too large initially, and (iii) $\sup_x(\partial_x \hat{v})$ large enough initially, *then* a classical solution does not exist for all $t > 0$, and hence $\sup_x |\partial_x \hat{v}| \to \infty$ in finite time.

Here, as in the theory of acceleration waves, one expects, but cannot as yet prove,[12] that the blow up of $|\partial_x v|$ or $|\partial_t v|$ in finite time implies the formation of a shear-shock.

Acknowledgments. I am grateful to William J. Hrusa and Daniel C. Newman for help and suggestions. While this article was in preparation my research was supported by the National Science Foundation through Grant DMS-88-15924 to Rutgers University.

---

[11] [26, pp. 67–69].

[12] Although it is strongly suggested by numerical studies. See, in particular, the study of a viscoelastic solid [obeying a special case of (5.3)] performed by Markowich and Renardy [31]. The participants in this workshop will find of particular interest Plohr's recent numerical study [35] of channel flows of a Johnson-Segalman fluid that does not obey several of the hypotheses made here, such as monotonicity of the shear-stress function τ and positivity of the modulus E_1, and gives rise to dynamical equations that change type.

REFERENCES

[1] B.D. Coleman and M.E. Gurtin, *J. Fluid. Mech.*, 33 (1968), pp. 165–181.
[2] H.K. Nason, *J. Appl. Phys.*, 16 (1945), pp. 338–343.
[3] R.S. Spencer and R.E. Dillon, *J. Colloid Sci.*, 4 (1949), pp. 241–255.
[4] J.P. Tordella, *J. Appl. Phys.*, 27 (1956), pp. 454–458.
[5] J.P. Tordella, *Trans. Soc. Rheology*, 1 (1957), pp. 203–212.
[6] J.P. Tordella, *J. Appl. Polymer Sci.*, 7 (1963), pp. 215–229.
[7] J.P. Tordella, in *Rheology, Vol. 5*, F.R. Eirich, ed., pp. 57–92, Academic Press, New York, 1969.
[8] E.B. Bagley, *J. Appl. Phys.*, 28 (1957), pp. 624–627.
[9] E.B. Bagley, *J. Appl. Polymer Sci.*, 7 (1963), pp. S7–S8.
[10] J.J. Benbow, R.V. Charley, and P. Lamb, *Nature, London*, 192 (1961), pp. 223–224.
[11] G.V. Vinogradov, A. Ya. Malkin, Yu. G. Yanooskii, E.K. Borisenkova, B.V. Yarlykov, and G.V. Berezhnaya, *J. Poly. Sci. A2*, 10 (1972), pp. 1061–1084.
[12] E.B. Bagley, I.M. Cabott, and D.C. West, *J. Appl. Phys.*, 29 (1958), pp. 109–110.
[13] W. Noll, *Arch. Rational Mech. Anal.*, 2 (1958/9), pp. 197–226.
[14] B.D. Coleman and W. Noll, *Ann. N.Y. Acad. Sci.*, 89 (1961), pp. 672–714.
[15] B.D. Coleman and W. Noll, *Arch. Rational Mech. Anal.*, 3 (1959), pp. 289–303.
[16] B.D. Coleman, *Arch. Rational Mech. Anal.*, 9 (1962), pp. 273–300.
[17] B.D. Coleman, H. Markowitz, and W. Noll, *Viscometric Flows of Non-Newtonian Fluids*, Springer, Berlin, Heidelberg, New York, 1966.
[18] B.D. Coleman and W. Noll, *Arch. Rational Mech. Anal.*, 6 (1960), pp. 355–370.
[19] B.D. Coleman and W. Noll, *Rev. Mod. Phys.*, 33 (1961), pp. 239–249; *Errata, ibid.*, 36 (1964), p. 239.
[20] B.D. Coleman and V.J. Mizel, *Arch. Rational Mech. Anal.*, 23 (1966), pp. 87–123.
[21] B.D. Coleman, M.E. Gurtin, and I. Herrera R., *Arch. Rational Mech. Anal.*, 19 (1965), pp. 1–19.
[22] B.D. Coleman and M.E. Gurtin, *Arch. Rational Mech. Anal.*, 19 (1965), pp. 239–265.
[23] B.D. Coleman, J.M. Greenberg, and M.E. Gurtin, *Arch. Rational Mech. Anal.*, 22 (1966), pp. 333–354.
[24] B. Bernstein, E.A. Kearseley, and L.J. Zapas, *Trans. Soc. Rheology*, 7 (1963), pp. 391–410.
[25] B.D. Coleman and L.J. Zapas, *J. Rheology*, 33 (1989), pp. 501–516.
[26] M. Renardy, W.J. Hrusa, and J.A. Nohel, *Mathematical Problems in Viscoelasticity*, Longman, Harlow, Essex, and Wiley, New York, 1987.
[27] W.J. Hrusa, J.A. Nohel, and M. Renardy, *Appl. Mech. Rev.*, 41 (1988), pp. 371–378.
[28] M. Slemrod, *Arch. Rational Mech. Anal.*, 68 (1978), pp. 211–225.
[29] H. Hattori, *Quart. Appl. Math*, 40 (1982), pp. 112–127.
[30] G. Gripenberg, *SIAM J. Math.*, 13 (1982), pp. 954–961.
[31] P. Markowich and M. Renardy, *SIAM J. Numer. Anal.*, 21 (1984), pp. 24–51; *Corrigenda, ibid.*, 22 (1985), p. 204.
[32] C.M. Dafermos, *Arch. Rational Mech. Anal.*, 91 (1986), pp. 365–377.
[33] M. Rammaha, *Commun. in Partial Differential Equations*, 12 (1987), pp. 243–262.
[34] J.A. Nohel and M. Renardy, in *Amorphous Polymers and non-Newtonian Fluids*, C. Dafermos, J.L. Ericksen, and D. Kinderlehrer, eds.; *IMA Volumes in Mathematics and its Applications*, Vol. 6, Springer, New York, *etc.*, 1987, pp. 139–152.
[35] B.J. Plohr, *Instabilities in Shear Flow of Viscoelastic Fluids with Fading Memory*, CMS Report 89–13, U. Wisconsin, Madison (1989).

COMPOSITE TYPE, CHANGE OF TYPE, AND DEGENERACY IN FIRST ORDER SYSTEMS WITH APPLICATIONS TO VISCOELASTIC FLOWS

L. PAMELA COOK* AND G. SCHLEINIGER*† AND R. J. WEINACHT*

Abstract. In this paper we discuss some features of systems of partial differential equations of first order related to change of type, to composite type and to degeneracy. We are interested in these effects with respect to viscoelastic fluid flow and hence we focus on the properties of two particular models of differential type, the Upper Convected Maxwell model and the Bird-DeAguiar model. Friedrichs' theory of symmetric positive operators is discussed as a means for treating these indefinite type systems and its use is illustrated for two simple systems, one of composite type and one of degenerate elliptic type.

Key words. viscoelastic flow, steady flow, composite systems, mixed-type systems, degenerate systems

AMS(MOS) subject classifications. 35M05, 76A10

1. Introduction. In this paper we discuss some features of systems of partial differential equations of first order related to change of type, to composite type and to degeneracy. We are particularly interested in these effects with respect to the system of equations describing viscoelastic fluid flow with constitutive equations of differential type [14]. We focus on two models, namely the Upper Convected Maxwell model (UCMM) and the Bird-DeAguiar model. The first is actually contained within the second.

It is well known that the equations governing the steady flow of an upper convected Maxwell fluid can be of changing type, Joseph et al [7]. Moreover, in addition to the possible complication of changing type, the equations are always of composite type, i.e. there are both real and complex characteristics. In the case of two-dimensional flow the equations of conservation of mass and momentum together with the constitutive equations lead to a six-by-six system of coupled first order partial differential equations. If one looks at the eigenvalues of that system one finds that two are real (in fact there is one repeated eigenvalue associated with the streamlines), two are pure imaginary, and the other two can be complex or real depending on the speed of the flow (Mach number). Thus the system always has two real characteristics and may have four within the flow region. Thus the label that the system is of composite and changing type. More recently two new phenomena have been observed: first, the system of equations governing the Bird-DeAguiar model can change type yet again to become fully hyperbolic and second, the linearized UCMM is actually degenerate at boundaries where the velocity components are zero. These two phenomena will be discussed in this paper. Both are of special importance in discussions of the well-posedness of the boundary value problems

*Department of Mathematical Sciences, University of Delaware, Newark, DE 19716. This work was supported by the National Science Foundation under grant # DMS-8714152.

†Work supported by ARO grant # DAAL03-88-K-0132

describing these flows. Both are also of importance in the numerical simulation of these flows.

Although some work has been carried out on reducing the original system for the UCMM model to one of coupled but higher order equations in terms of the vorticity and the streamfunction, we prefer in this work to retain the primitive variables and the first order system form. The advantage of this is that more complicated systems, e.g. the Bird-DeAguiar system, can hopefully be treated more simply and in a straightforward manner.

First we review some recent analytical results on the Bird-DeAguiar model. These have been reported in part in Calderer, Cook and Schleiniger [2] and in Schleiniger, Calderer and Cook [16]. Then we discuss a method of attack to deal with the questions of appropriate boundary conditions and well-posedness of systems which may be degenerate, of composite type and of changing type. In particular we discuss the application of Friedrichs' theory for symmetric positive linear operators [4] and we apply it to several prototype problems.

2. The Bird-DeAguiar model. The constitutive equation obtained by Bird and DeAguiar was derived from molecular theory using a phase-space kinetic theory for concentrated solutions and melts [1],[3]. It models the polymer molecules as finitely extensible nonlinear elastic dumbbells in which both the Brownian motion and the hydrodynamic forces are made anisotropic. Thus there are three parameters in the model that can be varied to cover the range from isotropic linear forces (a Maxwell model) to various degrees of nonlinearity in the spring (as measured by b), of anisotropy in the Stokes law (measured by the parameter s_0) and in the Brownian motion (measured by the parameter β). Other models which show aspects of these anisotropic molecular effects are integral models, namely the Doi-Edwards model and the Curtiss-Bird model. None allow for smooth variation in both Brownian motion and hydrodynamic drag.

We consider the model in the absence of a solvent so that only the polymer contribution is considered, analogous to the UCMM. The constitutive equation, in dimensionless form, is then

$$\left\{\frac{Z(s_0\beta+J)}{Y+J} - \mathrm{We}\frac{D}{Dt}\ln Z\right\}\mathbf{T} + \mathrm{We}\mathbf{T}_{(1)} = \frac{\mathrm{We}Y}{3}\dot{\boldsymbol{\gamma}} + \left\{\frac{(s_0\beta - Y)ZJ}{3(Y+J)} + \frac{\mathrm{We}Y}{3}\frac{D}{Dt}\ln Z\right\}\boldsymbol{\delta}. \tag{2.1}$$

Here $\dot{\boldsymbol{\gamma}} = \nabla\mathbf{u} + (\nabla\mathbf{u})^{\mathsf{T}}$ is (two times) the rate of deformation tensor, $\boldsymbol{\delta}$ is the identity tensor and subscript $_{(1)}$ means the upper convected derivative. Details of the nondimensionalization appear in [16]. In particular the dimensionless parameters are the Weissenberg number $\mathrm{We} = \lambda U/L$ and the Reynolds number $\mathrm{Re} = \rho U^2/(3n\kappa T_m)$ where U is a typical flow velocity, L is a typical length scale, ρ is the density of the fluid, κ is Boltzmann's constant, n is the number density of dumbbells, λ is a time constant, and T_m is the absolute mean temperature. The time and stress scales are L/U and $3n\kappa T_m$ respectively. Also we have that

$$W = \frac{\alpha+2\beta}{3}, \qquad Y = \frac{4\beta-\alpha}{3}, \qquad Z = \left\{1 + \frac{3(W+J)}{b}\right\}\frac{(Y+J)}{(W+J)}$$

and

$$J = \text{trace}(\mathbf{T}).$$

Note that we have changed notation slightly from the Bird-DeAguiar paper in that our $\mathbf{T}$ is their $-\mathbf{T}$ so that we can write the total stress $\boldsymbol{\pi}$ as

$$\boldsymbol{\pi} = -p\boldsymbol{\delta} + \mathbf{T}$$

where p is the pressure.

Note that if $\alpha = \beta = s_0 = 1$ then all forces are isotropic and the Tanner [17] constitutive model is obtained. If $\alpha = \beta = s_0 = 1$ and $b \to \infty$ (a Hookean spring) then the upper-convected Maxwell model is obtained

$$\mathbf{T} + \lambda \mathbf{T}_{(1)} = \eta \dot{\boldsymbol{\gamma}}. \tag{2.2}$$

The system is completed by adding the equations of conservation of mass and conservation of linear momentum:

$$\nabla \cdot \mathbf{u} = 0 \tag{2.3.a}$$

$$\text{Re}\{\mathbf{u}_t + \mathbf{u} \cdot \nabla \mathbf{u}\} = -\nabla p + \nabla \cdot \mathbf{T}. \tag{2.3.b}$$

Several base states for steady flow of the models have been analyzed corresponding to two-dimensional flows: 1) shear flow: $u_0 = y$, $v_0 = 0$; 2) extensional flow: $u_0 = x$, $v_0 = -y$; and 3) Poiseuille (or channel) flow: $u_0 = u(y)$, $v_0 = 0$. In addition axisymmetric pipe flow: $u_0 = u(r)$, $v_0 = 0$ was analyzed. In the first two cases base stresses which were constant were considered, while in the last two cases the stresses had some spatial dependence. In all cases one and only one base state was found which was evolutionary. Other base states were found but these suffered Hadamard instabilities.

Of particular interest was the result of linearization of the equations for two-dimensional perturbations about the base states mentioned above. The resulting system of equations in all four cases is a seven-by-seven first order system of partial differential equations of the form

$$AU_x + BU_y + CU = 0 \tag{2.4}$$

where $U = (u, v, p, \tau_{11}, \tau_{12}, \tau_{22}, \tau_{33})^{\mathsf{T}}$. This system was analyzed for its eigenvalues α defined by $\det(\alpha A + B) = 0$ and its corresponding characteristics, $dy/dx = \alpha$. What occurred in all cases was that not only did we find that the model in general was of composite and possibly of mixed type as the UCMM model, but also that it could be fully hyperbolic depending on the flow parameters. This is the first time that a region where the steady system is fully hyperbolic has been observed for viscoelastic flows [2],[16] (In a later paper, by Verdier and Joseph [19], similar behavior was shown to occur for the White-Metzner model.). In particular the system of equations, 7×7 in this case, has three real eigenvalues (associated with the streamlines) and the other four eigenvalues can switch from four complex to two complex and two real, to all real. Moreover there are, in the latter case, seven linearly independent eigenvectors, so that the system is in fact hyperbolic. This behavior is illustrated in the figures below. Note that this behavior occurs even for the simplified case of the Bird-DeAguiar model: $\alpha = \beta = s_0 = 1$ which is the Tanner model. Note also that in the case of shear flow the coordinates x, y can be scaled by the Reynolds number Re so that the boundary of the flow, $y = 1$, can occur at any horizontal level in Fig. 1, thus the flow can become fully hyperbolic at the wall.

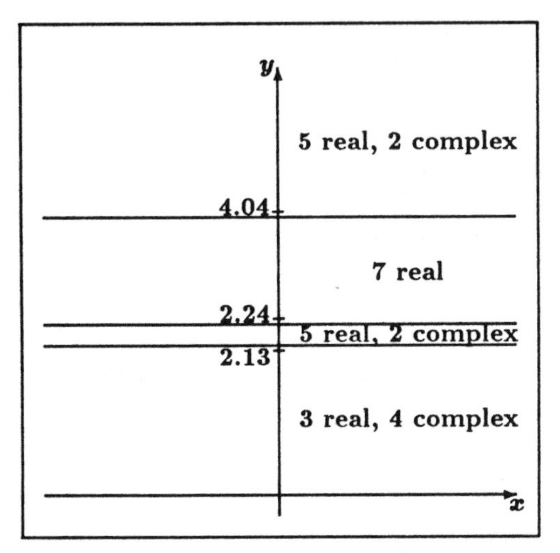

Fig. 1. *Distribution of eigenvalues: shear flow,* Re $= 1$, $\alpha = 1$, $\beta = 2.56$, $b = 2$, $s_0 = 0.6$, We $= 11.7$

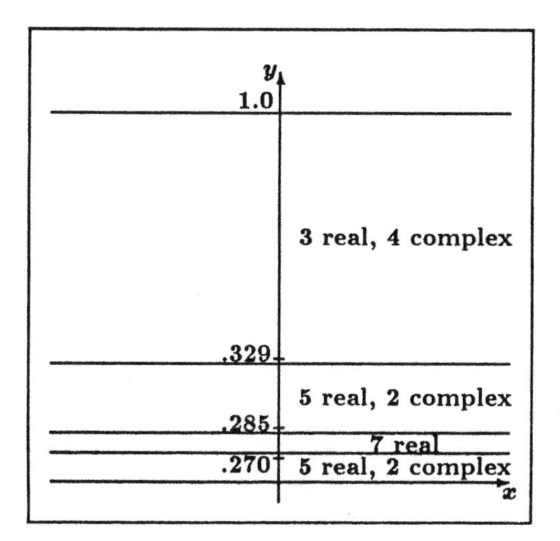

Fig. 2. *Distribution of eigenvalues: channel flow,* Re $= 1$, $\alpha = 1$, $\beta = 1$, $b = 2$, $s_0 = 1$, We $= 10$

One feature has been glossed over in the above discussion. The characteristic polynomial has a factor $\alpha u + v$, and in fact this factor occurs with multiplicity three. Thus all the comments above were for regions where u is not zero. However, if $v = u = 0$, then the characteristic polynomial is always zero, that is any α will do. In particular for channel flow if we are to assume that the base flow obeys no slip at the walls, then the system is degenerate at the walls. This introduces new questions about the correct boundary conditions to be prescribed. An analogous situation is studied for a degenerate elliptic system in section 5.

3. Symmetric positive systems. Friedrichs [4] introduced his theory of Symmetric Positive Systems specifically to treat systems which are not of one type, most notably of mixed elliptic-hyperbolic type. It is attractive therefore as a method that might also be useful for systems which are of composite type and degenerate type and of composite-mixed type which might arise, for example, in problems of viscoelastic flows.

In this section we outline certain aspects of the theory of Symmetric Positive Systems in a more or less self-contained way but only to the extent needed for its application in the next sections. There we show by some simple examples how correct boundary conditions for well-posed problems can be discovered through its use. The examples are chosen to exhibit certain features which are present in viscoelasticity.

Note that Friedrichs' theory is for *linear* systems of partial differential equations (however see [6]). Thus it can apply directly only to linearized equations (of, say, viscoelasticity) but one hopes that the results will suggest correct boundary conditions for the nonlinear case.

Consider then the system of m first order equations

$$Lu := A(x,y)u_x + B(x,y)u_y + C(x,y)u = f(x,y) \tag{3.1}$$

in a bounded region Ω in R^2, where the unknown u is an m-vector. Although the theory is valid for n independent variables, we consider the case $n = 2$. The m-by-m matrices A and B are assumed to be C^1 on $\bar{\Omega}$ with C belonging to C^0 on $\bar{\Omega}$. The Symmetry Condition is that

$$A \text{ and } B \text{ are symmetric on } \bar{\Omega}. \tag{3.2.a}$$

The Positivity Condition is that the symmetric matrix

$$H := C + C^\top - A_x - B_y \text{ is positive definite on } \bar{\Omega}. \tag{3.2.b}$$

This condition arises in a natural way as will be seen below. Systems satisfying (3.2.a) and (3.2.b) are called Symmetric Positive Systems.

Assume that Ω has a piecewise smooth boundary so that at all but a finite number of points of the boundary there is a well defined outward unit normal $n = (n_x, n_y)$. Application of the divergence theorem for smooth (say $C^1(\bar{\Omega})$) m-vectors u and v yields the Green's Identity

$$(v, Lu) = (u, L^*v) + \langle v, \beta u \rangle \tag{3.3}$$

where the $L^2(\Omega)$ inner product is denoted by $(\cdot,\cdot)$ with the corresponding norm $\|\cdot\|$; and $\langle\cdot,\cdot\rangle$ denotes the $L^2(\partial\Omega)$ inner product. The formal adjoint L^* of L is given by

$$L^* v := -(Av)_x - (Bv)_y + C^\top v \tag{3.4}$$

with $C^\top$ the transpose of C. The "boundary matrix" β is defined at all smooth points of $\partial\Omega$ by

$$\beta := n_x A + n_y B.$$

For ease of terminology we will discuss β as if it were defined at all points of $\partial\Omega$.

Taking $v = u$ in (3.3), and using the fact that $L^* = -L + H$ gives the identity

$$2(u, Lu) = (u, Hu) + \langle u, \beta u\rangle \tag{3.5}$$

with H given in (3.2.b). At each (smooth) point $P = (x, y)$ of $\partial\Omega$ let $\mathcal{M}(P)$ be a subspace of R^m. If the function u satisfies $u(P) \in \mathcal{M}(P)$ and β is positive semi-definite on $\mathcal{M}(P)$ then

$$2\|u\|\|Lu\| \geq 2(u, Lu) \geq (u, Hu) \geq 2h_0\|u\|^2, \quad h_0 > 0 \tag{3.6}$$

by virtue of (3.5) so that

$$\|Lu\| \geq h_0\|u\| \tag{3.7}$$

which is Friedrichs' "basic inequality". From this follows uniqueness of smooth solutions of the boundary value problem

$$Lu = f \text{ in } \Omega$$

$$u(P) \in \mathcal{M}(P) \text{ for } P \in \partial\Omega$$

provided that $\beta(P)$ is positive semi-definite on $\mathcal{M}(P)$. Clearly $\mathcal{M}(P)$ may be too restrictive a subspace for existence of solutions; for example the choice $\mathcal{M}(P) = \{0\}$ for all P on $\partial\Omega$ is too restrictive for existence of solutions for the nonhomogeneous degenerate elliptic system considered in section 5.

Suppose the subspace $\mathcal{M}(P)$ is given through the boundary condition

$$M(P)u(P) = 0$$

by means of a matrix $M = M(P)$. Then (3.3) can be rewritten as Friedrichs' first identity,

$$(v, Lu) + \langle v, Mu\rangle = (u, L^*v) + \langle u, M^*v\rangle \tag{3.8}$$

where $M^* = \beta + M^\top$. Using again $L^* = -L + H$ and putting $v = u$ one obtains

$$2(u, Lu) + \langle u, Mu\rangle = (u, Hu) + \langle u, M^*u\rangle. \tag{3.9}$$

Now if $Mu = 0$ on $\partial\Omega$ then again the basic inequality (3.7) results from using (3.9) provided that

$$(3.10) \qquad (M + M^*) \geq 0 \text{ on } \partial\Omega.$$

The condition (3.10) is Friedrichs' condition of semi-admissibility for the boundary condition $Mu = 0$ and yields uniqueness of smooth solutions of the boundary value problem

$$(3.11.a) \qquad Lu = f \text{ in } \Omega$$

$$(3.11.b) \qquad Mu = 0 \text{ on } \partial\Omega.$$

By putting $u = v$ in (3.8) to obtain

$$(3.12) \qquad (v, Lv) + \langle v, Mv \rangle = (v, L^*v) + \langle v, M^*v \rangle$$

one likewise has, for $M^*v = 0$ on $\partial\Omega$

$$(3.13) \qquad \|L^*v\| \geq h_0\|v\|$$

provided that the condition of semi-admissibility (3.10) is satisfied. Thus (3.10) also yields uniqueness of smooth solutions of the adjoint boundary value problem

$$(3.14.a) \qquad L^*v = g \text{ in } \Omega$$

$$(3.14.b) \qquad M^*v = 0 \text{ on } \partial\Omega.$$

It is expected from this latter result that there will be existence of an appropriately defined weak solution of (3.11). Indeed from (3.8) one is led to define a weak solution of (3.11) as an $L^2(\Omega)$ function u which satisfies

$$(3.15) \qquad (u, L^*v) = (f, v)$$

for all $C^1(\bar{\Omega})$ functions v satisfying $M^*v = 0$ on $\partial\Omega$. Then, in a familiar way, Friedrichs obtains the existence of such weak solutions provided the semi-admissibility criterion (3.10) is satisfied.

But for the *uniqueness* of *weak* solutions and their regularity as well as *existence* of *strong* solutions more is required than semi-admissibility of the boundary condition. Friedrichs introduced an admissibility criterion for the case of nonsingular β, i.e. the case of non-characteristic boundary. For nonsingular β this is equivalent to the *maximality condition* of Lax [8] which we follow here: a subspace $\mathcal{M}(P)$ of R^m is maximal non-negative with respect to L if for each $\xi \in \mathcal{M}(P)$

$$(3.16) \qquad \xi \cdot \beta(P)\xi \geq 0$$

and, moreover, there is no subspace of R^m which contains $\mathcal{M}(P)$ and on which (3.16) holds. Clearly $\mathcal{M}(P)$ determines a (linear homogeneous) boundary condition at each (smooth) point of $\partial\Omega$. Such a boundary condition is called admissible for L if $\mathcal{M}(P)$ is maximal non-negative on $\partial\Omega$.

In the examples treated below the partial differential equations are homogeneous while appropriate boundary conditions will be non-homogeneous. For the simple geometries considered there one can explicitly make a reduction to the case of homogeneous boundary conditions. For simplicity this reduction will not be carried out and only homogeneous admissible boundary conditions will be considered as in the discussion of this section. Of course, for questions of uniqueness alone such a reduction is superfluous. It should also be emphasized that other boundary conditions than those treated below are of interest and approachable by the theory of symmetric positive operators but the exposition would then be complicated considerably.

We should point out that some of the results of the theory of symmetric positive operators, particularly with regard to regularity of solutions ("weak=strong"), apply only to the case of noncharacteristic boundary points (i.e. where the boundary matrix β is non-singular) or when $\mathcal{M}(P)$ is of constant dimension on the boundary. In contrast, many problems of viscoelastic flow and the examples treated below do not satisfy these conditions. Also note that the regions that we consider below have corners. These points require special treatment for "weak = strong" results [4], [5], [8], [10], [15], [18]. We note also that Morawetz [9] treated a system of equations which includes Tricomi's equation (written as a 2-by-2 first order system) and obtained a weak solution for certain standard boundary value problems in regions including the parabolic line. She showed that for certain boundary geometries the theory of Symmetric Positive Systems does not apply directly so that some modification is needed to fit that case in Friedrichs' framework.

4. The Euler equations for uniform flow. The steady two-dimensional incompressible Euler equations linearized about the uniform flow $u_0 = 1, v_0 = 0$ and $p_0 =$ constant are given by

$$u_x + v_y = 0 \tag{4.1.a}$$

$$u_x + p_x = 0 \tag{4.1.b}$$

$$v_x + p_y = 0 \tag{4.1.c}$$

where u, v, give the x and y components of the velocity field and p is the pressure. The region under consideration is the channel $|y| < 1$ with an inlet at $x = 0$. This system, like its nonlinear version, is easily seen to be composite with one family of real characteristics y =constant and a family of complex characteristics $y \pm ix$ =constant. Since the pressure is determined only up to an additive constant by this system, in order to hope for uniqueness of a solution it is necessary to include the components of the gradient of p as separate dependent variables and thus to write the system in the augmented form

$$u_x + v_y = 0 \tag{4.2.a}$$

$$u_x + w = 0 \tag{4.2.b}$$

$$v_x + z = 0 \tag{4.2.c}$$

$$w_y - z_x = 0 \tag{4.2.d}$$

where we have put $w = p_x$, $z = p_y$. The system is not in symmetric form. We achieve a symmetric form by premultiplication by a symmetrizing matrix Z. The most general Z is

$$Z = \begin{pmatrix} 0 & a & b+c & 0 \\ b & c & l & d \\ d & -d & 0 & 0 \\ 0 & 0 & -d & 0 \end{pmatrix} \tag{4.3}$$

where a, b, c, d, l are arbitrary smooth functions of (x, y) to be chosen so as to satisfy $\det Z = ad^3 \neq 0$. The resulting symmetric system, obtained by multiplying the system on the left by Z is

$$AU_x + BU_y + CU = 0 \tag{4.4}$$

with $U = (u, v, w, z)^\top$ and

$$A = \begin{pmatrix} a & b+c & 0 & 0 \\ b+c & l & 0 & -d \\ 0 & 0 & 0 & 0 \\ 0 & -d & 0 & 0 \end{pmatrix}, B = \begin{pmatrix} 0 & 0 & 0 & 0 \\ 0 & b & d & 0 \\ 0 & d & 0 & 0 \\ 0 & 0 & 0 & 0 \end{pmatrix}, C = \begin{pmatrix} 0 & 0 & a & b+c \\ 0 & 0 & c & l \\ 0 & 0 & -d & 0 \\ 0 & 0 & 0 & -d \end{pmatrix}.$$

We note that A, B, C are everywhere singular so that the boundary matrix $\beta = n_x A + n_y B$ is singular everywhere on $\partial\Omega$: the boundary is everywhere characteristic. The positivity condition (3.2.b) requires that the symmetric matrix

$$H = \begin{pmatrix} -a_x & -(b_x + c_x) & a & b+c \\ -(b_x + c_x) & -l_x - b_y & c - d_y & l + d_x \\ a & c - d_y & -2d & 0 \\ b+c & l + d_x & 0 & -2d \end{pmatrix} \tag{4.5}$$

be positive definite. It is easy to see that this is the case if, for example, we choose $a = l = 2e^{-x}$, $b = c = 0$ and $-d$ a constant which is > 1 on $\bar{\Omega}$ (where we assume $x \geq 0$). With such a choice the quadratic form for the boundary matrix β on $y = \pm 1$ is

$$U \cdot \beta U = \pm 2dvw$$

and so the subspace determined by $v = 0$ (or $w = 0$) is maximal non-negative. Similarly on the inlet $x = 0$

$$U \cdot \beta U = -(au^2 + lv^2 - 2dvz)$$

so that the subspace determined by $u = v = 0$ is maximal non-negative. Note also that downstream on $x = 1$

$$U \cdot \beta U = au^2 + lv^2 - 2dvz$$

so that $v = 0$ (or $z = 0$) is maximal non-negative.

Thus we arrive at the expected result that admissible boundary conditions for the channel include the specification of v alone on $y = \pm 1$ and the specification of u and v on the inlet; only v need be specified at a downstream outlet. The components of the pressure gradient p_x and p_y are then uniquely determined. For the semi-infinite channel one would expect a limit condition that v would tend to zero at infinity.

5. A degenerate elliptic system: GASPT. A significant example of a degenerate elliptic equation is the equation of Weinstein's [20],[21] Generalized Axisymmetric Potential Theory (GASPT)

$$y(\phi_{xx} + \phi_{yy}) + k\phi_y = 0, \tag{5.1}$$

where k is a real constant. The equation is elliptic for $y > 0$ but for k not 0 the ellipticity degenerates on $y = 0$. For $k = 1/3$ equation (5.1) is the canonical form of Tricomi's equation

$$\sigma\psi_{\theta\theta} + \psi_{\sigma\sigma} = 0$$

in the elliptic (subsonic) half-plane $\sigma > 0$. For k not zero (5.1) is a prototype for elliptic partial differential equations which are singular or degenerate on a portion of the boundary of the region in question.

In consonance with several of the flow problems considered above we consider regions contained in the strip $0 < y < 1$. For definiteness let Ω denote the rectangular region $(0,1) \times (0,1)$ in the x, y plane.

A first order system obtainable from (5.1) is, with $\phi_x = u$, $\phi_y = v$,

$$\begin{aligned} u_y - v_x &= 0 \\ yu_x + yv_y + kv &= 0 \end{aligned} \tag{5.2}$$

which is of the form $\tilde{A}U_x + \tilde{B}U_y + \tilde{C}U = 0$ as in (3.1) but here with

$$\tilde{A} = \begin{pmatrix} 0 & -1 \\ y & 0 \end{pmatrix}, \tilde{B} = \begin{pmatrix} 1 & 0 \\ 0 & y \end{pmatrix}, \tilde{C} = \begin{pmatrix} 0 & 0 \\ 0 & k \end{pmatrix}.$$

Since $\det(\alpha\tilde{A} + \tilde{B}) = y(\alpha^2 + 1)$ the characteristics of (5.2) (and of course of (5.1)) are non-real for y not zero but for $y = 0$ every direction is characteristic, analogous to the UCMM or Bird-DeAguiar behavior for channel flow. By use of the symmetrizer

$$Z = \begin{pmatrix} -ya & b \\ yb & a \end{pmatrix}$$

the system assumes the symmetric form

$$AU_x + BU_y + CU = 0 \tag{5.3}$$

with

$$A = \begin{pmatrix} by & ay \\ ay & -by \end{pmatrix}, B = \begin{pmatrix} -ay & by \\ by & ay \end{pmatrix}, C = \begin{pmatrix} 0 & kb \\ 0 & ka \end{pmatrix}.$$

Note that Z is singular on $y = 0$.

The choice of $a = y^s$ and $b = 0$ gives for H in (3.2.b)

$$H = y^s \begin{pmatrix} s+1 & 0 \\ 0 & 2k-(s+1) \end{pmatrix} \tag{5.4}$$

which for $-1 < s < 2k-1$ is uniformly positive definite on sets bounded away from $y = 0$. Such an s exists only for $k > 0$. More relevant here is the fact that $H = y^s H_0$ with the constant matrix H_0 positive definite for such a choice of s. Indeed, it then follows that

$$(U, HU) \geq h_0(y^s U, U), \qquad h_0 > 0 \tag{5.5}$$

so that the theory of symmetric positive operators applies for a weighted-$L^2(\Omega)$ space rather than the usual (unweighted) space $L^2(\Omega)$. For the case of uniqueness this is obvious. For existence of weak solutions a more extensive analysis is needed but is valid at least for $k > 1$.

Note that as k tends to zero the system (5.2) has the Cauchy-Riemann system as the limit for which one obtains uniqueness of v only up to an additive constant if u is specified on the entire boundary. This is reflected in the fact that H can not be made positive definite in the domain for $k = 0$. The system must be augmented to include u or v explicitly in order to obtain uniqueness. With the positivity condition (3.2.b) considered in the modified way (5.5) we now examine possible admissible boundary conditions. On $y = 1$

$$U \cdot \beta U = y^{s+1}(v^2 - u^2)$$

so that $u = 0$ is an admissible boundary condition while on $y = 0$ no boundary condition is to be specified, provided that u and v remain bounded near $y = 0$ (in fact no boundary condition is to be specified as long as $y^{s+1}u^2$ and $y^{s+1}v^2$ tend to zero as y tends to zero). On $x = 0$ or 1

$$U \cdot \beta U = \mp 2y^{s+1}uv$$

so that $u = 0$ (or $v = 0$) is an admissible boundary condition there. Thus, for $k > 1/2$ (choosing $s > 0$) a correctly posed problem is to give u on $x = 0, 1$ and on $y = 1$ with no data given on $y = 0$, but only boundedness of u and v there.

A few remarks should be made for those familiar with the results of GASPT. For $k \geq 1$ a well-posed Dirichlet problem for (5.1) consists in specifying ϕ only on those parts of the boundary for which $y > 0$, provided merely that ϕ remains bounded near $y = 0$ (in fact, merely that $y^{k-1}\phi$ tends to zero for $k > 1$ [$(\log y)^{-1}\phi$ tends to zero for $k = 1$] as y tends to zero), whereas for $k < 1$ ϕ must be specified all around. In treating the system (5.2) we require that u and v (i.e. ϕ_x and ϕ_y) remain bounded near $y = 0$. This explains why we don't specify any conditions on u and v at $y = 0$ for $1/2 < k < 1$.

It is interesting to also consider an augmented version of the system (5.2) similar to the example of the Euler equations above. The point here is to be able to also obtain results for the Cauchy-Riemann system which was not possible through the corresponding two-by-two system. Letting $w = u_x$ and $z = u_y$ in (5.2) we arrive at a 3-by-3 system of the form (3.1), namely

$$\begin{aligned} v_x - z &= 0 \\ y(w + v_y) + kv &= 0 \\ w_y - z_x &= 0. \end{aligned} \tag{5.6}$$

By using the symmetrizer

$$Z = \begin{pmatrix} b & a & yc \\ 0 & c & 0 \\ -yc & 0 & 0 \end{pmatrix}$$

we obtain a corresponding symmetric form for $U = (v, w, z)^\top$ with

$$A = \begin{pmatrix} b & 0 & -yc \\ 0 & 0 & 0 \\ -yc & 0 & 0 \end{pmatrix}, B = \begin{pmatrix} ya & yc & 0 \\ yc & 0 & 0 \\ 0 & 0 & 0 \end{pmatrix}, C = \begin{pmatrix} ka & ya & -b \\ kc & yc & 0 \\ 0 & 0 & yc \end{pmatrix}.$$

Here again every direction is characteristic for this system.

The choices $b = 0$, $a = \alpha y^k$, and $c = \gamma y^{k-1}$ with α and γ positive for $k > 1$ yields

$$H = \begin{pmatrix} \alpha(k-1)y^k & \alpha y^{k+1} & 0 \\ \alpha y^{k+1} & 2\gamma y^k & 0 \\ 0 & 0 & 2\gamma y^k \end{pmatrix}$$

which can be written as $H = y^k H_0(y)$ with H_0 (uniformly) positive definite on $\bar{\Omega}$ provided that $2\gamma(k-1) > \alpha$ for $k > 1$. Thus a modified positivity condition like (5.5) is satisfied.

With such a choice of parameters we find that on $y =$constant

$$U \cdot \beta U = n_y y^k (\alpha y v^2 + 2\gamma v w)$$

so that $v = 0$ is an admissible boundary condition on $y = 1$ while on $y = 0$ no boundary condition need be specified provided that $k > 1$ and v and w remain bounded there (or merely that $y^{k+1}v^2$ and $y^k vw$ tend to zero as y tends to zero). On $x =$constant

$$U \cdot \beta U = -2\gamma y^k n_x v z$$

so that $v = 0$ (or $z = 0$) is an admissible boundary condition on $x = 0, 1$. Thus the situation for this augmented degenerate system is essentially the same as that for (5.2).

For the non-degenerate case, $k = 0$, this is not true. As k tends to zero in (5.6) (dividing first by y) one obtains an augmented Cauchy-Riemann system. For its symmetric form we have

$$A = \begin{pmatrix} b & 0 & -c \\ 0 & 0 & 0 \\ -c & 0 & 0 \end{pmatrix}, B = \begin{pmatrix} a & c & 0 \\ c & 0 & 0 \\ 0 & 0 & 0 \end{pmatrix}, C = \begin{pmatrix} 0 & a & -b \\ 0 & c & 0 \\ 0 & 0 & c \end{pmatrix}$$

so that the choice $b = 0$, $a = -\alpha y$ and $c = \gamma$ $(\alpha > 0, \quad \gamma > 0)$ yields

$$H = \begin{pmatrix} \alpha & -\alpha y & 0 \\ -\alpha y & 2\gamma & 0 \\ 0 & 0 & 2\gamma \end{pmatrix}$$

which is positive definite on Ω under a restriction similar to that above: $\alpha < 2\gamma$. On y =constant

$$U \cdot \beta U = n_y(2\gamma v w - \alpha y v^2)$$

so that the boundary condition $v = 0$ is admissible on $y = 0, 1$. On x =constant

$$U \cdot \beta U = -2 n_x \gamma v z$$

so that the boundary condition $v = 0$ (or $z = 0$) is admissible also on $x = 0, 1$. Thus by augmenting the Cauchy-Riemann system we obtain, via the theory of symmetric positive operators, the uniqueness of the solution $(v, w, z)^\top$ as compared to uniqueness up to an additive constant for $(u, v)^\top$ in (5.2). It should be mentioned that the Cauchy-Riemann system was also treated in the original paper of Friedrichs [4] along with other examples including the Tricomi system (see also Morawetz [9]).

6. Conclusion.

It is clear that the differential models for viscoelastic fluid flow contain a wide variety of interesting features. In particular the equations form a coupled system of partial differential equations of first order which is not of standard type. The systems are composite, may be of changing type and may be degenerate at flow boundaries. This has been indicated in particular for the Bird-DeAguiar model where the system can also become fully hyperbolic in the flow region. For such systems the coupling can be of great significance, as was discussed for a two-by-two degenerate elliptic system in section 5. As the parameter k changes (which couples the system through lower order terms) the boundary conditions for a well-posed problem also change. We have applied Friedrichs' theory of Symmetric Positive Systems to several examples to show how it may be useful for attacking the more complicated problems.

In a recent paper [13] Renardy considered small perturbations of plane Poiseuille flow of Maxwell type fluids at small shear rates with inflow and outflow boundaries.

Assuming no-slip (zero velocity at solid walls), he gave an existence and uniqueness proof for small enough data (in some appropriate norm). A well-posed problem in that context is obtained if, in addition to the no-slip condition at the walls, inflow and outflow velocities as well as normal stresses at the inlet are prescribed. These inflow data must satisfy certain compatibility conditions which rule out singularities at the corners where inflow boundaries meet the walls (see also [11],[12]). Thus apparently the degeneracy has no effect on the boundary conditions for this particular coupling. It will be interesting to see if similar results hold for the more complex nonlinear model, the Bird-DeAguiar model. We are presently applying the Friedrichs' method to various differential constitutive models governing viscoelastic fluid flow.

Acknowledgments. Gilberto Schleiniger was on leave at the Institute for Mathematics and its Applications at the University of Minnesota (IMA) when part of this work was in progress and he thanks the IMA and the Minnesota Supercomputer Institute for their support.

REFERENCES

[1] R. B. Bird and J. R. DeAguiar, *An encapsulated dumbbell model for concentrated polymer solutions and melts I. Theoretical development and constitutive equation*, J. Non-Newtonian Fluid. Mech., 13 (1983), pp. 149–160.

[2] M. C. Calderer, L. Pamela Cook and G. Schleiniger, *An analysis of the Bird-DeAguiar model for polymer melts*, J. Non-Newtonian Fluid. Mech., 31 (1989), pp. 209–225.

[3] J. R. DeAguiar, *An encapsulated dumbbell model for concentrated polymer melts II. Calculation of material functions and experimental comparisons*, J. Non-Newtonian Fluid. Mech., 13 (1983), pp. 161–179.

[4] K. O. Friedrichs, *Symmetric positive linear differential equations*, Comm. Pure Appl. Math., 11 (1958), pp. 333–418.

[5] K. O. Friedrichs and P. D. Lax, *Boundary value problems for first order operators*, Comm. Pure Appl. Math., 18 (1965), pp. 355–388.

[6] S. Hahn-Goldberg, *Generalized linear and quasilinear accretive systems of partial differential equations*, Comm. in Partial Differential Equations, 2 (1977), pp. 165–191.

[7] D. D. Joseph, M. Renardy and J.-C. Saut, *Hyperbolicity and change of type in the flow of viscoelastic fluids*, Arch. Rational Mech. Anal., 87 (1985), pp. 213–251.

[8] P. D. Lax and R. S. Phillips, *Local boundary conditions for dissipative symmetric linear differential operators*, Comm. Pure Appl. Math, 13 (1960), pp. 427–455.

[9] C. S. Morawetz, *A weak solution for a system of equations of elliptic-hyperbolic type*, Comm. Pure Appl. Math., 11 (1958), pp. 315–331.

[10] R. S. Phillips and L. Sarason, *Singular symmetric positive first order differential operators*, J. Math. Mech., 15 (1966), pp. 235–271.

[11] M. Renardy, *Inflow boundary conditions for steady flows of viscoelastic fluids with differential constitutive laws*, Rocky Mt. J. Math., 18 (1988), pp. 445–453.

[12] ————, *A well-posed boundary value problem for supercritical flow of viscoelastic fluids of Maxwell type*, These proceedings.

[13] ————, *Compatibility conditions at corners between walls and inflow boundaries for fluids of Maxwell type*, preprint, May 1989.

[14] M. Renardy, W. J. Hrusa and J. A. Nohel, *Mathematical Problems in Viscoelasticity*, Longman Scientific and Technical (with John Wiley and Sons. Inc), New York, 1987.

[15] L. Sarason, *On weak and strong solutions of boundary value problems*, Comm. Pure Appl. Math, 15 (1962), pp. 237–288.

[16] G. Schleiniger, M. C. Calderer and L. Pamela Cook, *Embedded hyperbolic regions in a nonlinear model for viscoelastic flow*, to appear in the AMS Contemporary Mathematics Series, Proceedings of the 1988 Joint Summer Research Conference on Current Progress in Hyperbolic Equations: Riemann Problems and Computations, Bowdoin, Maine.

[17] R. I. Tanner, *Stresses in dilute solutions of bead-nonlinear-spring macromolecules. II Unsteady flows and approximate constitutive relations*, Trans. Soc. Rheol., 19 (1975), pp. 37–65.
[18] D. S. Tartakoff, *Regularity of solutions to boundary value problems for first order systems*, Indiana J., 21 (1972), pp. 1113–1129.
[19] C. Verdier and D. D. Joseph, *Change of type and loss of evolution of the White-Metzner model*, to appear, J. Non-Newtonian Fluid Mech..
[20] A. Weinstein, *Generalized axially symmetric potential theory*, Bull. Amer. Math. Soc., 59 (1953), pp. 20–38.
[21] ———, *Singular partial differential equations and their applications*, in *Fluid Dynamics and Applied Mathematics (Proc. Sympos., Univ. of Maryland, 1961)*, edited by Diaz and Pai, Gordon and Breach, New York, New York, 1962, pp. 29–49.

NUMERICAL SIMULATION OF INERTIAL VISCOELASTIC FLOW WITH CHANGE OF TYPE

M.J. CROCHET and V. DELVAUX*

SUMMARY. We examine plane inertial flows of viscoelastic fluids with an instantaneous elastic response. In such flows, the vorticity equation can change type when the velocity of the fluid exceeds the speed of shear waves. We use a finite element algorithm which has been developed for calculating highly viscoelastic flows.

The algorithm is tested for supercritical flow regimes on the problem of the flow through a wavy channel. We next consider the problem of the flow of a Maxwell fluid around a circular cylinder for various flow regimes. We compare creeping flows, Newtonian flows and supercritical viscoelastic flows. We show that the flow kinematics is affected by supercritical flow conditions. In particular, a vorticity shock forms ahead of the cylindrical body.

1. Introduction. Earlier papers on the numerical simulation of viscoelastic flow have been essentially devoted to creeping flows. The major reason is that most rheological and experimental data relate to very viscous fluids in flows where the Reynolds number is much smaller than one. For such flows, inelastic calculations show that inertia effects have little influence on the creeping flow field ($Re = 0$). Another reason is that viscoelastic flow calculations already are quite difficult for creeping flows; the coupling between viscoelastic and inertia effects, to be discussed below, amounts to an additional numerical complexity. Still, it is now recognized that the smallness of the Reynolds number is not a sufficient reason for neglecting the inertia terms in the momentum equations. The crucial parameter is the product M^2 of the Reynolds times the Weissenberg number; M is the viscoelastic Mach number, or the ratio of a characteristic velocity of the fluid to the speed of shear waves. In this chapter, we study steady inertial flows of viscoelastic fluids with an instantaneous elasticity.

The mathematical analysis of the partial differential equations governing such flows was first addressed by Rutkevich [1, 2]. More recently, Joseph, Renardy and Saut [3] have shown that, under some flow conditions, the vorticity equation in steady flow changes type. The flow domain contains subcritical (elliptic) and supercritical (hyperbolic) regions for the vorticity, which can become discontinuous across real characteristics in the hyperbolic region. Other mathematical features of the governing equations of viscoelastic flow, such as the loss of evolution, may be found in [4–6].

Several interesting flow phenomena show features which are related to change of type. Significant examples are the sink flow considered by Metzner et al [7], the delayed die swell phenomenon evidenced by Joseph et al. [8] and the flow around

*Unité de Mécanique Appliquée, Université Catholique de Louvain, 1348 Louvain-la-Neuve, Belgium. The work of V. Delvaux is supported by the "Service de Programmation de la Politique Scientifique".

a circular cylinder studied by James and Acosta [9]. In the latter case, it was shown that, under supercritical conditions, the Nusselt number characterizing heat transfer from the cylinder to the surrounding fluid reaches an asymptotic value for velocities exceeding the wave velocity. A simplified analysis of that flow was given in [10].

In the present paper, we wish to analyze the kinematics of the flow of a Maxwell fluid around a circular cylinder under supercritical conditions; further results on the drag and the transport properties will be examined in a later paper. Our goal is to solve the full set of governing equations in the non-linear regime, without resorting to a linearized analysis [11]. We make use of a finite element analysis of viscoelastic flow, which has known considerable progress over the last few years. Recent review on the state of the art may be found in [12, 13], where many references are available. It is interesting to note that one of the difficulties affecting the numerical calculation of viscoelastic flow, even creeping flow, is the inaccuracy of the calculated stress field which can produce artificial change of type in the discretized system of governing equations [14]; with standard Galerkin methods, numerical errors then propagate throughout the flow domain.

Earlier numerical attempts at calculating viscoelastic flow in the supercritical regime [15–17] have failed to simulate the flow far beyond the onset of criticality. In this chapter, we use the mixed finite element algorithm of Marchal and Crochet [18] which is characterized by bilinear sub-elements for the stresses and the use of streamline-upwinding for discretizing the constitutive equations. A recent investigation on the mesh convergence properties of such an algorithm may be found in [19].

In section 2, we examine the basic equations and the associated non-dimensional numbers, and we explain the numerical method. In section 3, we analyze the supercritical flow of a Maxwell fluid through a wavy channel, for which an exact analysis for small amplitude has been given by Yoo and Joseph [20]. We compare the performance of several numerical methods, and find that the method developed in [18] is satisfactory. Finally, we analyze in section 4 the flow of a Maxwell fluid around a circular cylinder. We find that the conjectures made by Joseph [11] are verified by our numerical analysis of the problem.

2. Basic Equations and Numerical Method. Let $\underline{\sigma}$ denote the Cauchy stress tensor which is decomposed as follows

$$\underline{\sigma} = -p\underline{I} + \underline{T} \ , \tag{2.1}$$

where p is the pressure and $\underline{T}$ the extra-stress tensor. For the upper-convected Maxwell fluid, which we study below, $\underline{T}$ is given by the constitutive equation

$$\underline{T} + \lambda \overset{\nabla}{\underline{T}} = 2\eta \ \underline{d} \ , \tag{2.2}$$

where λ is the relaxation time, η the shear viscosity and $\underline{d}$ the rate of deformation tensor. The triangular superscript in (2.2) denotes the upper-convected derivative,

$$\overset{\nabla}{\underline{T}} = D\underline{T}/Dt - (\nabla \mathbf{v})^T \underline{T} - \underline{T} \nabla \mathbf{v} \ , \tag{2.3}$$

where $\mathbf{v}$ is the velocity vector and D/Dt stands for the material derivative. The momentum and incompressibility equations are given as follows

$$-\nabla p + \nabla \bullet \underline{T} + \mathbf{f} = \rho \mathbf{a} , \tag{2.4}$$

$$\nabla \bullet \mathbf{v} = 0 , \tag{2.5}$$

where $\mathbf{f}$ is the body force, ρ the density and $\mathbf{a}$ the acceleration. In later sections, we assume that $\mathbf{f}$ vanishes but, by contrast with earlier papers, we take inertia effects into account.

We limit ourselves to steady-state problems. We assume that the fluid does not slip along rigid walls. The flow in the entry sections is fully developed; we impose $\mathbf{v}$ and $\underline{T}$ in these sections. In exit sections where the flow is assumed to be fully developed, we impose $\mathbf{v}$ as a boundary condition.

Let V and L denote a characteristic velocity and a characteristic length of the flow. The dimensionless quantities associated with the flow are the Weissenberg number We,

$$We = \lambda V/L , \tag{2.6}$$

and the Reynolds number Re,

$$Re = \rho V L/\eta . \tag{2.7}$$

The velocity of the shear waves for a Maxwell fluid is given by

$$c = (\eta/\lambda\rho)^{1/2}; \tag{2.8}$$

the viscoelastic Mach number M is defined as

$$M = V/c = (Re\ We)^{1/2}. \tag{2.9}$$

The elasticity number, defined as

$$E = \eta\lambda/\rho L^2 , \tag{2.10}$$

is also of interest; when the flow rate or a characteristic velocity of the flow increases in a given problem, one finds that We, Re and M increase linearly with V while E, which depends upon the fluid and the geometry, remains a constant.

It is worthwhile to introduce two additional tensorial quantities $\underline{T}_A$ and $\underline{T}_B$ which are defined as follows:

$$\begin{aligned} \underline{T}_A &= \underline{T} + \eta/\lambda \underline{I} , \\ \underline{T}_B &= \underline{T}_A - \rho\ \mathbf{v}\ \mathbf{v} . \end{aligned} \tag{2.11}$$

If one considers the integral form of the constitutive equation (2.2), it is easy to show that the tensor $\underline{T}_A$ should always be positive definite [14]. For creeping flows, $\underline{T}_B = \underline{T}_A$. However, when inertia is taken into account, $\underline{T}_B$ can lose its positiveness; it was shown by Joseph, Renardy and Saut [3] that, in regions where $\underline{T}_B$ is non-positive, the vorticity equation is hyperbolic. These two tensors are important in our applications, because the lack of positiveness of $\underline{T}_A$ is a sign of numerical inaccuracy, while the loss of positiveness of $\underline{T}_B$ signals the presence of a hyperbolic region for the vorticity equation.

Let Ω denote the flow domain on which we wish to solve the system of eqns. (2.1-5). In view of the implicit character of (2.2), we select as primitive unknowns the extra-stress tensor $\underline{T}$, the velocity field $\mathbf{v}$ and the pressure p which belong to their respective spaces T, V and P. The flow domain is discretized by means of finite elements which cover a domain Ω^h. The elements are characterized by their shape, their nodes, the nodal values and the interpolating functions. The finite element representations of $\underline{T}, \mathbf{v}$ and p are denoted by $\underline{T}^h, \mathbf{v}^h$ and p^h which belong to their respective discrete spaces T^h, V^h and P^h.

A straightforward approach for calculating $\underline{T}^h, \mathbf{v}^h$ and p^h is to substitute for (2.2), (2.4) and (2.5) their weak formulation which may be expressed as follows [21]:

find $\underline{T}^h \in T^h, \mathbf{v}^h \in V^h$ and $p^h \in P^h$ such that

$$(2.12) \qquad \langle \underline{T}^h + \lambda \overset{\nabla}{\underline{T}}{}^h - \eta(\nabla \mathbf{v}^h + \nabla \mathbf{v}^{hT}); \underline{S}^h \rangle = 0, \quad \forall\, \underline{S}^h \in T^h ,$$

$$(2.13) \qquad \langle -p^h \underline{I} + \underline{T}^h + \rho \mathbf{a}^h; \nabla \mathbf{u}^h + \nabla \mathbf{u}^{hT} \rangle = 2 \ll \mathbf{t};\ \mathbf{u}^h \gg, \ \forall \mathbf{u}^h \in V^h ,$$

$$(2.14) \qquad \langle \nabla \bullet \mathbf{v}^h; q^h \rangle = 0, \ \forall q^h \in P^h .$$

The brackets $\langle\ \rangle$ and $\ll\ \gg$ denote the L^2 scalar product over Ω^h and its boundary, respectively, $\mathbf{t}$ denotes the surface force vector on the boundary, and expressions such as $\underline{A}; \underline{B}$ and $\mathbf{a}$; $\mathbf{b}$ denote respectively $tr\underline{A}\underline{B}^T$ and $\mathbf{a} \bullet \mathbf{b}$.

The finite element algorithm called MIX1 in [21], which has been used in many earlier publications, is based on the formulation (2.12–14) together with a $P^2 - C^0$ interpolation for T^h and V^h and $P^1 - C^0$ interpolation for P^h. Such an algorithm is able to calculate creeping flows of viscoelastic fluids for low values of We. When We increases, a first sign of numerical inaccuracy is observed when $\underline{T}_A$ loses its positiveness [14]. This early symptom is soon followed by the loss of convergence of Newton's method for solving the non-linear algebraic system of equations resulting from (2.12–14). The loss of positiveness of $\underline{T}_A$ causes in fact an *artificial change of type* of the discretized version of the field and constitutive equations.

Much higher values of the Weissenberg number have been reached by means of a new mixed method introduced by Marchal and Crochet [18]. The method rests on two important improvements of the earlier mixed methods.

First, an analysis of the structure of the mixed method for the stress-velocity-pressure formulation reveals that the stresses are the primary variables; the stress equations of motion act as a constraint on the stress components, while the incompressibility condition acts as a constraint on the velocity field. The spaces

T^h, V^h, P^h must satisfy the Babuska-Brezzi conditions of stability [22-23]; it is not the case with the MIX1 element [24]. The element developed in [18] consists of a $P^2 - C^0$ representation for the velocity field and a $P^1 - C^0$ representation for the pressure. For the stresses, Marchal and Crochet [18] subdivide the quadrilaterals for the velocity field into 4×4 sub-elements with bilinear shape functions. It has been shown in [24] that such a mixed element converges for Stokes flow.

The second ingredient of the method introduced in [18] has to do with the hyperbolic character of the constitutive equation (2.2), where the streamlines are characteristics. It is well-known that the classical Galerkin (or Bubnov-Galerkin) method is unsuitable for solving advection dominated problems, unless one resorts to elements of small and sometimes unpractical size. In [18] Marchal and Crochet make use of the streamline-upwind method developed by Brooks and Hughes [25] with a special emphasis on bilinear quadrilateral elements. Let us consider in Fig. 1 an isoparametric bilinear element together with the deformed ξ and η axes of the parent element and let v_ξ, v_η denote the scalar product of the velocity vector $\mathbf{v}$ and the vectors $\mathbf{h}_\xi$ and $\mathbf{h}_\eta$ indicated on Fig. 1. Let $\overline{k}$ and $\mathbf{w}$ be defined as follows,

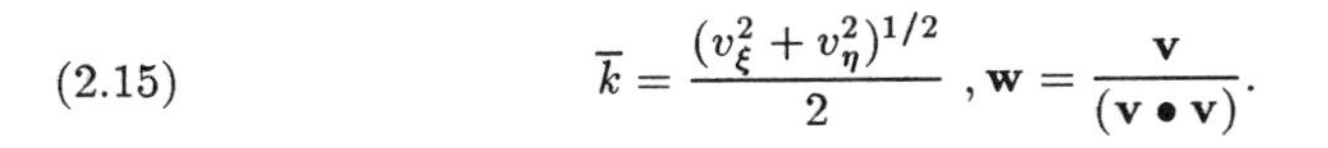

$$(2.15) \qquad \overline{k} = \frac{(v_\xi^2 + v_\eta^2)^{1/2}}{2}, \mathbf{w} = \frac{\mathbf{v}}{(\mathbf{v} \bullet \mathbf{v})}.$$

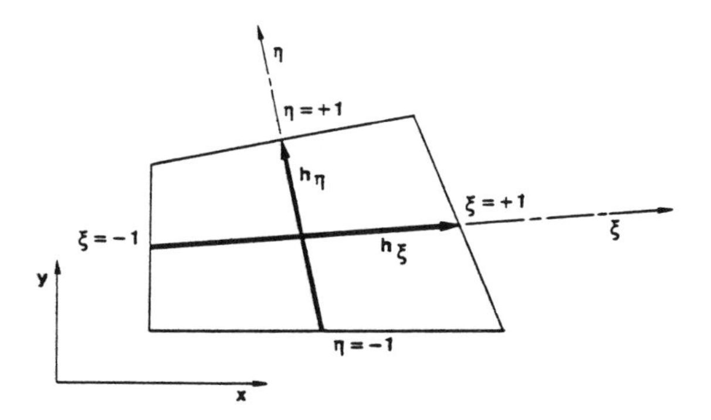

Fig. 1. Isoparametric bilinear element for the streamline-upwind scheme.

Instead of (2.12), we adopt the following form for the weak formulation of the constitutive equation,

$$(2.16) \quad \langle \underline{T}^h + \lambda \overset{\nabla}{\underline{T}}{}^h - \eta(\nabla \mathbf{v}^h + \nabla \mathbf{v}^{hT}); \underline{S}^h \rangle + \langle \lambda \mathbf{v} \bullet \nabla \underline{T}^h ; \overline{k} \mathbf{w} \bullet \nabla \underline{S}^h \rangle = 0, \quad \forall \underline{S}^h \in T^h .$$

The streamline-upwind scheme introduces an artificial stress diffusivity along the streamlines. However, it is easy to show [19] that the diffusivity vanishes when the size h of the element goes to zero.

The method has been successfully applied to the calculation of several creeping flows. We show below that it can also be used for problems where inertia cannot be neglected, and, in particular, when the vorticity equation changes type in some parts of the flow.

3. Viscoelastic Flow Through Channels. Yoo and Joseph [20] have examined the consequences of viscoelasticity and inertia on the flow of a Maxwell fluid through a planar channel with wavy walls. The vorticity equation changes type when the velocity in the center of the channel is larger than the velocity of propagation of shear waves. The geometry of the problem is shown in Fig. 2; in a Cartesian system of coordinates, the location of the wall is given by the equation

$$y = H(1 + \varepsilon \sin\ 2\pi x/L), \tag{3.1}$$

where H is the half width of the channel, L is the wave length of the perturbation, and ε a small parameter. In what follows, we select a ratio $H/L = 0.5$. Let V_0 denote the unperturbed velocity along the plane of symmetry. The Reynolds and Weissenberg number are defined as follows,

$$Re = \rho V_0 H/\eta, \qquad We = \lambda V_0/H\ , \tag{3.2}$$

while the viscoelastic Mach number is

$$M = V_0/(\eta/\lambda\rho)^{1/2}\ . \tag{3.3}$$

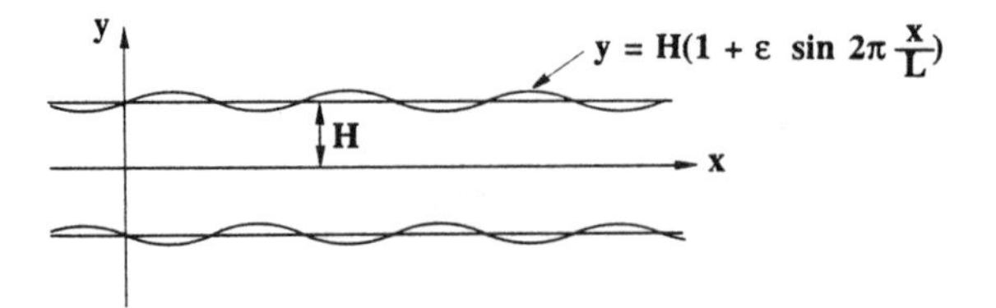

Fig. 2. Geometry of the channel with wavy walls.

Yoo and Joseph [20] solve the vorticity equation by means of a domain perturbation analysis. Through a separation of variables, they reduce the problem to a set of two fourth order differential equations which are solved by a standard shooting technique employing a fourth order Runge-Kutta method. Their results are given in terms of the first order perturbation of the vorticity and the streamfunction. More precisely, let ω and ψ denote respectively the vorticity and the streamfunction, and let ω_0, ψ_0 denote their values in the unperturbed flow, that is the Poiseuille flow. The first order perturbation ω_1, ψ_1 is defined as

$$\omega = \omega_0 + \varepsilon\omega_1 + \ldots,\ \psi = \psi_0 + \varepsilon\psi_1 + \ldots\ . \tag{3.4}$$

Several plots of ω_1 and ψ_1 are presented in [20] for various values of We and Re.

We have performed a similar analysis by means of the numerical method explained in section 2. We have selected a small value of $\varepsilon = 0.001$, and obtained numerical results $\tilde{\omega}$ and $\tilde{\psi}$ for the vorticity and streamfunction. The first order perturbation ω^*, ψ^* is then deduced as follows

$$\omega^* = (\tilde{\omega} - \omega_0)/\varepsilon, \qquad \psi^* = (\tilde{\psi} - \psi_0)/\varepsilon \; ; \tag{3.5}$$

these functions can be readily compared with ω_1 and ψ_1.

The finite element mesh covers one wavelength of the wavy channel. It consists of a uniform distribution of eight elements in the flow direction and twenty four in the width. We impose the flow rate through the channel and apply periodic boundary conditions. No-slip conditions are imposed along the wavy wall. The primitive variables of the finite element calculation are the extra stress and velocity components and the pressure. The streamfunction is calculated a posteriori by solving the equation

$$\Delta\tilde{\psi} = \frac{\partial \tilde{u}}{\partial y} - \frac{\partial \tilde{v}}{\partial x} \tag{3.6}$$

while, for the vorticity, we calculate a continuous representation $\tilde{\omega}$ by means of a least square projection of the discontinuous function

$$\omega^+ = \frac{\partial \tilde{v}}{\partial x} - \frac{\partial \tilde{u}}{\partial y} \; . \tag{3.7}$$

We have analyzed three situations with three numerical methods. The mixed method MIX1 does not use stress sub-elements and streamline upwinding. The 4×4 algorithm is a mixed method with stress sub-elements introduced in [18] with no streamline upwinding. Finally, the $SU4\times4$ algorithm is the method introduced by Marchal and Crochet in [18].

In Figs 3 to 5, we show contour lines of ψ^* on the left and ω^* on the right. The wavy wall is on top of the figures; the flow goes from right to left. In Fig. 3, we show a subcritical case at $Re = 1$, $We = .01$ and $M = .1$. This is an easy problem where we find an excellent correspondence between Yoo and Joseph's analytical results and the numerical results with all three algorithms. In Fig. 4, we consider a supercritical case (which was not analyzed by Yoo and Joseph) where $Re = 5$, $We = .5$ and $M = 1.58$. Although the Weissenberg number is low, we find that MIX1 is unable to produce smooth vorticity contours. The situation improves with 4×4 (in the absence of upwinding), while smooth contourlines are obtained with $SU4\times4$. It is interesting to note that the streamlines are the same with all three algorithms. In Fig. 5, we analyze the situation $Re = 50$, $We = .5$ and $M = 5$ and compare the results with those of Yoo and Joseph. Again, we find that the streamline pattern is good with all three algorithms. The MIX1 method is unable to provide smooth vorticity contours. The situation improves with 4×4 while smooth lines are obtained with $SU4\times4$. The comparison with the analytical results is excellent. On the right hand side of the figures, we note that $SU4\times4$ slightly planes some of the contours. This is attributed to the artificial stress diffusivity which is proper to the streamline upwind method [19].

We conclude from these examples that, at least in these simple cases, the $SU4\times4$ algorithm is able to calculate solutions where the vorticity equation changes type.

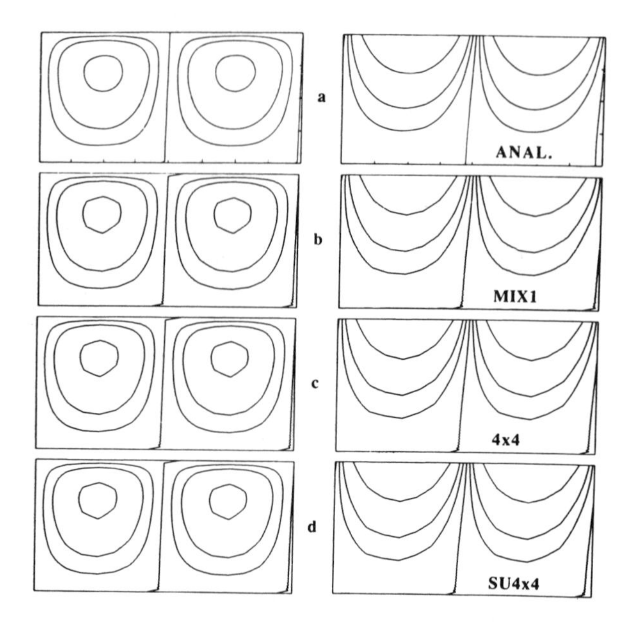

Fig. 3. Streamlines and vorticity contours at $Re = 1$, $We = .01$ and $M = .1$. The corresponding values of the streamfunction are $0, \pm.05, \pm.1$ and $\pm.2$. The values of the vorticity are $0, \pm1, \pm2$ and ±5.

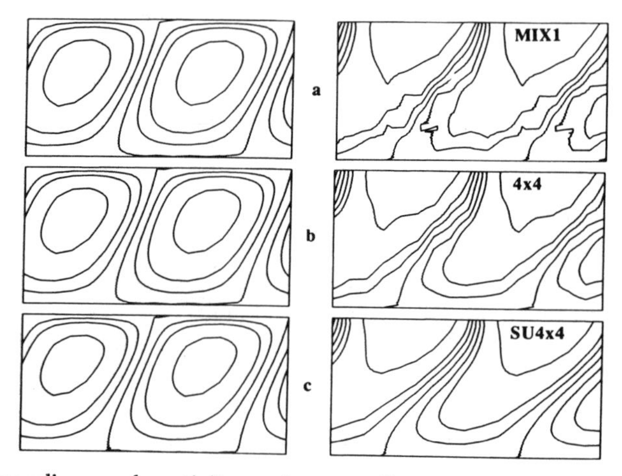

Fig. 4. Streamlines and vorticity contours at $Re = 5$, $We = .5$ and $M = 1.58$. The associated values are the same as in Fig. 3.

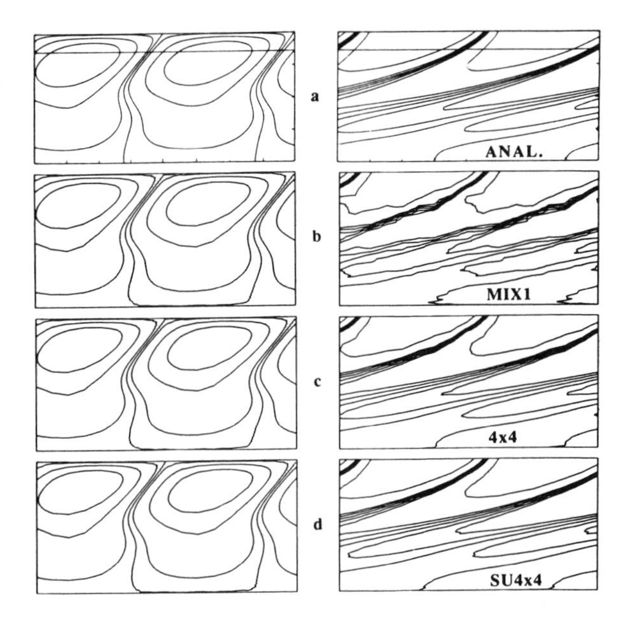

Fig. 5. Streamlines and vorticity contours at $Re = 50$, $We = .5$ and $M = 5$. The corresponding values of the streamfunction are $0, \pm.01, \pm.05$ and $\pm.1$. The values of the vorticity are $0, \pm.5, \pm1$ and ±5.

4. Flow around a circular cylinder. Let us consider the combined effect of viscoelasticity and inertia on the flow around a circular cylinder. Fig. 6 shows the general expectation as explained by Joseph in his recent book [11], for a supercritical condition. "Vorticity is created at the surface of the body due to the no-slip condition, but this vorticity cannot propagate upstream into the region of uniform flow where the velocity is larger than the speed c of vorticity waves into rest. In a linear theory, the first changes in the vorticity would occur across the leading characteristic surface which, in the axisymmetric case, forms a cone like the Mach cone of gas dynamics. The undisturbed region in front of the cone is like the region of silence in gas dynamics." With our numerical method, we are not constrained by the linearized theory. We are able to consider a flow which is supercritical away from the body and subcritical near the body.

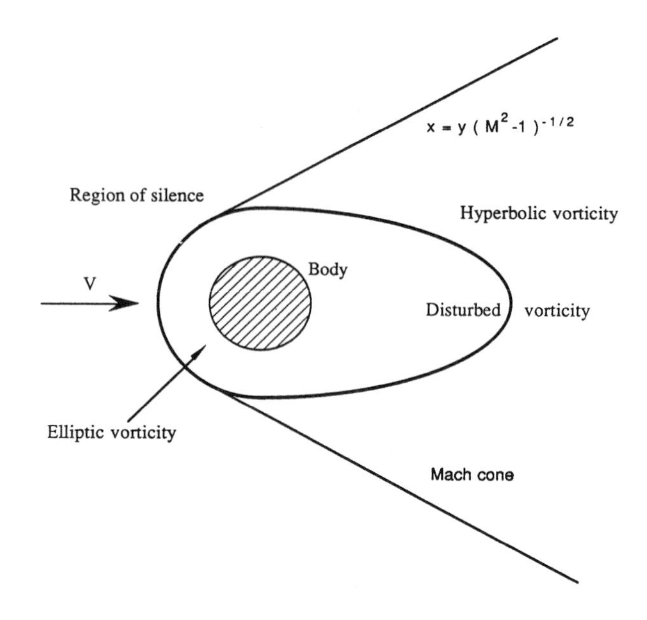

Fig. 6. Expected distribution of the vorticity in a perturbed uniform flow around a circular cylinder.

In the present chapter, we limit ourselves to the effect of change of type upon the flow kinematics and in particular upon the vorticity. In a later paper [26], we will analyze the consequences of the modified kinematics upon drag and heat transfer, which are indeed dramatic.

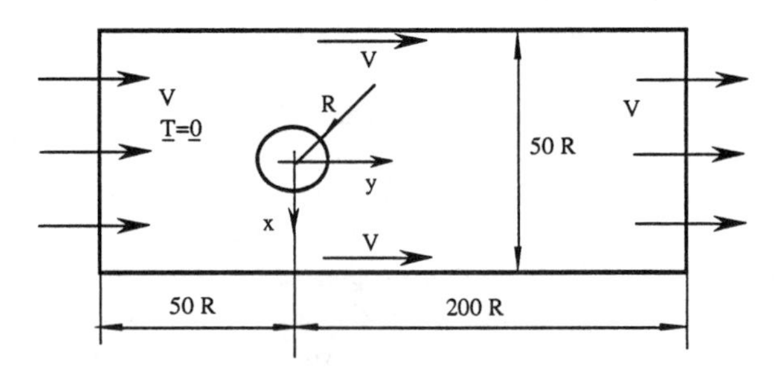

Fig. 7. Geometry and boundary conditions for the flow around a circular cylinder.

The geometry and boundary conditions of the problem are shown in Fig. 7. For economical purposes, we consider the flow of a Maxwell fluid around a circular cylinder located between two parallel walls. The distance between the walls is $50R$, where R is the radius of the cylinder. Preliminary calculations have shown that, for the range of Re and We that we wish to consider, we need upstream and downstream regions which are respectively $50R$ and $200R$ long. We impose a uniform velocity profile in the entry and exit sections, and a uniform velocity V along the walls. The velocity vanishes on the cylindrical wall. Additionally, we impose vanishing stresses in the entry section. For Re, We and M we use the definitions

$$Re = \rho V R/\eta, \qquad We = \lambda V/R, \qquad M = V/(\eta/\lambda\rho)^{1/2} . \tag{4.1}$$

The numerical method is $SU4\times4$. On the basis of the velocity field, we calculate the streamfunction ψ and the vorticity ω as in section 3. Another quantity of interest for the present problem is the sign of $\det \underline{T}_B$, defined by (2.11). A negative value of $\det \underline{T}_B$ indicates the extent of the region where the vorticity equation is hyperbolic.

A partial view of the finite element mesh used for the results of the present section is shown in Fig. 8. In our later paper [26], we will show that our results are confirmed by a coarser and a finer grid. The mesh contains 532 elements, 2233 nodes and 31214 degrees of freedom. We note that very thin elements have been selected near the cylindrical wall with a view to a good definition of the boundary layer.

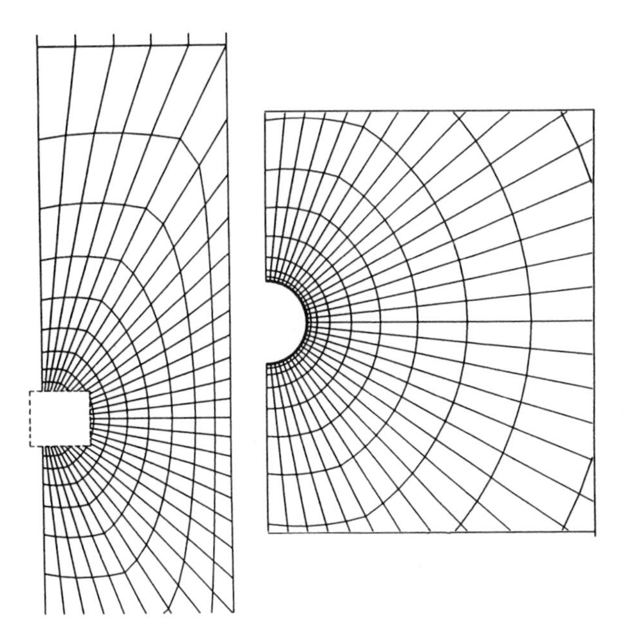

Fig. 8. Partial view of the finite element mesh together with a zoom near the cylindrical wall.

Let us first analyze the results obtained for creeping flow, i.e. for the purely viscoelastic case. We have calculated solutions up to $We = 10$, at which value a more refined mesh is already needed near the cylindrical boundary. In Fig. 9, we show streamlines near the cylinder for respective values of We equal to 0, 5 and 10. The step between the streamlines is the same for all three values. We observe that their spacing increases when We increases; this indicates a tendency to stagnation near the cylinder when viscoelastic effects increase. We also detect a slight downstream shift of the streamlines. In Fig. 10, we analyze the axial velocity profile and the vorticity on the axis $y = 0$ indicated in Fig. 7. We note a hump in the velocity profile, since the flow rate is the same for all sections. As we expect from Fig. 9, we find that the velocity gradient is less steep at some distance of the cylinder for high values of We. We find however that the vorticity is steeper near the cylinder when We increases. The wiggles at $We = 10$ indicate that the finite element mesh is too coarse for an accurate resolution of the boundary layer. The vorticity contours are shown in Fig. 11. We find that, at $We = 5$, the contourlines are almost symmetric with respect to the line $y = 0$. The wiggly behavior at $We = 10$ again indicates the coarseness of the mesh.

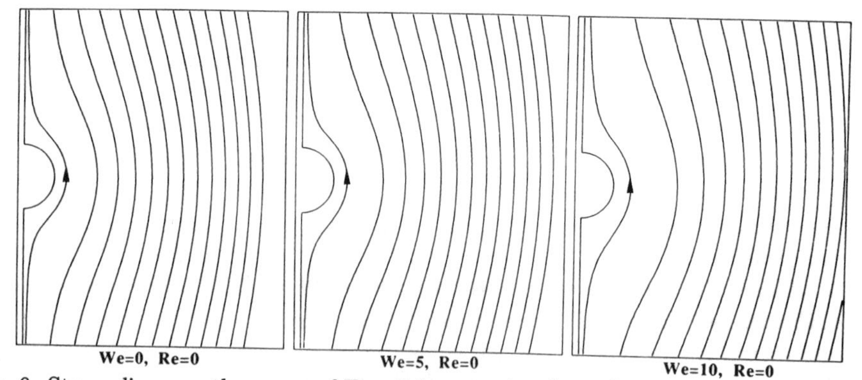

Fig. 9. Streamlines on the zoom of Fig. 8 for creeping flow. Streamfunction values from left to right are given by $(24.95 - n\ 0.4)/VR$, with $n = 0, \ldots, 12$.

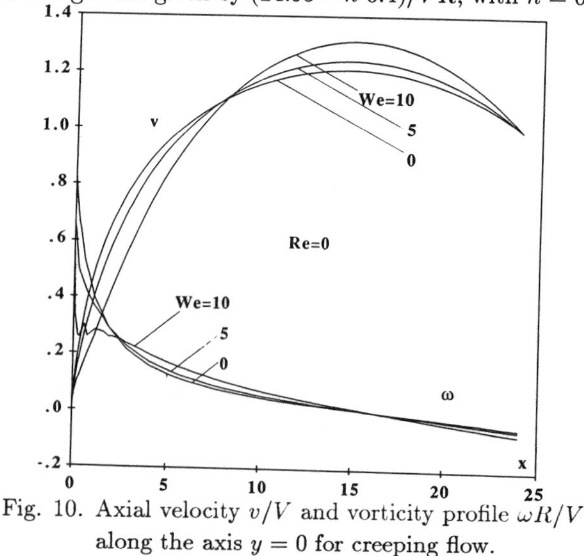

Fig. 10. Axial velocity v/V and vorticity profile $\omega R/V$ along the axis $y = 0$ for creeping flow.

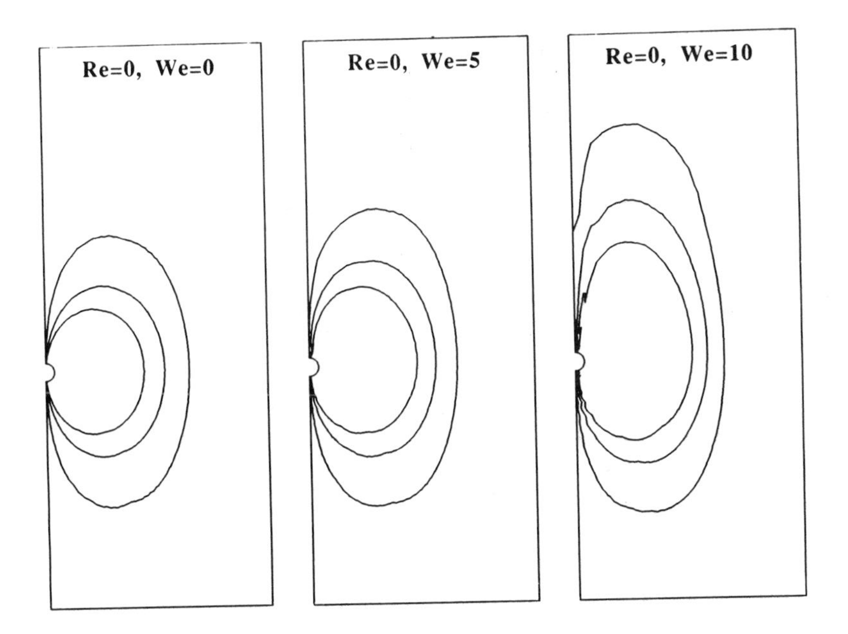

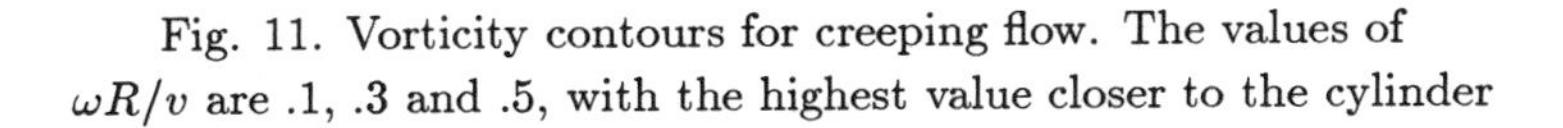
Fig. 11. Vorticity contours for creeping flow. The values of $\omega R/v$ are .1, .3 and .5, with the highest value closer to the cylinder

These relatively high values of We allow us to consider small values of Re. Indeed, at $We = 10$, criticality is reached at $Re = 0.1$. Before examining the combined effects of inertia and viscoelasticity, let us consider the purely Newtonian situation with inertia. Fig. 12 shows the streamlines at $Re = 0.3$ and $Re = 0.6$. The effect of inertia is clearly a downstream shift of the streamlines with little effect on the extent of the stagnant region. This is confirmed in Fig. 13, where we also find little change in the vorticity when Re goes from 0.3 to 0.6. However, the effect of inertia on vorticity, even at such low values of Re, is clearly seen in Fig. 14; we find that the contourlines are shifted in the downstream direction. Let us observe in particular the spacing between the isovorticity lines in the upstream region, which will be modified once the vorticity equation changes type.

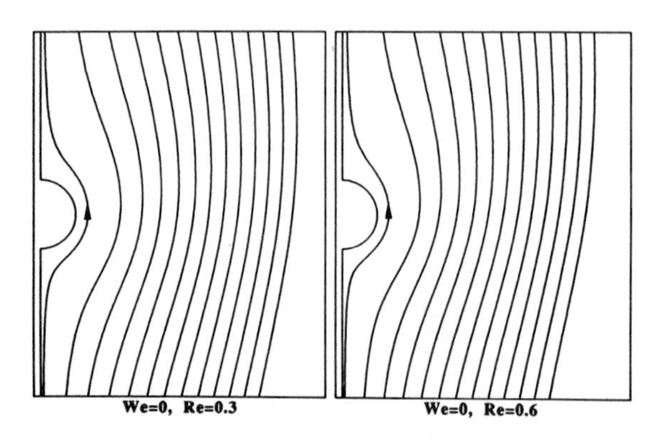

Fig. 12. Streamlines on the zoom of Fig. 8 for inelastic flow. The values are the same as in Fig. 9.

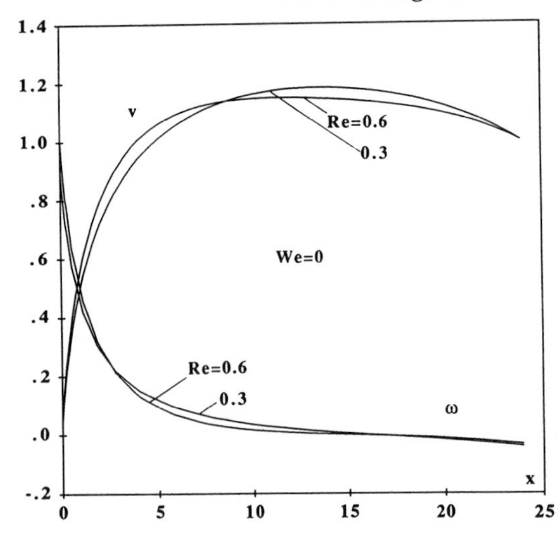

Fig. 13. Axial velocity v/V and vorticity profile $\omega R/V$ along the axis $y = 0$ for inelastic flow.

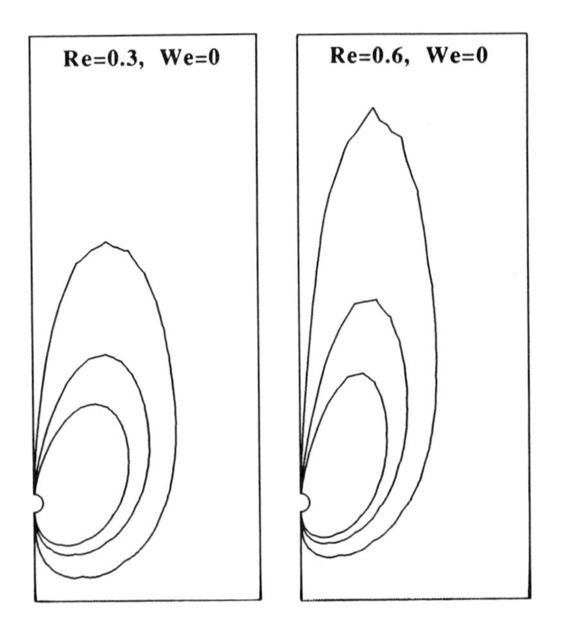

Fig. 14. Vorticity contours for inelastic flow. The values are the same as in Fig. 11.

Let us now analyze the coupled effect of viscoelasticity and inertia at $We = 5$. In Fig. 15, we show the extent of the hyperbolic region when Re goes from 0.18 to 0.6. At $Re = 0.18$, a hyperbolic region has developed between the cylinder and the wall, in view of the hump in the velocity profile detected in Fig. 10, although the Mach number, based on the velocity on the wall, is still below 1. We find that, at $M = 1.5$ and $M = 1.75$, the cylinder is still surrounded by an elliptic region. The effect on the streamlines is shown in Fig. 16 which, compared with Fig. 9, shows that the downstream shift is accentuated, as we might actually expect from Fig. 12. One of the most striking features is shown in Fig. 17, where we show the axial velocity and vorticity profiles along the line $y = 0$. The major difference between the solution at $M = 0.95$ on one hand and $M = 1.5$ and 1.73 on the other is that the velocity profile is affected by an essentially discontinuous slope. The location of the discontinuity corresponds to the shock of vorticity observed on the same figure. The isovorticity lines of Fig. 18, to be compared with those of Fig. 14, confirm the existence of such a shock. On Fig. 18, we have also indicated the slope of the Mach cone, parallel to the line $x = y\ (M^2 - 1)^{-1/2}$. Quite clearly, the calculations exhibit the features of Fig. 6. The wiggles on the outer contour line are due to fact that ω is quite flat in the immediate neighborhood of the shock, and that the mesh is coarse in the radial direction at such a distance from the cylinder (see Fig. 8).

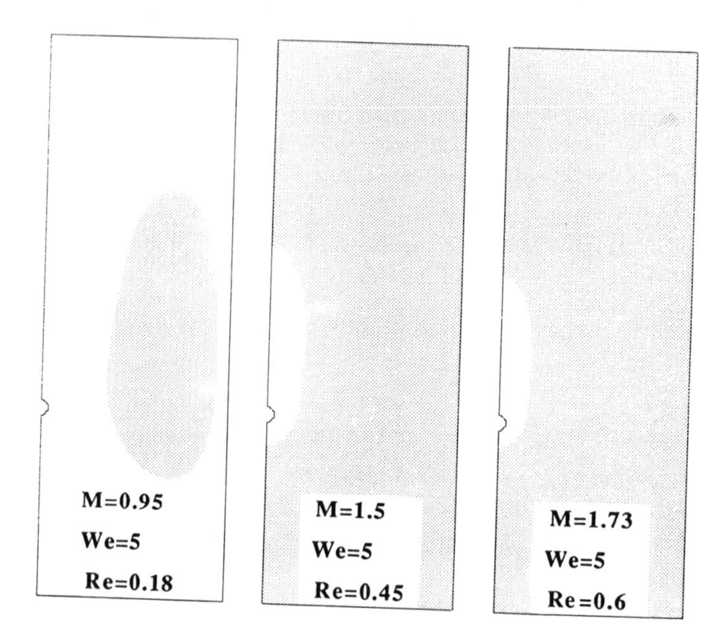

Fig. 15. Region of hyperbolic vorticity indicated by the shaded area.

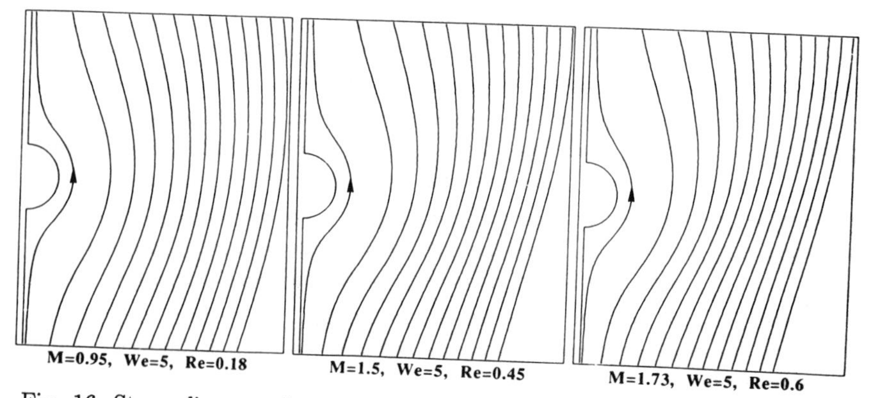

Fig. 16. Streamlines on the zoom of Fig. 8 for viscoelastic flow, $We = 5$. The values are the same as in Fig. 9.

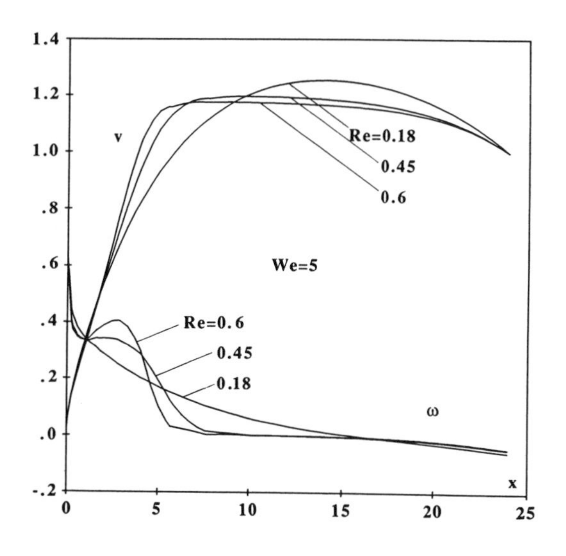

Fig. 17. Axial velocity v/V and vorticity profile $\omega R/V$ along the axis $y = 0$ for viscoelastic flow, $We = 5$.

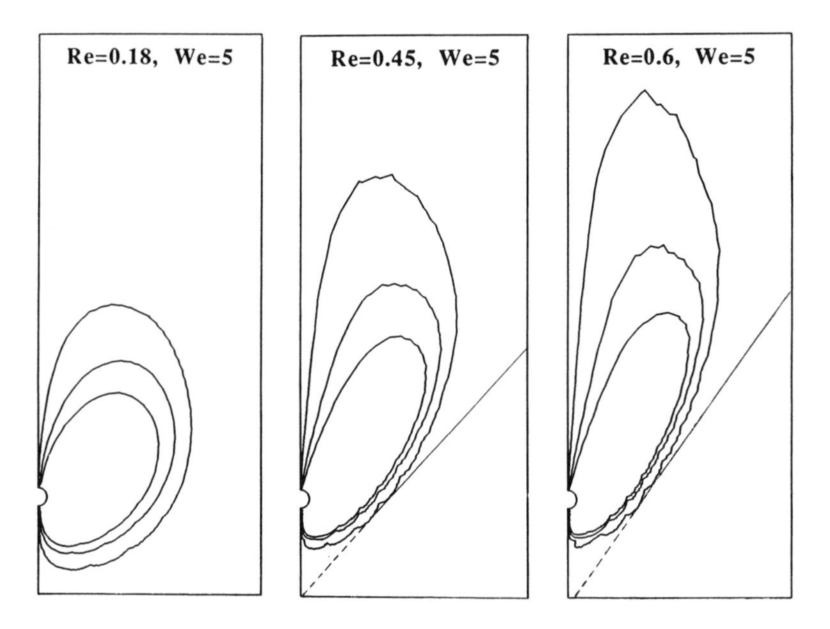

Fig. 18. Vorticity contours for viscoelastic flow, $We = 5$. The values are the same as in Fig. 11.

The problem has also been studied for the same values of M at a higher value of We. The elasticity We/Re is now four times as high. In Fig. 19, we observe that the extent of the downstream elliptic region is now definitely larger, as anticipated by Joseph [11]. Again, Fig. 20 demonstrates the abrupt change of the vorticity.

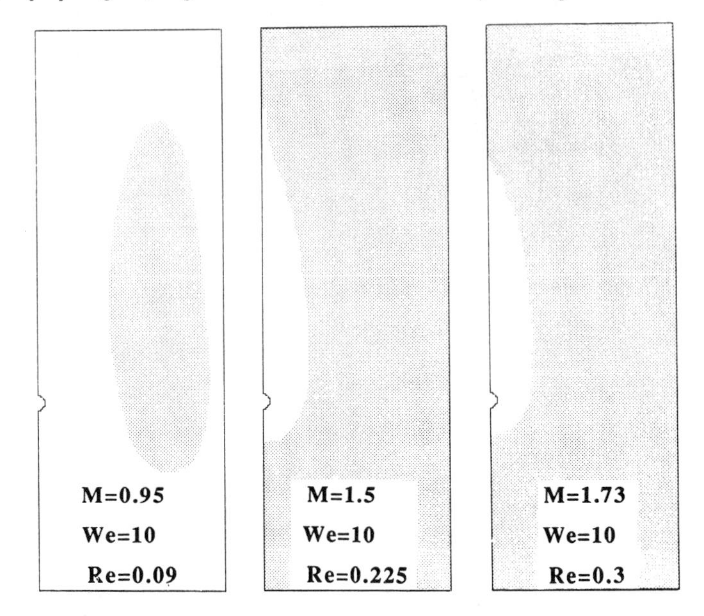

Fig. 19. Region of hyperbolic vorticity indicated by the shaded area.

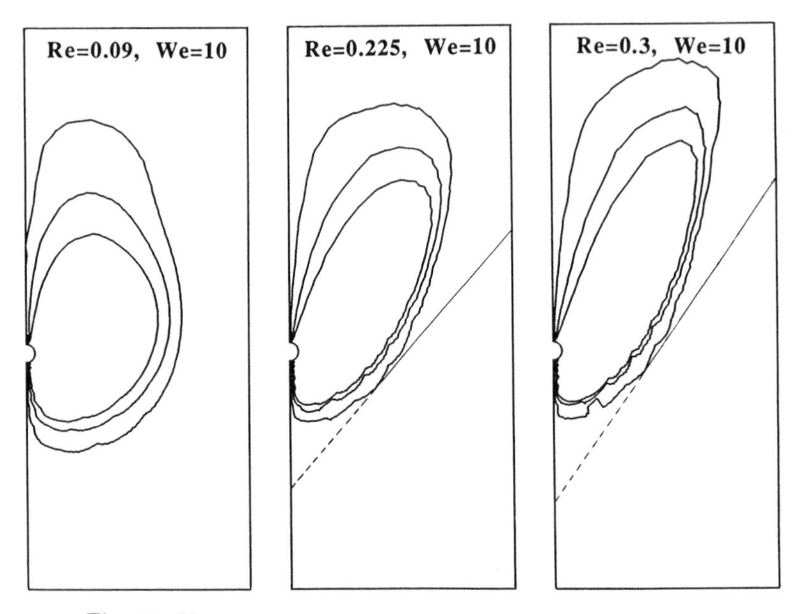

Fig. 20. Vorticity contours for viscoelastic flow, $We = 10$.
The values are the same as in Fig. 11.

Conclusions. We have shown that a mixed finite element method with a proper integration of the constitutive equations allows one to calculate flows of viscoelastic fluids with instantaneous elasticity in the supercritical regime. The validity of the algorithm has been tested on the flow through a wavy channel which has been calculated analytically in [20]. The flow of a Maxwell fluid around a cylindrical body shows that the coupling between viscoelasticity and inertia can produce dramatic effects at fairly low values of the Reynolds number. In particular, we have observed the shock transition between regions of vanishing and active vorticity. In a later paper [26], we will analyze the consequences of such flows on heat and momentum transfer properties.

REFERENCES

[1] I.M. Rutkevich, *The propagation of small perturbations in a viscoelastic fluid*, J. Appl. Math. Mech., 34 (1970), 35–50.

[2] I.M. Rutkevich, *On the thermodynamic interpretation of the evolutionary conditions of the equation of the mechanics of finitely deformable viscoelastic media of Maxwell type*, J. Appl. Math. Mech., 36 (1972), 283–295.

[3] D.D. Joseph, M. Renardy and J.C. Saut, *Hyperbolicity and change of type in the flow of viscoelastic fluids*, Arch. Ration. Mech. Anal., 87 (1985), 213–251.

[4] D.D. Joseph and J.C. Saut, *Change of type and loss of evolution in the flow of viscoelastic fluids*, J. Non-Newtonian Fluid Mech., 20 (1986), 117–141.

[5] F. Dupret and J.M. Marchal, *Sur le signe des valeurs propres du tenseur d' extra-contraintes dans un écoulement de fluide de Maxwell*, J. de Mécanique théorique et appliquée, 5 (1986), 403–427.

[6] F. Dupret and J.M. Marchal, *Loss of evolution in the flow of viscoelastic fluids*, J. Non-Newtonian Fluid Mech., 20 (1986), 143–171.

[7] A.B. Metzner, E.A. Uebler and C.F.C.M. Fong, *Converging flows of viscoelastic materials*, Am. Inst. Ch. Eng. J., 15 (1969), 750–758.

[8] D.D. Joseph, J.E. Matta and K. Chen, *Delayed die swell*, J. Non-Newtonian Fluid Mech., 24 (1987), 31–65.

[9] D.F. James and A.J. Acosta, *The laminar flow of dilute polymer solutions around circular cylinders*, J. Fluid Mech., 42 (1970), 269–288.

[10] J.S. Ultman and M.M. Denn, *Anomalous heat transfer and a wave phenomenon in dilute polymer solutions*, Trans. Soc. Rheol., 14 (1970), 307–317.

[11] D.D. Joseph, *Fluid dynamics of viscoelastic liquids*, Springer-Verlag (1989).

[12] M.J. Crochet, *Numerical simulation of viscoelastic flow: A review*, Rubber Chemistry and Technology - Rubber Reviews for 1989.

[13] R. Keunings, *Simulation of viscoelastic flow*, in C.L. Tucker III (ed). Fundamentals of computer modeling for polymer processing (1988).

[14] F. Dupret, J.M. Marchal and M.J. Crochet, *On the consequence of discretization errors in the numerical calculation of viscoelastic flow*, J. Non-Newtonian Fluid Mech., 18 (1985) 173–186.

[15] R.A. Brown, R.C. Armstrong, A.N. Beris and P.W. Yeh, *Galerkin finite element analysis of complex viscoelastic flows*, Comp. Meth. Appl. Mech. Eng., 58 (1986), 201–226.

[16] J.H. Song and J.Y. Yoo, *Numerical simulation of viscoelastic flow through sudden contraction using type dependent method*, J. Non-Newtonian Fluid Mech., 24 (1987), 221–243.

[17] R.E. Gaidos and R. Darby, *Numerical simulation and change in type in the developing flow of a non linear viscoelastic fluid*, J. Non-Newtonian Fluid Mech., 29 (1988), 59–79.

[18] J.M. Marchal and M.J. Crochet, *A new mixed finite element for calculating viscoelastic flow*, J. Non-Newtonian Fluid Mech., 26 (1987) 77–114.

[19] M.J. Crochet, V. Delvaux and J.M. Marchal, *On the convergence of the streamline-upwind mixed finite element*, J. Non-Newtonian Fluid Mech., submitted (1989).

[20] J.Y. Yoo and D.D. Joseph, *Hyperbolicity and change of type in the flow of viscoelastic fluids through channels*, J. Non-Newtonian Fluid Mech., 19 (1985) 15–41.

[21] M.J. Crochet, A.R. Davies and K. Walters, *Numerical simulation of Non-Newtonian flow*, Elsevier, Amsterdam (1984).

[22] I. Babuska, *The finite element method with Lagrangian multipliers*, Numer. Math., 20 (1973) 179–192.

[23] F. Brezzi, *On the existence, uniqueness and approximation of saddle-point problems arising from Lagrange multipliers*, RAIRO Num. Analysis, 8-R2 (1974), 129–151.

[24] M. Fortin and R. Pierre, *On the convergence of the mixed method of Crochet and Marchal for viscoelastic flows*, to appear in Com. Meth. Appl. Mech. Eng. (1989).

[25] A.N. Brooks and T.J.R. Hughes, *Streamline-upwind/Petrov-Galerkin formulations for convection dominated flows with particular emphasis on the incompressible Navier-Stokes equations*, Comp. Meth. Appl. Mech. Eng., 32 (1982) 199–259.

[26] V. Delvaux and M.J. Crochet, in preparation.

SOME QUALITATIVE PROPERTIES OF 2×2 SYSTEMS OF CONSERVATION LAWS OF MIXED TYPE

H. HOLDEN* AND L. HOLDEN† AND N. H. RISEBRO‡

Abstract. We study qualitative features of the initial value problem $z_t + F(z)_x = 0$, $z(x,0) = z_0(x)$, $x \in \mathbf{R}$, where $z(x,t) \in \mathbf{R}^2$, with Riemann inital data, viz. $z_0(x) = z_l$ if $x < 0$ and $z_0(x) = z_r$ if $x > 0$. In particular we are interested in the case when the system changes type when the eigenvalues of the Jacobian dF become complex. It is proved that if z_l and z_r are in the elliptic region, and the elliptic region is convex, then part of the solution has to be outside the elliptic region. If both z_l and z_r are in the hyperbolic region, then the solution will not enter the elliptic region. We show with an explicit example that the latter property is not true for general Cauchy data. This example is investigated numerically.

Key words. conservation laws, mixed type, Riemann problems

AMS(MOS) subject classifications. 35L65,35M05,76T05

1. Introduction. In this note we analyze certain qualitative properties of the 2×2 system of partial differential equations in one dimension of the form

$$\frac{\partial}{\partial t}\begin{pmatrix} u \\ v \end{pmatrix} + \frac{\partial}{\partial x}\begin{pmatrix} f(u,v) \\ g(u,v) \end{pmatrix} = 0 \tag{1.1}$$

with $u = u(x,t)$, $v = v(x,t)$, $x \in \mathbf{R}$. In particular we are interested in the initial value problem with Riemann initial data, i.e.

$$\begin{pmatrix} u(x,0) \\ v(x,0) \end{pmatrix} = \begin{cases} \binom{u_l}{v_l}, & \text{for } x < 0 \\ \binom{u_r}{v_r}, & \text{for } x > 0 \end{cases} \tag{1.2}$$

where u_l, u_r, v_l, v_r are constants.

The system (1.1),(1.2) arises as a model for a diverse range of physical phenomena from traffic flow [2] to three–phase flow in porous media [1]. Common for these applications is that one obtains from very general physical assumptions a system of mixed type, i.e. there is a region $E \subset \mathbf{R}^2$ of phase space where the 2×2 matrix

$$dF = \begin{pmatrix} f_u(u,v) & f_v(u,v) \\ g_u(u,v) & g_v(u,v) \end{pmatrix} \tag{1.3}$$

has no real eigenvalues. The system is then called elliptic in E.

---

*Division of Physics, Mathematics and Astronomy, California Institute of Technology, Pasadena, CA 91125, USA. On leave from Institute of Mathematics, University of Trondheim, N–7034 Trondheim–NTH, Norway. Supported in part by Vista, NAVF and NSF grant DMS-8801918. H.H. would like to thank Barbara Keyfitz, Michael Shearer and the Institute for Mathematics and its Applications for organizing a very stimulating workshop and for the invitation to present these results.

†Norwegian Computing Center, P.O. Box 114, Blindern, N–0314 Oslo 3, Norway. Supported in part by Vista and NTNF.

‡Department of Mathematics, University of Oslo, P.O. Box 1053, Blindern, N–0316 Oslo 3, Norway. Supported by Vista and NTNF.

Consider e.g. the case of three–phase flow in porous media where the unknown functions u and v denote saturations, i.e. relative volume fractions, of two of the phases, e.g. oil and water respectively. A recent numerical study [1] gave as a result with realistic physical data that there in fact is a small compact region E in phase space, and quite surprisingly the Riemann problem (1.1),(1.2) turned out to be rather well–behaved numerically in this situation.

Subsequent mathematical analysis [25], [9], [16], [27] showed that one in general has to expect mixed type behavior in this case. Also in applications to elastic bars and van der Waal fluids [14], [28], [22], [23], [24] there is mixed type behavior. See also [20], [10], [11], [12], [13], [15], [17], [18], [19].

Parallel to this development there has been a detailed study of certain model problems with very simple flux functions (f, g) with elliptic behavior in a compact region E which has revealed a very complicated structure of the solution to the Riemann problem [7], [8]. In general one must expect nonuniqueness of the solution for Riemann problems, see [5].

We prove two theorems for general 2×2 conservation laws of mixed type. Specifically the flux function is *not* assumed to be quadratic. The first theorem states that if z_l is in the elliptic region E, then z_l is the only point on the Hugoniot locus of z_l inside E provided E is convex. In the second theorem we show that one cannot connect a left state outside E via an intermediate state inside E to a right state outside E if we only allow shocks with viscous profiles as defined by (2.21).

This latter theorem has also been proved independently by Azevedo and Marchesin (private communication).

Combining these two theorems we see that if $z_l, z_r \notin \bar{E}$ then also the solution $z(x,t) \notin E$ for all $x \in \mathbf{R}, t > 0$. Finally we explicitly show that this property is *not* valid for the general Cauchy problem. The consequences of this for the Glimm's scheme is discussed by Pego and Serre[21] and Gilquin[3]. For the most recent result on conservation laws of mixed type we refer to the other contributions to these proceedings.

2. Qualitative properties. We write (1.1) as

$$z_t + F(z)_x = 0 \tag{2.1}$$

where $z = \binom{u}{v}$ and $F = \binom{f}{g}$, with Riemann initial data

$$z(x,0) = \begin{cases} z_l, & \text{for } x < 0 \\ z_r, & \text{for } x > 0. \end{cases} \tag{2.2}$$

We assume that f and g are real differentiable functions such that the Jacobian dF has real eigenvalues exept in components of $\mathbf{R}^2$, each of which are convex. Let

$$E = \Big\{ z \in \mathbf{R}^2 | \lambda_j(z) \notin \mathbf{R} \Big\}. \tag{2.3}$$

A shock solution is a solution of the form

$$z(x,t) = \begin{cases} z_l, & \text{for } x < st \\ z_r, & \text{for } x > st. \end{cases} \tag{2.4}$$

where the shock speed s must satisfy the Rankine–Hugoniot relation [29]

$$s(z_l - z_r) = F(z_l) - F(z_r). \tag{2.5}$$

The Hugoniot locus of z_l is the set of points satisfying

$$H_{z_l} = \left\{ z \in \mathbf{R}^2 | \exists s \in \mathbf{R}, s(z_l - z) = F(z_l) - F(z) \right\}. \tag{2.6}$$

For $z \in E$ we let E_z denote the convex component of E containing z. Then we have

THEOREM 2.1. *Let $z_l \in E$ and assume that E_{z_l} is convex, then*

$$H_{z_l} \cap E_{z_l} = \{z_l\} \tag{2.7}$$

and if $z_r \in E$ and E_{z_r} is convex, then

$$H_{z_r} \cap E_{z_r} = \{z_r\}. \tag{2.8}$$

Proof. We will show (2.7), (2.8) then follows by symmetry. Let $z_r \in H_{z_l}$ and assume that

$$z_r \in E_{z_l}. \tag{2.9}$$

Then the straight line connecting z_l and z_r is contained in E_{z_l}, viz.

$$\alpha(t) = tz_r + (1-t)z_l \in E_{z_l} \tag{2.10}$$

for $t \in [0,1]$ by convexity. Let

$$\beta(t) = F\big(\alpha(t)\big). \tag{2.11}$$

Then

$$\beta'(t) = dF\big(\alpha(t)\big)(z_r - z_l). \tag{2.12}$$

We want to show the existence of $k \in \mathbf{R}$ and of $\tilde{t} \in [0,1]$ such that

$$\beta'(\tilde{t}) = k(z_r - z_l). \tag{2.13}$$

Assuming (2.13) for the moment we obtain by combining (2.12) and (2.13)

$$\Big[dF\big(\alpha(\tilde{t})\big) - k\Big](z_r - z_l) = 0 \tag{2.14}$$

which contradicts (2.10).

To prove (2.13) we consider the straight line passing through $F(z_l)$ in the direction $z_l - z_r$. By assumption

$$s(z_r - z_l) = F(z_r) - F(z_l). \tag{2.15}$$

Using this we see that this line passes through $F(z_r)$ and that there is a $\tilde{t} \in [0,1]$ such that $\beta'(\tilde{t}) || (z_r - z_l)$ proving (2.13). □

This implies that if $z_l \in E$ and $\{z_l, z_r\}$ are the initial values of a Riemann problem, then, the state immediately adjacent to z_l (z_r) in the solution will be outside of E_{z_l} (E_{z_r}). This is so since this state must either be a point on a rarefaction or a shock. Rarefaction curves do not enter E, and we have just shown that neither does the Hugoniot locus.

This result cannot be extended to an arbitrary elliptic region consisting of more than one component, as the following construction shows. Let $z_l \neq z_r$ with $z_r \in H_{z_l}$ and consider two neighborhoods of z_r, viz. $z_r \in \Omega_1 \subset \Omega_2$ such that $z_l \notin \Omega_2$. Let $G(z)$ be a smooth function such that dG is elliptic in Ω_1 and $G(z_r) = F(z_r)$. Now we use G to modify F near z_r

$$\tilde{F}(z) = \begin{cases} F(z) & \text{if} \quad z \notin \Omega_2 \\ G(z) & \text{if} \quad z \in \Omega_1, \end{cases} \tag{2.16}$$

and we make $\tilde{F}$ smooth everywhere. If now z_l is in the elliptic region of F, then both z_l and z_r are in elliptic regions for $\tilde{F}$, and $z_r \in H_{z_l}$ for $\tilde{F}$.

The other basic ingredient in the solution of the Riemann problem is rarefaction waves. These are smooth solutions of the form $z = z(x/t)$ that satisfy (2.1). The value $z(\xi)$ must be an integral curve of r_j, $j = 1, 2$ where r_j is a right eigenvector of dF corresponding to λ_j. ξ is the speed of the wave; $\xi = \lambda_j(z(x/t))$, therefore λ_j has to increase with ξ as z moves from left to right in the solution of the Riemann problem. Note that no rarefaction wave can intersect E since the eigenvectors are not defined there.

For a system of non–strictly hyperbolic conservation laws, the Riemann problem does not in general possess a unique solution, and by making the entropy condition sufficiently lax in order to obtain existence of a solution, one risks losing uniqueness. It is believed, see however [6], that the correct entropy condition which singles out the right physical solution is that the shock should be the limit as $\epsilon \to 0$ of the solution of the associated parabolic equation

$$z_t^\epsilon + F(z^\epsilon)_x = \epsilon z_{xx}^\epsilon \qquad \epsilon > 0.$$

We then say that the shock has a viscous profile. Let now z_l, z_r be two states that can be connected with a shock of speed s. We seek solutions of the form

$$z^\epsilon = z^\epsilon\left(\frac{x - st}{\epsilon}\right) = z^\epsilon(\xi) \tag{2.17}$$

and then obtain

$$-s\frac{d}{d\xi}z^{\epsilon}+\frac{d}{d\xi}F(z^{\epsilon})=\frac{d^2}{d\xi^2}z^{\epsilon} \tag{2.18}$$

which can be integrated to give

$$\frac{d}{d\xi}z^{\epsilon}=F(z^{\epsilon})-sz^{\epsilon}+A \tag{2.19}$$

where A is a constant of integration. If $z^{\epsilon}(\xi)$ converges to the correct solution we must have

$$\lim_{\xi\to-\infty} z^{\epsilon}(\xi)=z_l \qquad \lim_{\xi\to\infty} z^{\epsilon}(\xi)=z_r \tag{2.20}$$

(provided the derivatives decay sufficiently fast) which implies

$$\frac{d}{d\xi}z^{\epsilon}=\big(F(z^{\epsilon})-F(z_l)\big)-s(z^{\epsilon}-z_l). \tag{2.21}$$

We see that z_l and z_r are fixpoints for this field, and if it admits an orbit *from* z_l *to* z_r we say that the shock is *admissible* and has a *viscous profile*. The associated eigenvalues of this field are

$$\lambda_j(z)-s \qquad j=1,2. \tag{2.22}$$

THEOREM 2.2. *Assume that we have two admissible shocks, one connecting the left state* z_l *with a state* z_m *with speed* s_l *and one connecting* z_m *with* z_r *having speed* s_r. *If* z_l *and* z_r *are in the hyperbolic region, i.e.* $z_l, z_r \notin \bar{E}$, *then*

$$z_m \notin \bar{E}. \tag{2.23}$$

Proof. Assume that $z_m \in \bar{E}$. In $\bar{E}$ the eigenvalues constitute a pair of complex conjugates. z_m is a source (sink) if $\mathrm{Re}(\lambda_j(z_m))-s_l>0$ $(\mathrm{Re}(\lambda_j(z_m))-s_r<0)$, hence we obtain

$$s_l \geq \mathrm{Re}(\lambda_j(z_m)) \geq s_r \tag{2.24}$$

which contradicts the fact that z_r is to the right of z_l unless $s_l=s_r$ in which case there is no z_m. □

Combining Theorem 2.1 and Theorem 2.2 we obtain

COROLLARY 2.3. *Consider an admissible solution* $z=z(x,t)$ *of (2.1) with Riemann initial data (2.2). Assume that* E *is convex. Then*

(1) *If* $z_l, z_r \notin \bar{E}$, *then also* $z(x,t)\notin E$ *for all* $x\in\mathbf{R}$, $t>0$.
(2) *If* $z_l\in E$ *or* $z_r \in E$ *and* $z(\tilde{x},\tilde{t})\in E$ *for some* $\tilde{x},\tilde{t}$, *then* $z(\tilde{x},\tilde{t})\in\{z_l,z_r\}$.

The corollary states that if the initial values in a Riemann problem are inside the convex elliptic region, then the solution will contain values outside this region if the entropy condition is based on the "vanishing viscosity" approach. Furthermore if the initial values are outside the convex elliptic region, then the solution will not enter this region.

3. The Cauchy problem — a counterexample. Based on the results of the Riemann problem in the previous section it is natural to ask whether the same property is true for the general Cauchy problem: If

$$z_t + F(z)_x = 0 \qquad z(x,0) = z_0(x) \tag{3.1}$$

and for all $x \in \mathbf{R}$

$$z_0(x) \notin E \tag{3.2}$$

is

$$z(x,t) \notin \bar{E} \tag{3.3}$$

for all $x \in \mathbf{R}$ and $t > 0$?

The following example shows this not always to be the case. Let

$$f(u,v) = \frac{1}{2}\Big(\frac{u^2}{2} + v^2\Big) + v \qquad g(u,v) = uv. \tag{3.4}$$

Then

$$E = \Big\{(u,v) \in \mathbf{R}^2 \;\Big|\; \frac{u^2}{16} + \big(v + \frac{1}{2}\big)^2 < \frac{1}{4}\Big\}. \tag{3.5}$$

Making the *ansatz*

$$u(x,t) = \alpha(x)\beta(t) \qquad v(x,t) = \gamma(t) \tag{3.6}$$

we easily find

$$u(x,t) = \frac{2c_1 x + c_2}{c_1 t + c_3}, \qquad v(x,t) = \frac{c_4}{(c_1 t + c_3)^2}, \tag{3.7}$$

for constants $c_i \in \mathbf{R}$, $i = 1, \dots, 4$. Choosing

$$c_1 = c_3 = 1, \qquad c_2 = 0, \qquad c_4 = -2, \tag{3.8}$$

we find

$$u_0(x) = 2x, \qquad v_0(x) = -2, \tag{3.9}$$

and

$$u(x,t) = \frac{2x}{t+1}, \qquad v(x,t) = \frac{-2}{(t+1)^2}. \tag{3.10}$$

For this choice (3.2) is valid but (3.3) fails for some $x \in \mathbf{R}$ for $t > \sqrt{2} - 1$, see figures 1 and 2. This and other [21] examples of solutions entering the elliptic region do however have the property that the solutions u and v are also solutions to the viscous equations since $u_{xx} = v_{xx} = 0$, as well as to hyperbolic equations without any elliptic regions since we have that $v_x = 0$.

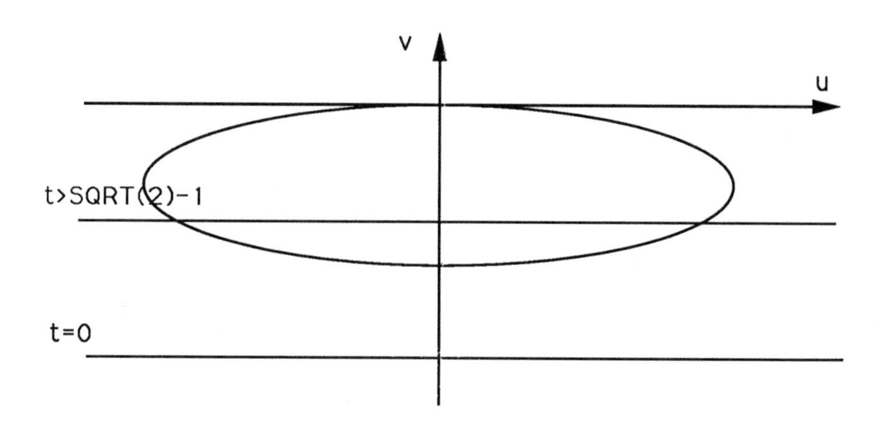

Figure 1. The solution at $t = 0$ and $t > SQRT(2) - 1$ in the z-plane

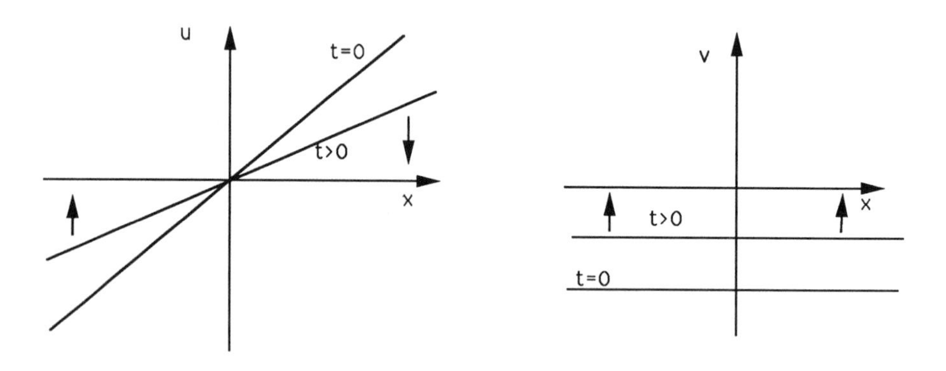

Figure 2. The solution for $t = 0$ and for $t > 0$.

Comparing the general properties of the Riemann problem and the example just presented, it is clear that the Glimm's scheme [4] will be highly unstable when the system is of mixed type since one in this scheme replaces the general Cauchy problem by a series of Riemann problems. This has recently been discussed by Pego and Serre [21], where another counterexample is provided and by Gilquin [3].

It was found that difference schemes also exhibit instabilities in this mixed type problem. The scheme used for the numerical examples was itself a mixed scheme: If both eigenvalues had positive (negative) real part a upwind (downwind) scheme was used, else a Lax–Friedrichs scheme was used. A pure Lax–Friedrichs scheme will have the same kind of oscillations, but they appear at a much smaller Δx.

In figures 3–5 we see the numerical solution to the initial value problem

$$u_0(x) = \frac{2x}{10}, \qquad v_0(x) = \frac{-11}{10} \tag{3.11}$$

at times $t = 0.0$, $t = 1.5$ and $t = 3.0$ respectively. This is (3.7) with $c_1 = 1$, $c_2 = 0$, $c_3 = 10$ and $c_4 = -110$, and the exact solution enters the elliptic region at $t = \sqrt{110} - 10 \approx 0.4881$. In all the examples $\Delta x = 0.01$ and $\Delta t = 0.002$. In figures 6–7 we see numerical solutions to initial value problems with perturbations of these initial values at $t = 2.0$.

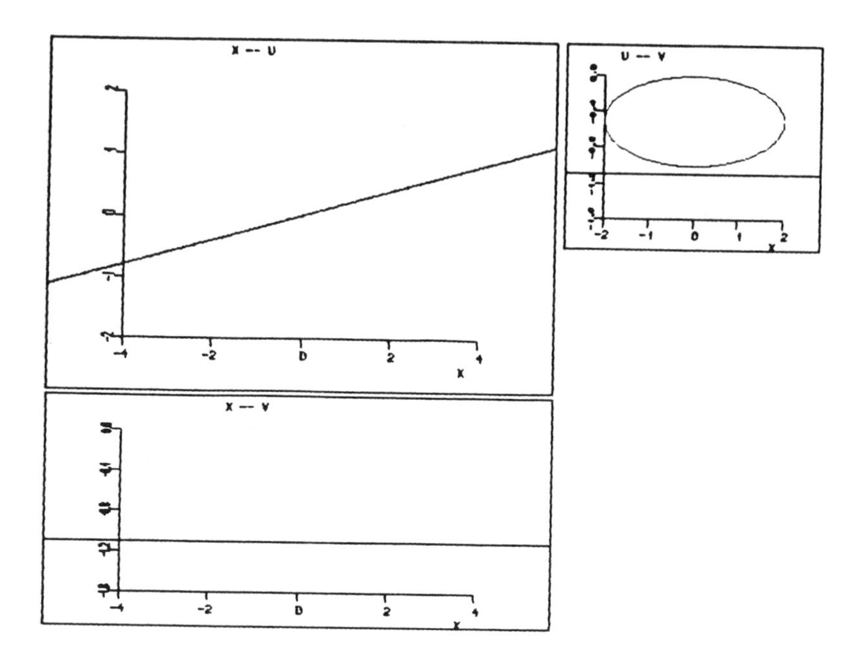

Figure 3.

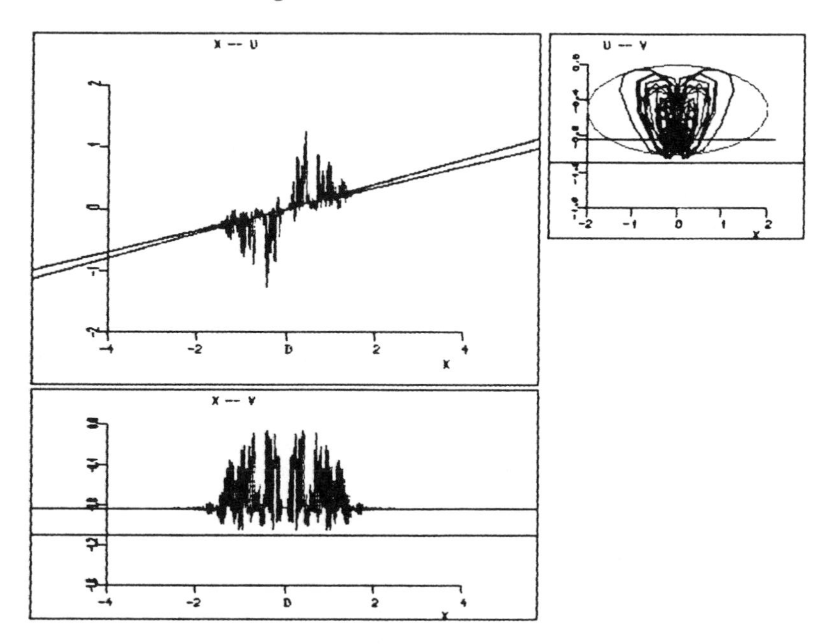

Figure 4.

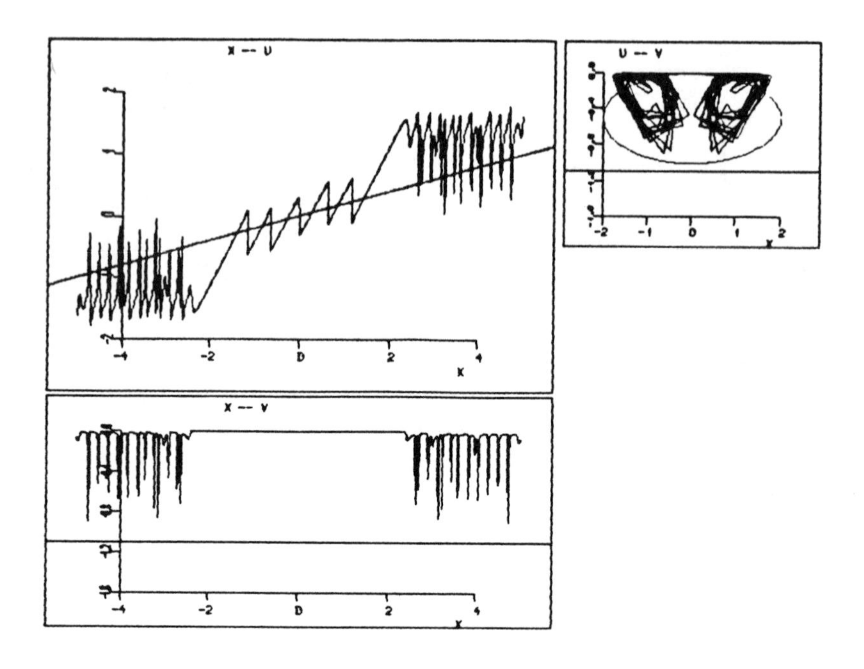

Figure 5.

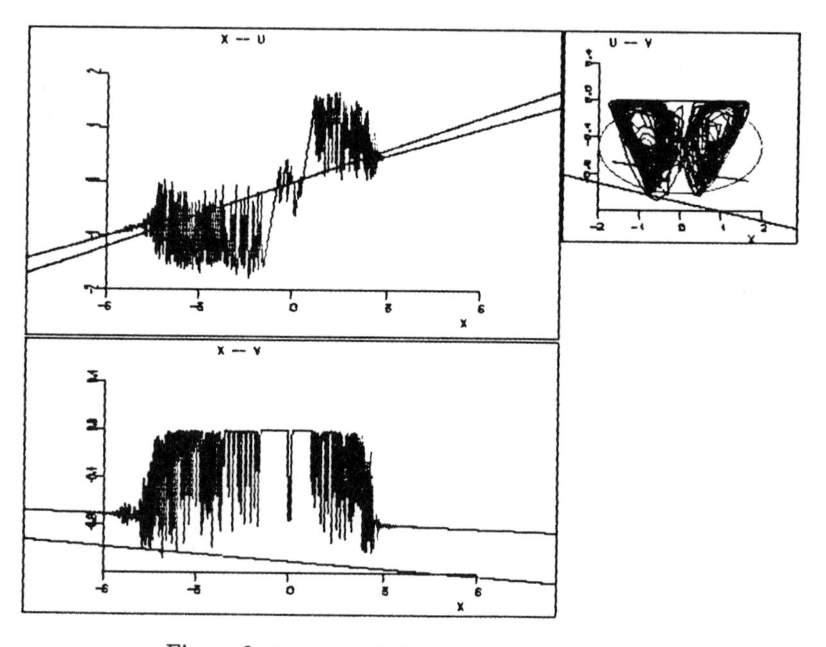

Figure 6. $\bar{v}_0 = v_0 - 0.02x$, $\bar{u}_0 = u_0$.

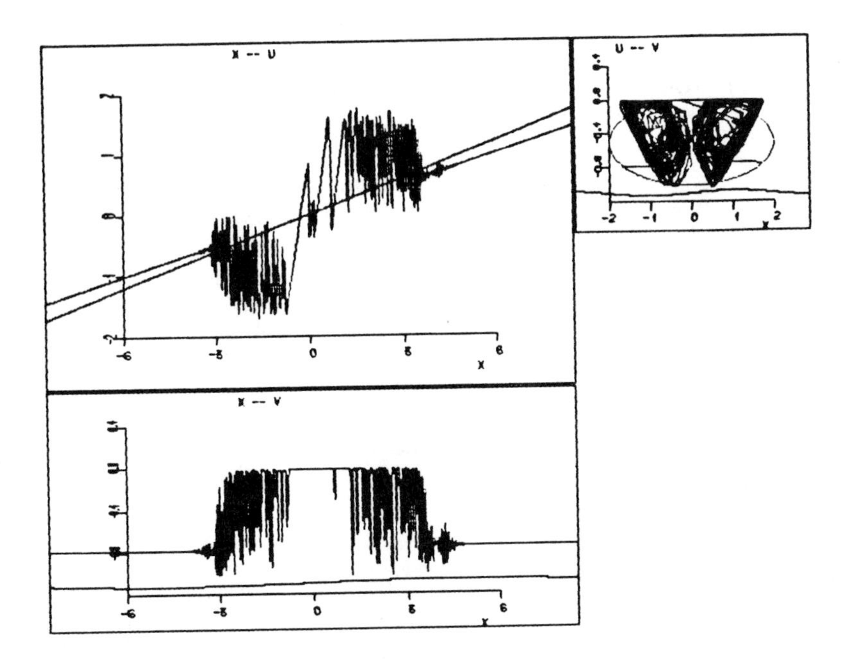

Figure 7. $\bar{v}_0 = v_0 + 0.03 \sin \frac{\pi}{10} x$, $\bar{u}_0 = u_0$.

These examples indicate that the solutions do enter the elliptic region, but this is difficult to determine due to the oscillations.

REFERENCES

[1] J.B. Bell, J.A. Trangenstein, G.R. Shubin, *Conservation laws of mixed type describing three–phase flow in porous media*, SIAM J. Appl. Math, 46 (1986), pp. 1000–1017.

[2] J.H. Bick, G.F. Newell, *A continuum model for two–directional traffic flow*, Quart. J. Appl. Math., 18 (1961), pp. 191–204.

[3] H. Gilquin, *Glimm's scheme and conservation laws of mixed type*, SIAM J. Sci. Stat. Comp., 10 (1989), pp. 133–153.

[4] J. Glimm, *Solutions in the large for nonlinear hyperbolic systems of equations*, Comm. Pure. Appl. Math., 18 (1965), pp. 697–715.

[5] J. Glimm, *Nonuniqueness of solutions for Riemann problems*, New York Univ. Preprint (1988).

[6] M.E.S. Gomes, *The viscous profile entropy condition is incomplete for realistic flows*, New York Univ. Preprint (1988).

[7] H. Holden, *On the Riemann problem for a prototype of a mixed type conservation law*, Comm. Pure and Appl. Math., 40 (1987), pp. 229–264.

[8] H. Holden, L. Holden, *On the Riemann problem for a prototype of a mixed type conservation law II*, in *Contemporary Mathematics*, Edited by W. B. Lindquist (to appear).

[9] L. Holden, *On the strict hyperbolicity of the Buckley–Leverett equations for three–phase flow in a porous medium*, Norwegian Computing Centre, Preprint (1988).

[10] L. Hsiao, *Admissible weak solution for nonlinear system of conservation laws of mixed type*, J. Part. Diff. Eqns. (to appear).

[11] L. Hsiao, *Nonuniqueness and uniqueness of admissible solutions of Riemann problem for system of conservation laws of mixed type*, Indiana University Preprint (1988).

[12] L. HSIAO, *Admissible weak solution of the Riemann problem for nonisothermal motion in mixed type system of conservation laws*, Indiana University Preprint (to appear).
[13] E.L. ISAACSON, B. MARCHESIN, B.J. PLOHR, *Transitional waves for conservation laws*, University of Wisconsin Preprint (1988).
[14] R.D. JAMES, *The propagation of phase boundaries in elastic bars*, Arch. Rat. Mech. Anal., 13 (1980), pp. 125–158.
[15] B.L. KEYFITZ, *The Riemann problem for nonmonotone stress–strain functions: A "hysteresis" approach*, Lectures in Appl. Math., 23 (1986), pp. 379–395.
[16] B.L. KEYFITZ, *An analytical model for change of type in three–phase flow*, in *Numerical Simuation in Oil Recovery*, Edited by M.F. Wheeler, Springer–Verlag, New York–Berlin–Heidelberg, 1988, pp. 149–160.
[17] B.L. KEYFITZ, *A criterion for certain wave structures in systems that change type*, in *Contemporary Mathematics*, Edited by W.B. Lindquist (to appear).
[18] B.L. KEYFITZ, *Admissibility conditions for shocks in conservation laws that change type*, Proceedings of GAMM International Conference on Problems Involving Change of Type (to appear).
[19] B.L. KEYFITZ, *Change of type in three–phase flow: A simple analogue*, J. Diff. Eqn. (to appear).
[20] M.S. MOCK, *Systems of conservations laws of mixed type*, J. Diff. Eqn., 37 (1980), pp. 70–88.
[21] R.L. PEGO, D. SERRE, *Instabilities in Glimm's scheme for two systems of mixed type*, SIAM J. Numer. Anal., 25 (1988), pp. 965–988.
[22] M. SHEARER, *The Riemann problem for a class of conservation laws of mixed type*, J. Diff. Eqn., 46 (1982), pp. 426–445.
[23] M. SHEARER, *Admissibility criteria for shock wave solutions of a system of conservation laws of mixed type*, Proc. Roy. Soc. (Edinburgh), 93A (1983), pp. 223–224.
[24] M. SHEARER, *Non–uniqueness of admissible solutions of Riemann initial value problems for a system of conservation laws of mixed type*, Arch. Rat. Mech. Anal., 93 (1986), pp. 45–59.
[25] M. SHEARER, *Loss of strict hyperbolicity for the Buckley–Leverett equations of three–phase flow in a porous medium*, in *Numerical Simuation in Oil Recovery*, Edited by M.F. Wheeler, Springer–Verlag, New York–Berlin–Heidelberg, 1988, pp. 263–283.
[26] M. SHEARER, D.G. SCHAEFFER, *The classification of 2×2 systems of non–strictly hyperbolic conservation laws, with application to oil recovery*, Comm. Pure. Appl. Math., 40 (1987), pp. 141–178.
[27] M. SHEARER, J. TRANGENSTEIN, *Change of type in conservation laws for three–phase flow in porous media*, North Carolina State Univ., Preprint (1988).
[28] M. SLEMROD, *Admissibility criteria for propagating phase boundaries in a van der Waals fluid*, Arch. Rat. Mech. Anal., 81 (1983), pp. 301–315.
[29] J. SMOLLER, *Shock Waves and Reaction–Diffusion Equations*, Springer, New York, 1983.

ON THE STRICT HYPERBOLICITY OF THE BUCKLEY-LEVERETT EQUATIONS FOR THREE-PHASE FLOW

LARS HOLDEN†

Abstract. It is proved that the standard assumptions on the Buckley-Leverett equations for three-phase flow imply that the equation system is not strictly hyperbolic. Therefore, the solution of the Buckley-Leverett equations for three-phase flow is very complicated. We also discuss four different models for the relative permeability. It is stated that Stone's model almost always gives (an) elliptic region(s). Furthermore, it is proved that Marchesin's model is hyperbolic under very weak assumptions. The triangular model is hyperbolic and the solution is well-defined and depends L_1-continuously upon the initial values in the Riemann problem.

1. Introduction. The Buckley-Leverett equations describe the flow of three phases in a porous medium neglecting capillary effects. If the system is not strictly hyperbolic, the solution is much more complicated. If the system is not hyperbolic, then the solution is not stable. Therefore it is important to know whether the system is strictly hyperbolic, hyperbolic or not hyperbolic.

In the last years there have been several papers on the strict hyperbolicity of the Buckley-Leverett equations. Bell, Trangenstein and Shubin [1] showed that Stone's model, which is the most used model for three-phase relative permeabilities, may give an elliptic region. Shearer [9] proved that if two interaction conditions between the relative permeabilities are satisfied, then strict hyperbolicity fails. Shearer and Trangenstein [10] propose some other interaction conditions which also imply that strict hyperbolicity fails. They also discuss several alternatives to Stone's model. In a recent paper, Trangenstein [12] proves that a large class of models have elliptic regions when gravity is included. Fayers [2] examines the elliptic region(s) in Stone's model for different choices of the residual oil parameter. The properties of the solution near an elliptic region are not known. See [5] and [8].

In the following section the Buckley-Leverett equations are defined and the standard assumptions are listed. In the third section we will make an additional assumption called Stone's assumption. It is stated that with this assumption the Buckley-Leverett equations fail to be strictly hyperbolic. In the final section four different models are discussed.

2. The Model. Let u, v and $w = 1-u-v$ be the water, gas and oil saturations. Let $f(u,v)$, $g(u,v)$ and $h(u,v)$ be the relative permeabilities divided by the viscosity for water, gas and oil respectively. Then the Buckley-Leverett equations for three-phase flow are

$$(1)\qquad \begin{aligned} u_t + \left(\frac{f(u,v)}{f(u,v)+g(u,v)+h(u,v)}\right)_x &= 0 \\ v_t + \left(\frac{g(u,v)}{f(u,v)+g(u,v)+h(u,v)}\right)_x &= 0. \end{aligned}$$

†Norwegian Computing Center, P.O.Box 114 Blindern, 0314 Oslo 3, Norway.

The equations are defined in

$$\Omega = \{(u,v); 0 < u, v < u+v < 1\}. \tag{2}$$

We will assume that the functions f, g and h satisfy

$$\begin{aligned} f(0,v) &= 0,\\ g(u,0) &= 0,\\ h(u,1-u) &= 0, \end{aligned} \tag{3}$$

$$\begin{aligned} f_u(0,v) &= 0,\\ g_v(u,0) &= 0,\\ -h_u(u,1-u) - h_v(u,1-u) &= 0, \end{aligned} \tag{4}$$

$$\begin{aligned} &f_u(u,v) > 0 \quad \text{and} \quad f_u(u,v) - f_v(u,v) > 0 \quad \text{for } (u,v) \in \Omega,\\ &g_v(u,v) > 0 \quad \text{and} \quad g_v(u,v) - g_u(u,v) > 0 \quad \text{for } (u,v) \in \Omega,\\ &h_u(u,v) < 0 \quad \text{and} \quad h_v(u,v) < 0 \quad \text{for } (u,v) \in \Omega, \end{aligned} \tag{5}$$

and

(6)

$$\begin{aligned} &f_{uu}(u,v) > 0 \quad \text{and} \quad f_{uu}(u,v) - 2f_{uv}(u,v) + f_{vv}(u,v) > 0 \quad \text{for } (u,v) \in \Omega,\\ &g_{vv}(u,v) > 0 \quad \text{and} \quad g_{vv}(u,v) - 2g_{vu}(u,v) + g_{uu}(u,v) > 0 \quad \text{for } (u,v) \in \Omega,\\ &h_{uu}(u,v) > 0 \quad \text{and} \quad h_{vv}(u,v) > 0 \quad \text{for } (u,v) \in \Omega. \end{aligned}$$

(3) implies that the phase is not moving if it is not present. (4) states that the speed of the phase vanishes when the saturation vanishes. (5) and (6) state that the rate and the speed increase when the saturation increases and one of the two other phases has constant saturation. These assumptions are widely accepted. It is possible to give physical arguments for these properties. With an additional assumption at the corners of Ω, it is possible to prove that the system is not strictly hyperbolic, see [7].

3. Stone's assumption. It is also usual to assume that the relative permeability of water and gas depends only on the water and gas saturation, namely

$$f(u) = f(u,v) \text{ and } g(v) = g(u,v). \tag{7}$$

This assumption is called Stone's assumption, see [11]. It uses the fact that water is usually wetting both in contact with oil and gas, and gas is usually not wetting either in contact with water or oil.

Some experiments indicate that the isoperms of the relative permeability of oil are concave, see [11]. See figure 1 for an illustration of concave isoperms. Then we may state the following theorem.

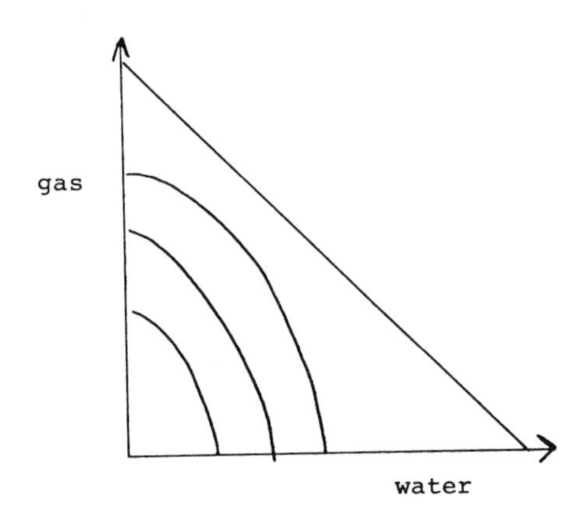

Figure 1. Concave isoperms of the relative permeability of oil

THEOREM 1. *Assume (3), (4), (5), (6) and (7) are satisfied. Then (1) is not strictly hyperbolic in* Ω. *If the three curves* $f' + h_u = 0$, $g' + h_v = 0$ *and* $f' + g' = 0$ *intersect at a point* P *in* Ω, *then (1) is not strictly hyperbolic at* P. *If the three curves intersect and the isoperms of* h *are concave, then the system is strictly hyperbolic except at* P. *If the three curves do not intersect in* Ω, *then there is at least one elliptic region in* Ω.

This theorem is proved in [7]. See figure 2 for an illustration.

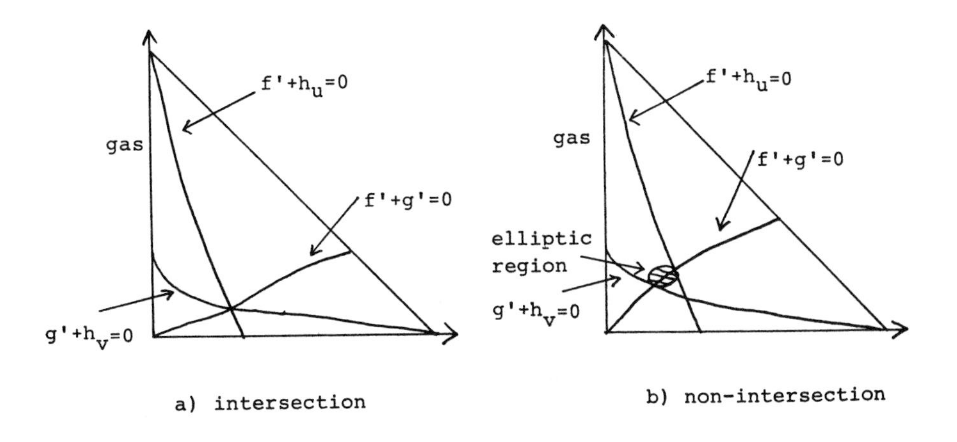

Figure 2. The three curves $f' + h_u = 0, g' + h_v = 0$ and $f' + g' = 0$.

Since the gas viscosity is much smaller than the viscosity of oil and water, the intersection point between $g' + h_v = 0$ and $f' + g = 0$ has very small gas saturation. Therefore, the elliptic region which is caused by the non-intersection of the three curves has very small gas saturation. The elliptic region reported in [1] is another elliptic region. In their example there is also another elliptic region which is much smaller with smaller gas saturation.

4. Four different models. Since only two-phase relative permeabilities are measured, the formulas for three-phase relative permeabilities are usually only interpolation formulas in the three-phase region. In the following section we will discuss four different models for the relative permeabilities.

4.1 Stone's model. This is the most commonly used expression for the three-phase relative permeabilities. See [11]. It assumes that the relative permeabilities satisfy (7) and in addition that

$$h(u,v) = \frac{(1-u-v)\ h(u,0)\ h(0,v)}{(1-u)(1-v)}. \tag{8}$$

$f(u)$, $g(v)$, $h(u,0)$ and $h(0,v)$ are usually found from experiments. With reasonable assumptions on the two-phase relative permeabilities, Stone's model satisfies assumptions (3), (5), (6). (4) is satisfied for the water and gas relative permeabilities and only in the corners for the relative permeability of oil. Therefore, it is not possible to use Theorem 1 directly. If we make the following minor modification on the model

$$h(u,v) = \frac{(1-u-v)^{1+\epsilon}\ h(u,0)\ h(0,v)}{(1-u)^{1+\epsilon}(1-v)^{1+\epsilon}} \tag{9}$$

for $\epsilon > 0$, then (4) is also satisfied for the relative permeability of oil. Then we may use Theorem 1 and state that the modified Stone's model is not strictly hyperbolic. Since the modified Stone's model is almost identical to Stone's model, this indicates that except for very special two-phase data, Stone's model (8) is not strictly hyperbolic.

4.2 Marchesin's model. In Marchesin's model [9] it is assumed that the relative permeability for each phase only depends on its own saturation, i.e.

$$h(1-u-v) = h(u,v) \tag{10}$$

in addition to (7). Therefore, we may use Theorem 1 making the additional assumptions (3), (4), (5) and (6). It is trivial to see that the three curves in Theorem 1 intersect at a common point. We will instead prove a theorem with much weaker assumptions on the relative permeabilities. Let us first define the following

$$\begin{aligned} f_D &= \{y \in R; f'(u) = y \ \text{ for } \ 0 < u < 1\}, \\ g_D &= \{y \in R; g'(u) = y \ \text{ for } \ 0 < u < 1\}, \\ h_D &= \{y \in R; h'(u) = y \ \text{ for } \ 0 < u < 1\}, \\ H &= f_D \cap g_D \cap h_D \qquad \text{and} \\ K &= \{(u,v) \in \Omega; f'(u) \in H,\ g'(u) \in H, \ \text{ and } \ h'(u) \in H\}. \end{aligned}$$

We may then formulate the following theorem.

THEOREM 2. *Assume that (7) and (10) are satisfied. Then the system (1) is hyperbolic. If K is empty, then (1) is strictly hyperbolic. If K is not empty and (5) is satisfied, then (1) is strictly monotone except at a unique point P.*

Proof. It is straightforward to prove that (1) with the assumption (7) and (9) is strictly hyperbolic except if, and only if,

$$A = (B - C + E)^2 + 4BC > 0$$

where

$$B = g(f' + h_u),\ C = f(g' + h_v) \text{ and } E = h(f' - g').$$

The system is hyperbolic if, and only if, $A \geq 0$. If $BC > 0$, then obviously $A > 0$.

Assume $B \geq 0$ and $C \leq 0$. Since $f > 0$, $g > 0$ and $h_u = h_v$, $E > 0$. This implies that

$$A = (B - C + E)^2 + 4BC >$$
$$(B - C)^2 + 4BC = (B + C)^2.$$

The argument is similar for $B \leq 0$ and $C \geq 0$. Thus we have proved that A is non-negative and vanishes if, and only if $f' = g' = h_u = h_v$. If K is empty, then obviously (1) is strict hyperbolic. If K is not empty and (5) is satisfied, it is easy to prove that (1) is strictly hyperbolic except at a unique point. □

4.3 The triangular model. The fractional flow function of gas is

$$G(u,v) = \frac{g(u,v)}{f(u,v) + g(u,v) + h(u,v)}.$$

In [6] it is proposed to let the fractional flow depend only on the gas saturation i.e.

$$G(v) = G(u,v). \tag{11}$$

Since the gas viscosity is much smaller than the oil and water viscosity, this is a good approximation. This results in a model

$$u_{i_t} + f_i(u_1, ..., u_i)_x = 0 \text{ for } i = 1, ..., n. \tag{12}$$

This model was studied in [6], [3] and [4] with Riemann initial data i.e.

$$u_i(x,0) = \begin{cases} u_{i_-} & \text{for } x < 0 \\ u_{i_+} & \text{for } x > 0 \end{cases} \quad \text{for } i = 1, ..., n. \tag{13}$$

The following theorem was proved in [6].

THEOREM 3. *Assume that f_i is continuous, piecewise smooth and that $\frac{\partial f_i}{\partial u_i}$ increases or decreases faster than linear when u_i increases to ∞ or decreases to $-\infty$ for i=1,...,n. Then there is a solution of (12) and (13). The solution is unique except for $f \in M$ and for each f, $(u_{1_-}, ..., u_{n_-}, u_{1_+}, ..., u_{n_+}) \in M_f$. M has measure zero in the supremum norm and M_f has Lebesgue measure zero for all f not in M. There is always uniqueness for $n < 3$.*

For the Buckley-Leverett equations we may state the following theorem.

THEOREM 4. *Assume (11) is satisfied. Then (1) is strictly hyperbolic except at a curve from one corner to another corner where the system is hyperbolic. The solution of the Riemann problem exists uniquely, is well-defined in Ω and depends L_1-continuously on the initial data.*

The first part of this theorem is obvious. Gimse proves the second part of the theorem in [3] and [4].

4.4 The hyperbolic model. It is possible to find a model which satisfies (3), (4), (5), (6), (7) and in addition ensures that the three curves in Theorem 1 intersect. In [7] is given an example of such a model which always is hyperbolic. The disadvantage with such models is that they become technically difficult in order to ensure that the three curves in Theorem 1 intersect. It is proved in [12] that such models have elliptic regions when gravity is included.

Acknowledgement. The author wishes to thank Helge Holden for valuable discussions.

REFERENCES

[1] BELL, J. B., TRANGENSTEIN, J. A. AND SHUBIN, G. R., *Conservation laws of mixed type describing three-phase flow in porous media*, Siam J. of Appl. Math., 46 (1986), pp. 1000-1117.

[2] FAYERS, F. J., *Extensions of Stone's Method I and the Condition for real characteristics in three-phase flow*, Presented at SPE conference in Dallas, September 1987.

[3] GIMSE, T., *A numerical method for a system of equations modelling one-dimensional three-phase flow in a porous medium*, in Nonlinear hyperbolic equations - Theory, Computation Methods, and Applications, J. Ballmann and R. Jeltsch (Eds.), Vieweg, Braunschweig, Germany, 1989.

[4] GIMSE, T., Thesis, University of Oslo, Norway, 1989.

[5] HOLDEN, H., HOLDEN, L. AND RISEBRO, N. H., *Some qualitative properties of 2×2 systems of conservation laws of mixed type*, This proceeding.

[6] HOLDEN, L. AND HØEGH-KROHN, R., *A class of n nonlinear hyperbolic conservation laws*, J. Diff. Eq. (to appear).

[7] HOLDEN, L., *On the strict hyperbolicity of the Buckley-Leverett Equations for Three-Phase Flow in a Porous Medium*, Siam J. of Appl. Math. (to appear).

[8] PEGO R. L. AND SERRE D., *Instabilities in Glimm's scheme for two systems of mixed type*, Siam J. Num. Anal., 25 (1988), pp. 965-988.

[9] SHEARER, M., *Loss of strict hyperbolicity of the Buckley-Leverett Equations for three-phase flow in a porous medium, in Numerical Simulation in Oil Recovery*, Edited by M.F. Wheeler, Springer-Verlag, New York-Berlin-Heidelberg (1988), pp. 263-283.

[10] SHEARER, M. AND TRANGENSTEIN, J., *Change of type in conservation laws for three-phase flow in porous media*, Preprint.

[11] STONE, H. L., *Estimation of Three-Phase Relative Permeability and Residual Oil Data*, J. Cnd. Pet., 12 (1973), pp. 53-61.

[12] TRANGENSTEIN, J.A., *Three-phase flow with gravity*, J. Contemp. Math. (to appear).

ADMISSIBILITY CRITERIA AND ADMISSIBLE WEAK SOLUTIONS OF RIEMANN PROBLEMS FOR CONSERVATION LAWS OF MIXED TYPE: A SUMMARY*

L. HSIAO†

Consider the model system

$$\begin{cases} v_t + p(u)_x = 0 \\ u_t - v_x = 0 \end{cases} \tag{1.1}$$

which describes the one-dimensional isothermal motion of a compressible elastic fluid or solid in Lagrangian coordinate system. Here v denotes the velocity, u the specific volume for a fluid or displacement gradient for a solid, and $-p$ is the stress which is determined through a constitutive relation.

For many materials, $p(u)$ is a decreasing function of u, and the system (1.1) is strictly hyperbolic. However, (1.1) can be of mixed type when it is used for dynamics of a material exhibiting change of phase such as in a Van der Waals fluid [SL] (see Figure 1.1).

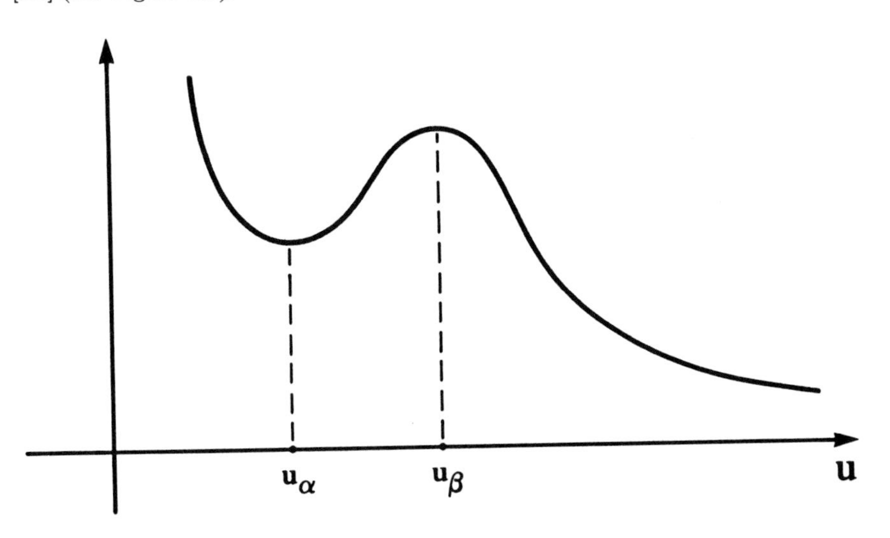

(Figure 1.1)

How should we generalize the admissibility criteria used for strictly hyperbolic system to the case of mixed type in order to define an admissible discontinuity and

*This research was supported in part by the Institute for Mathematics and its Applications with funds provided by the National Science Foundation.

†Department of Mathematics, Indiana University, Bloomington, IN 47405 and Institute of Mathematics, Academia Sinica, Beijing, P.R. China

an admissible weak solution for Riemann problems? Is it still possible to prove the existence and uniqueness of the admissible weak solution and determine whether or not the solution is continuously dependent on the initial data?

Many results are now available which generalize admissibility criteria from strictly hyperbolic systems to those of mixed type. (See [J], [K_1], [K_2], [HA_1], [HA_2], [HM], [HS_1], [HS_2], [HSL], [P], [SE_1], [SE_2], [SL].) We apply the generalization of shock (E) criterion, introduced in [HS_2] to obtain the result summarized here. See [HS_3] for the details.

For a strictly hyperbolic system of the form (1.1), the Oleinik-Liu criterion (so called shock (E) criterion) can be described in two different ways which are equivalent. A discontinuity $(\sigma; u_+, v_+; u_-, v_-)$ is called admissible according to shock (E) criterion, if $\sigma = \sigma_i(u_-, u_+)$ is such that

$$\text{(I)} \quad \begin{cases} (u_+, v_+) \in H_i(u_-, v_-), i = 1 \text{ or } 2, H_i(u_-, v_-) \text{ is the Hugoniot locus} \\ \text{determined by the Rankine–Hugoniot condition with } (u_-, v_-) \text{ given,} \\ \text{and } \sigma_i(u_-, u_+) \leq \sigma_i(u_-, u) \text{ for all } (u, v) \in H_i(u_-, v_-) \text{ with } u \\ \text{between } u_- \text{ and } u_+. \end{cases}$$

or

$$\text{(II)} \quad \begin{cases} (u_-, v_-) \in H_i(u_+, v_+), i = 1 \text{ or } 2, H_i(u_+, v_+) \text{ is the Hugoniot locus} \\ \text{determined by the Rankine–Hugoniot condition with } (u_+, v_+) \text{ given,} \\ \text{and } \sigma_i(u, u_+) \leq \sigma_i(u_-, u_+) \text{ for all } (u, v) \in H_i(u_+, v_+) \text{ with } u \\ \text{between } u_- \text{ and } u_+. \end{cases}$$

Corresponding to the two criteria, (I) and (II), we have two types, (I′) and (II′) of *generalized shock (E) criteria*;

$$\text{(I}'\text{)} \quad \begin{cases} (u_+, v_+) \in H_i(u_-, v_-), \sigma = \sigma_i(u_+, u_-), i = 1 \text{ or } 2 \\ \sigma_i(u; u_-) \geq \sigma_i(u_+; u_-) \quad \text{for any } u \text{ between } u_- \text{ and } u_+ \\ \text{where } \sigma_i(u; u_-) \text{ is defined as a real-valued function} \end{cases}$$

and similarly for (II′).

Now a function $(u(\xi), v(\xi))(\xi = x/t)$ is called an *admissible type I* solution to the Riemann problem for (1.1) with data

$$(1.2) \qquad (u, v)\Big|_{t=0} = \begin{cases} (u_-, v_-) & \text{for} \quad x < 0 \\ (u_+, v_+) & \text{for} \quad x > 0 \end{cases}$$

if

i) it satisfies the boundary condition $(u, v) \to (u_\mp, v_\mp)$ as $\xi \to \mp\infty$

ii) it is either a rarefaction wave or a constant state wherever it is smooth

iii) at any discontinuity it satisfies (I′).

Similarly we define *admissible type II* solutions. Finally we define an *admissible type III* solution to be a function $(u(\xi), v(\xi))$ that satisfies i), ii) and

iii) any discontinuity of the first family satisfies (I′) while any discontinuity of the second family satisfies (II′).

It was proved in [HS$_2$] that there is a unique admissible type I solution for any given Riemann data and the same is true for type II. Moreover, the two solutions are identical for any given Riemann data. It is proved in [HS$_3$] that an admissible type III weak solution may not exist for some choices of Riemann data; however, there is a unique type III solution if the Riemann data belong to the same phase (that is, $u_- \leq u_\alpha$ and $u_+ \leq u_\alpha$ or $u_- \geq u_\beta$, $u_+ \geq u_\beta$ as shown in Figure 1.1). Furthermore, the three types of admissible weak solutions are identical if the Riemann data belong to the same phase [HS$_3$].

As far as continuous dependence is concerned, it is to be expected in general that the solution is not continuously dependent on the initial data in problems of mixed type. However, some kind of stable behavior can still be expected for our admissible weak solution; this is discussed in [HS$_3$].

In conclusion, we see that the generalized shock (E) criterion supplies a reasonable generalization of an admissibility criterion for the mixed type system (1.1). Moreover, it can be shown that the solution constructed by Shearer in [SE$_1$] can be obtained as our type I or type II admissible weak solution, by using the approach introduced in [HS$_3$].

REFERENCES

[HA$_1$] H. Hattori, *The Riemann problem for a Van der Waals fluid with entropy rate admissibility criterion–Isothermal case*, Arch. Rational Mech. Anal., 92 (1986), 247–263.

[HA$_2$] H. Hattori, *The Riemann problem for a Van der Waals fluid with entropy rate admissibility criterion—Nonisothermal case*, J. differential Equations, 65, 2 (1986), 158–174.

[HM] L. Hsiao and P. de Mottoni, *Existence and uniqueness of Riemann problem for a nonlinear system of conservation laws of mixed type*, Trans. Amer. Soc. (to appear).

[HS$_1$] L. Hsiao, *Admissible weak solution for nonlinear system of conservation laws in mixed type*, P.D.E., 2, 1 (1989), 40–58.

[HS$_2$] L. Hsiao, *The uniqueness of admissible solutions of Riemann problem for system of conservation laws of mixed type*, (to appear in Jour Diff Equn.).

[HS$_3$] L. Hsiao, *Admissibility criteria and admissible weak solutions of Riemann problems for conservation laws of mixed type*, IMA Preprint Series (1990).

[HSL] R. Hagan and M. Slemrod, *The viscosity-capillarity criterion for shocks and phase transitions*, Arch. Rational Mech. Anal. 83 (1984), 333–361.

[J] R.D. James, *The propagation of phase boundaries in elastic bars*, Arch. Rational Mech. Anal. 73 (1980), 125–158.

[K$_1$] B.L. Keyfitz, *Change of type in three-phase flow: A single analogue*, J. Diff. E. (to appear).

[K$_2$] B.L. Keyfitz, *Admissibility conditions for shocks in conservation laws that change type*, (to appear).

[P] R. Pego, *Phase transitions: Stability and admissibility in one dimensional nonlinear viscoelasticity*, (to appear).

[SE_1] M. SHEARER, *The Riemann problem for a class of conservation laws of mixed type*, J. Diff. E., 46 (1982), 426–443.

[SE_2] M. SHEARER, *Admissibility criteria for shock wave solution of a system of conservation laws of mixed type*, Proc. Royal. Soc. Edinburgh 93 A (1983), 233–244.

[SL] M. SLEMROD, *Admissibility for propagating phase boundaries in a Van der Waals fluid*, Arch. Rational Mech. Anal. 81 (1983), 301–315.

SHOCKS NEAR THE SONIC LINE: A COMPARISON BETWEEN STEADY AND UNSTEADY MODELS FOR CHANGE OF TYPE*

BARBARA LEE KEYFITZ†

Abstract. We look at the structure of shocks for states near a locus where equations change type. Two basic models are considered: steady transonic flow, and models for unsteady change of type. Our result is that these two problems may be distinguished by the nature of the timelike directions and the forward light cone. This leads in a natural way to different candidates for admissible shocks in the two cases.

AMS(MOS) subject classifications. AMS (1980) Subject Classification Primary 35L65, 35M05 Secondary 76H05, 35A30

1. Introduction. Models for Change of Type. Systems of quasilinear equations which are of different type at different states arise in two different ways in applications. The first is typified by the pair of conservation laws which governs steady, inviscid, irrotational, isentropic flow [2]:

$$
\begin{aligned}
(\rho u)_x + (\rho v)_y &= 0 \\
v_x \quad - \quad u_y &= 0
\end{aligned}
\tag{1.1}
$$

Here $w = (u, v)$ represents the velocity of a flow in the xy plane; $\rho = \rho(|w|)$ is the density, given by Bernoulli's law as a function of the speed. The first equation expresses conservation of mass; the second irrotationality of the flow. Under the assumption that the medium is an ideal gas (or some other reasonable thermodynamics), there is a speed, c_*, the *sonic speed*, such that system (1.1) is hyperbolic if $|w| > c_*$ ("supersonic flow") and nonhyperbolic (elliptic) if $|w| < c_*$ ("subsonic"). The flow changes type along the curve $u^2 + v^2 = c_*^2$, which we shall call the *sonic line* in this paper. The classification of these equations, and their relation to other models for gas flow, were studied by Courant and Friedrichs [2]. Here we are interested in the following property of *weak solutions* of (1.1): for every supersonic state w_0, there is a one-parameter family of states, $w_1(\beta)$, such that the function

$$
w(x, y) = \begin{cases} w_0, & x\cos\beta + y\sin\beta < 0 \\ w_1, & x\cos\beta + y\sin\beta > 0 \end{cases}
\tag{1.2}
$$

is a weak solution of (1.1). The set $\{w_1(\beta)\}$, called the *shock polar* through w_0, has a self-intersection at w_0. The portion of the curve for which $|w_1| < |w_0|$, which we shall call the *Hugoniot loop*, $H(w_0)$, contains a subinterval in the elliptic region.

*This work was performed during an extended visit to the IMA for the Nonlinear Waves program, and many of the ideas in this paper emerged from discussions with other participants. I'd like to thank them, as well as the IMA for hosting us.

†Research supported by the Air Force Office of Scientific Research, Air Force Systems Command, USAF, under Grant Number AFOSR 86-0088. The U.S. Government is authorized to reproduce and distribute reprints for Governmental purposes notwithstanding any copyright notation thereon.

All weak solutions (1.2) for which $w_1 \in H(w_0)$ are said to be *admissible*. Courant and Friedrichs [2] discuss the fact that only such shocks are compressive; they also point out that there may be other restrictions, depending on the particular boundary value problem one is trying to solve. However, all compressive shocks are felt to be physically realizable in some context, and none that are expansive are ever physical. In [7], we discuss these admissibility conditions in the context of steady viscous perturbations and give more details of the calculations summarized above.

The situation described by (1.1), *transonic steady* (TS) flow, may be replicated in other physical situations: steady visco-elastic flows of *composite* type have flow regimes where a pair of characteristic speeds may change from real to complex - as if (1.1) were embedded in a larger system where all the remaining characteristic fields are of constant type [11]; change of type which may resemble this also occurs in steady flows of granular materials [12].

Change of type in systems of conservation laws modelling unsteady flows has also been observed. The list of flow models exhibiting this phenomenon includes several models for two-fluid flow [13], multi-phase porous media flow [1], and possibly some examples in granular flow [12]. Some models for unsteady visco-elastic flows exhibit change of type, but most likely not in flow regions where the models are valid. In addition, there are models for fluids undergoing phase transitions where the equations change type, but where it may be the case that additional, different physical mechanisms are necessary to model the fluid behavior. However, in two-phase nonreacting flows, such as are modelled in large-scale computations in nuclear reactor engineering, the standard model exhibits change of type in the fluid regime of interest [13]. These models have been in practical use for years.

A prototype model is given by a pair of equations for *unsteady* flow (US)

$$f(w)_x + w_t = 0 \tag{1.3}$$

with $w = (u, v), f = (f_1, f_2)$. In discussing the general structure of shock solutions, it is convenient to write (1.1) or (1.3) as

$$f(w)_x + g(w)_y = 0. \tag{1.4}$$

In the next section, we shall make precise what is meant by the change of type locus or *sonic line* for (1.4); and we shall prove that under some nondegeneracy conditions equivalent to genuine nonlinearity in the hyperbolic region, (1.4) admits a Hugoniot loop near the sonic line. In section three we shall review and abstract a shock classification based on a determination of the forward light cone for (1.4). This criterion, which says, in essense, that the shock surface is spacelike when viewed from the upstream direction, applies also when (1.4) is not hyperbolic at one state, and yields the well-known admissibility of compressive shocks for the transonic case, TS. For the unsteady case, US, the situation is more complicated. When we apply the same principle which leads to the standard conclusion in TS, then we can use the theory of local unfolding of vector field dynamics to show that the Hugoniot loop can be divided into three intervals, two of which represent admissible shocks. However, determination of the precise end points of these intervals has not yet been accomplished.

2. Systems of Conservation Laws near the Sonic Line.

We write (1.4) in quasilinear form

$$P(w, \partial)w \equiv A(w)\partial_x w + B(w)\partial_y w = 0 \tag{2.1}$$

where

$$A(w) = (a_{ij}) = df \quad B(w) = (b_{ij}) = dg \tag{2.2}$$

and form the linearized symbol (at a constant state w)

$$P(w, \zeta) = A(w)\xi + B(w)\eta \tag{2.3}$$

with $\zeta = (\xi, \eta) \neq 0$. We may write $\zeta = \rho(\cos\beta, \sin\beta)$ where

$$\rho \in \mathbf{R}\backslash\{0\} \text{ and } \beta \in [-\frac{\pi}{2}, \frac{\pi}{2}].$$

(We will justify this choice later). It will be sufficient to consider unit vectors. The *characteristic covectors* or "characteristics" (by an abuse of notation which is sufficiently confusing that we shall avoid it) are the solutions to

$$\det P(w, \zeta) \equiv \det(A(w)\cos\beta + B(w)\sin\beta) = 0. \tag{2.4}$$

System (2.1) is *strictly hyperbolic* at a state w if (2.4) has real distinct solutions $\beta_1 < \beta_2$, and *nonhyperbolic* if the roots of (2.4) are not real; we let $\mathcal{H} \subset \mathbf{R}^2$ be the set of states w where (2.1) is strictly hyperbolic, $\mathcal{E}$ the set of nonhyperbolic states. Define $\mathcal{B}$, the *sonic line* (by analogy with steady transonic flow) as the locus of states w where (2.4) has equal roots, β.

We introduce the notation B^* for the adjunct (transpose of the matrix of cofactors) of B and $q(w) = tr(B^*A)$. Note that $\det B^* = \det B$. Let

$$D(w) = q^2 - 4\det(B^*A). \tag{2.5}$$

The following proposition, which is a slight generalization of Proposition 2.1 of [5], characterizes the sonic line in a nondegenerate case.

PROPOSITION 2.1. *Assume $f, g \in C^2$ and that there is a point w_0 where $D = 0$ and $\nabla_w D \neq 0$. Then in a neighborhood $\mathcal{O}(w_0)$, $\mathcal{B} = \{w | D(w) = 0\}$ is a C^1 curve separating states $w \in \mathcal{H}$ from states $w \in \mathcal{E}$. Furthermore Rank $P(w, \zeta) \geq 1$ everywhere, including $\mathcal{B}$; at a point $w \in \mathcal{B}$, at least one of $\det A$ or $\det B$ is nonzero, and P has a nonzero eigenvalue there if and only if the other has precisely one nonzero eigenvalue.*

Proof. Define

$$p(w, \beta) = \det P(w, \zeta) = \det A\cos^2\beta + q\cos\beta\sin\beta + \det B\sin^2\beta. \tag{2.6}$$

This has real roots, β_i, if $D(w) \geq 0$. Now $w \in \mathcal{B}$ if (2.6) has a double root; since (2.6) is a smooth function of β, this implies

$$\frac{\partial p}{\partial \beta}(w, \beta) = 0. \tag{2.7}$$

By a calculation, this is equivalent to

$$D(w) = 0. \tag{2.8}$$

The nondegeneracy condition $\nabla_w D \neq 0$ guarantees, by the implicit function theorem, that there is a open ball $\mathcal{O} \in \mathbf{R}^2$ centered at w_0 in which solutions of (2.8) form a connected C^1 curve, and also that $D > 0$ on one side, $\mathcal{H}$, and $D < 0$ on the other, $\mathcal{E}$. Since

$$\nabla_w D = 2q \nabla_w q - 4 \det A \nabla(\det B^*) - 4 \det B^* \nabla(\det A), \tag{2.9}$$

and $\det(B^*A) = 0 \Leftrightarrow q = 0$ when (2.8) holds, we see that $\nabla D \neq 0$ only if at least one of $\det A, \det B$ is nonzero. Hence on $\mathcal{B}$, (2.6) is a nontrivial form and β is well defined:

$$\tan \beta = -(\det A / \det B)^{1/2}.$$

Now let $C = B^*A$, so $D = (trC)^2 - 4 \det C$ and, with $C = (c_{ij})$

$$\frac{1}{2} \nabla D = (c_{11} - c_{22}) \nabla c_{11} + (c_{22} - c_{11}) \nabla c_{22} + 2c_{12} \nabla c_{21} + 2c_{21} \nabla c_{12}. \tag{2.10}$$

If we suppose, without loss of generality, that $\det B(w_0) \neq 0$, then at the double root of (2.6),

$$B^* P(w, \zeta) = C \cos \beta + (\det B) I \sin \beta = \cos \beta (C - \lambda I)$$

where $\lambda = -(\det B) \tan \beta = (\det C)^{1/2} = \frac{1}{2} trC$. We note that $\det B \neq 0$ if and only if $\cos \beta \neq 0$ so that $P = 0$ if and only if

$$N = C - \left(\frac{1}{2} trC\right) I = \frac{1}{2} \begin{pmatrix} c_{11} - c_{22} & 2c_{12} \\ 2c_{21} & c_{22} - c_{11} \end{pmatrix}$$

is zero. But if every component of N is zero, then, from (2.10), ∇D is also zero, contrary to hypothesis. This shows that Rank $P \geq 1$ everywhere. Finally, still assuming $\det B \neq 0$, we find

$$P(w, \zeta) = \frac{\cos \beta}{|\det B|} BN \tag{2.11}$$

Now, the last statement of the theorem concerns whether $P(w_0, \zeta)$ is nilpotent: clearly $P^2 = 0 \Leftrightarrow (BN)^2 = 0$ and since B is invertible, we have $P = 0$ if and only if $NBN = 0$. Now $N = B^*A - \frac{1}{2} tr(B^*A) I = B^*A - (\det B^*A)^{1/2} I$, and so $NBN = N(NB + BB^*A - B^*AB) = |B|NA - NB^*AB$ and $NB^*A = N^2 + \sqrt{|AB|}N = \sqrt{|AB|}N$, so

$$NBN = N\left\{|B|A - \sqrt{|AB|}B\right\}.$$

If $\det A = 0$, then $N = B^*A$ so $NBN = |B|B^*A^2$ and this is zero precisely when $A^2 = 0$, i.e. A is also nilpotent. If $\det A \neq 0$, then $NBN = N\left\{A - \sqrt{\frac{|A|}{|B|}}N\right\} = \sqrt{abs|A|}N\left\{\tilde{A} - \tilde{B}\right\}$ where $\det \tilde{A} = \det \tilde{B} = 1$ and, since $P = BN, \det|\tilde{A} - \tilde{B}| = 0$. Since $\tilde{A}$ and $\tilde{B}$ have a common eigenvector $\tilde{r}$, we can write them in a basis $\{\tilde{r}, \tilde{s}\}$ as

$$\tilde{A} = \begin{pmatrix} 1 & 0 \\ \alpha & \beta \end{pmatrix} \quad \tilde{B} = \begin{pmatrix} 1 & 0 \\ \gamma & \delta \end{pmatrix}:$$

and now $\beta = \delta = 1$; hence $\tilde{A} - \tilde{B}$ is nilpotent. Thus we have proved the final assertion of the proposition, and we have also shown the useful fact that P has a nonzero eigenvalue only in the case that one of A or B is singular. □

From now on we will work in the ball $\mathcal{O}$ given by the proof of Proposition 2.1. It is not necessary to restrict $\mathcal{O}$ to be a ball, but we omit details of the extension to other geometries. We also note that the matter of P being nilpotent is not invariant under reasonable notions of equivalence, such as interchanging the order of writing the equations. The appropriate definition of equivalence will be given after we discuss shocks.

Requiring that (1.2) be a weak solution of (1.4) results in the equation

$$V(w_1, w_0, \beta) \equiv (f(w_1) - f(w_0))\cos\beta + (g(w_1) - g(w_0))\sin\beta = 0. \tag{2.12}$$

We consider this as a bifurcation equation with distinguished parameter β: for a fixed w_0 in $\mathcal{O}$, we are interested in the zero-set $w_1 = w_1(\beta)$ of $V(w_1, w_0\beta)$. This set (the bifurcation diagram) is unchanged if V is multiplied on the left by an invertible matrix, and equivalent bifurcation diagrams result under coordinate changes $\beta \longmapsto \beta'(\beta), w \longmapsto w'(w, \beta)$. The mapping V has the property $V(w_0, w_0, \beta) \equiv 0$ and so do equivalent mappings if both w_0 and w_1 are transformed by the same function. This type of contact equivalence is called t-equivalence in [3, p 129]. We use the singularity theory approach of [3] to prove the following theorem which extends [6].

THEOREM 2.2. *Under the further nondegeneracy condition*

$$l_0 \cdot (d^2 f \cdot r_0 r_0 \cos\beta_0 + d^2 g \cdot r_0 r_0 \sin\beta_0) \neq 0 \tag{2.13}$$

at (w_0, β_0), where $w_0 \in \mathcal{B}$ and β_0 is the solution of (2.4) and $l_0 P = P r_0 = 0$, there is a ball $\mathcal{O} \subset \mathbf{R}^2, w_0 \in \mathcal{O}$, in which the bifurcation problem

$$V(w, w_0, \beta) = 0 \tag{2.14}$$

is t-equivalent to the one-state-variable problem

$$h(x, \mu) = \varepsilon_1 x^2 + \varepsilon_2 x \mu^2 = x(\varepsilon_1 x + \varepsilon_2 \mu^2) \tag{2.15}$$

near $x = \mu = 0$. Here we may take $\mu = \beta - \beta_0$ and $x = w \cdot r_0$; ε_1 has the sign of the expression in (2.13) and ε_2 is also non-zero. Problem (2.15) has t-codimension one; an unfolding is given by

$$h_a(x, \mu) = x(\varepsilon_1 x + \varepsilon_2 \mu^2 + a) \tag{2.16}$$

where we may take $a = w_0 \cdot r_0, w_0 \notin \mathcal{B}$. The problem and its unfolding are shown in Figure 2.1, a-c.

Proof. We summarize the calculation. Fix $w_0 \in \mathcal{B}$, and suppose $\det B(w_0) \neq 0$. It can be shown that (2.13) implies that r_0 is not tangent to $\mathcal{B}$; we choose a direction for (r_0, l_0) so that r_0 is directed towards $\mathcal{H}$ and (2.13) is positive. We write

$$w = w_0 + xr_0 + W(x, \beta)r_0^{\perp}$$

and let

$$h(x, \beta) = l_0 V(w_0 + xr_0 + W(x, \beta)r_0^{\perp}, w_0, \beta) \tag{2.17}$$

where W is determined from $l_0^{\perp} V \equiv 0$. Performing the steps of the Liapunov Schmidt reduction recursively yields

$$h_x(0, \beta_0) = W_x(0, \beta_0) = 0$$

and

$$h_{xx}(0, \beta_0) = l_0 d^2 V(w_0, w_0, \beta_0) \cdot r_0 r_0 = \varepsilon_1 \neq 0.$$

Also $h_\beta = 0$ and $W_\beta = 0$ and

$$h_{x\beta}(0, \beta_0) = l_0 P_\beta(w_0, \beta_0) r_0; \tag{2.18}$$

using properties of P:

$$Pr_0 = Ar_0 \cos\beta + Br_0 \sin\beta = 0,$$

so, with $Br_0 = y, Ar_0 = -y \tan\beta$ and $P_\beta r_0 = y \sec\beta, l_0 P_\beta r_0 = l_0 B r_0 \sec\beta$. Now, $l_0 B$ is the left eigenvector of N (in (2.11)); hence it is orthogonal to the right eigenvector r_0. Hence in (2.18), $h_{x\beta} = 0$. Calculating $(l_0^{\perp} V)_{x\beta} = 0$:

$$l_0^{\perp} P_\beta r_0 + (l_0^{\perp} P r_0^{\perp}) W_{x\beta} = 0$$

and neither matrix vector product is zero, since

$$l_0^{\perp} P_\beta r_0 = a_2 (l_0 B)^{\perp} r_0 \sec\beta,$$

where $l_0^{\perp} B = a_1 l_0 B + a_2 (l_0 B)^{\perp}$ and $a_2 \neq 0$ since B is invertible; since $(l_0 B)^{\perp}$ is orthogonal to the left eigenvector of N it is not orthogonal to r_0. Finally,

$$l_0^{\perp} P r_0^{\perp} = \frac{\cos\beta}{|\det B|} l_0^{\perp} B \alpha r_0 = \frac{a_2 \alpha \cos\beta}{|\det B|} t^{\perp} r_0 \neq 0.$$

Now $h_{\beta\beta} = W_{\beta\beta} = 0$ as a consequence of t-equivalence, and similarly $h_{\beta\beta\beta} = W_{\beta\beta\beta} = 0$. The final calculation is,

$$h_{x\beta\beta} = l_0 P_{\beta\beta} r_0 + 2W_{x\beta} l_0 P_\beta r_0^\perp = 2W_{x\beta} l_0 P_\beta r_0^\perp, \tag{2.19}$$

for $l_0 P_{\beta\beta} r_0 = -l_0 P r_0 = 0$, and thus $h_{x\beta\beta} \neq 0$ by the calculations above. These defining and nondegeneracy conditions determine the singularity to be equivalent to (2.15); by our normalization, we may take $\varepsilon_1 = 1$. In the unsteady and transonic cases where we have done the calculation, the sign of ε_2 is -1; this means that the nontrivial solutions of (2.15) have $x > 0$ and hence $w \in \mathcal{H}$. If we unfold the singularity (2.15), we may do so by perturbing w_0 away from $\mathcal{B}$, say to $w_0 = ar_0 + w_0^*$, where the calculations above are performed at a fixed point $w_0^* \in \mathcal{B}$. Now we obtain the normal form

$$h_a(x, \beta) = x(x - \mu^2 + \varepsilon a),$$

where $\varepsilon > 0$ in both the unsteady and transonic cases. Again the case $a > 0$ corresponds to $w_0 \in \mathcal{H}$.

This concludes the proof. □

The qualitative shape of the Hugoniot locus and its representation as a bifurcation diagram are shown in Figures 2.1 and 2.2.

We remark that to each triple $\{w_1, w_0, \beta\}$ there correspond two different shock configurations: (1.2) is one and

$$w(x, y) = \begin{cases} w_1, & x\cos\beta + y\sin\beta < 0 \\ w_0, & x\cos\beta + y\sin\beta > 0 \end{cases} \tag{2.20}$$

(its mirror image) is the other.

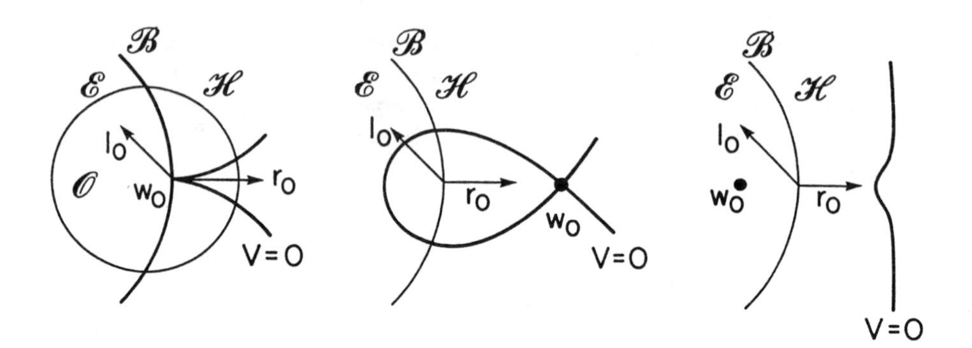

2.1 The Hugoniot locus in state space.

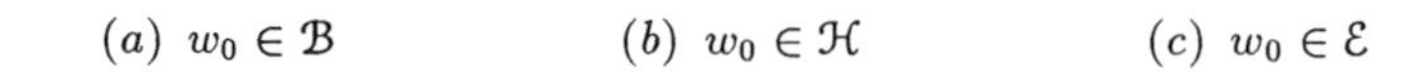

(a) $w_0 \in \mathcal{B}$ (b) $w_0 \in \mathcal{H}$ (c) $w_0 \in \mathcal{E}$

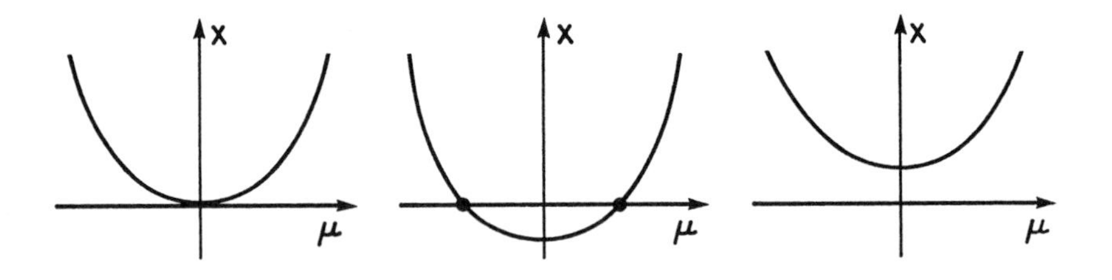

2.2 The bifurcation diagram for the Hugoniot locus.

(a) $h(x, \beta) = 0$ (b) $h_a, a > 0$ (c) $h_a, a < 0$

Our final use of the local theory in this section will be to obtain the relation between the shock angle β, given by (2.12), and the local characteristic angles $\beta_1(w) < \beta_2(w)$ obtained by solving (2.4) at w_0 or w_1, when these points are in $\mathcal{H}$. We observe the following in the case $w_0 \in \mathcal{H}$.

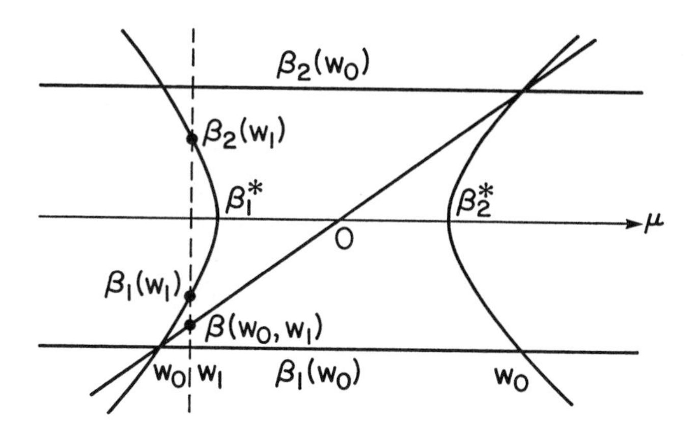

2.3 The characteristic angles along the loop.

PROPOSITION 2.3. *At the points $w_1 = w_0$, corresponding to $\mu = \pm\sqrt{a}$, we have $\beta = \beta_1(w_0)$ or $\beta_2(w_0)$; thus $\beta_1 = \beta_0 - \sqrt{a}$ and $\beta_2 = \beta_0 + \sqrt{a}$, approximately. The genuine nonlinearity of the system in $\mathcal{O} \cap \mathcal{H}$,(a consequence of (2.13)), and the fact that the loop crosses $\mathcal{B}$, show that the angles $\beta_i(w_1)$, calculated as functions of β along the locus, have the structure shown in Figure 2.3. For $\beta_1(w_0) \leq \beta < \beta_1^*, w_1 \in \mathcal{H}$ and $\beta(w_0, w_1) < \beta_1(w_1) < \beta_2(w_1)$; in the interval $\beta_1^* < \beta_2^*, w_1 \in \mathcal{E}$ and there are no real solutions to the characteristic equation. There is a second point β_2^* such that $w_1(\beta_2^*) \in \mathcal{B}$ and for $\beta \in (\beta_0^*, \beta_2(w_0)), \beta_1(w_1) < \beta_2(w_1) < \beta(w_0, w_1)$.*

This section has generalized the results in [6] and localized the picture in [7]: our main conclusion is that the nondegeneracy conditions of Proposition 2.1 and Theorem 2.2 are sufficient to ensure a shock polar or Hugoniot loop near the sonic line and to establish the direction of variation of the characteristic covectors along the loop. We now show how this local behavior is related to shock admissibility.

3. Spacelike curves and timelike conormals. We recall, using the notation of section 2, that the characteristic covectors $\zeta = (\cos\beta, \sin\beta)$ are normals (or conormals) to the *characteristic surfaces* $\phi_i(x, y) = 0$, defined by $\det P(w, \nabla_x \phi) = 0$. Linearizing around a constant state, we write the corresponding surfaces as

$$\phi_i(x, y) \equiv x \cos\beta_i + y \sin\beta_i = 0$$

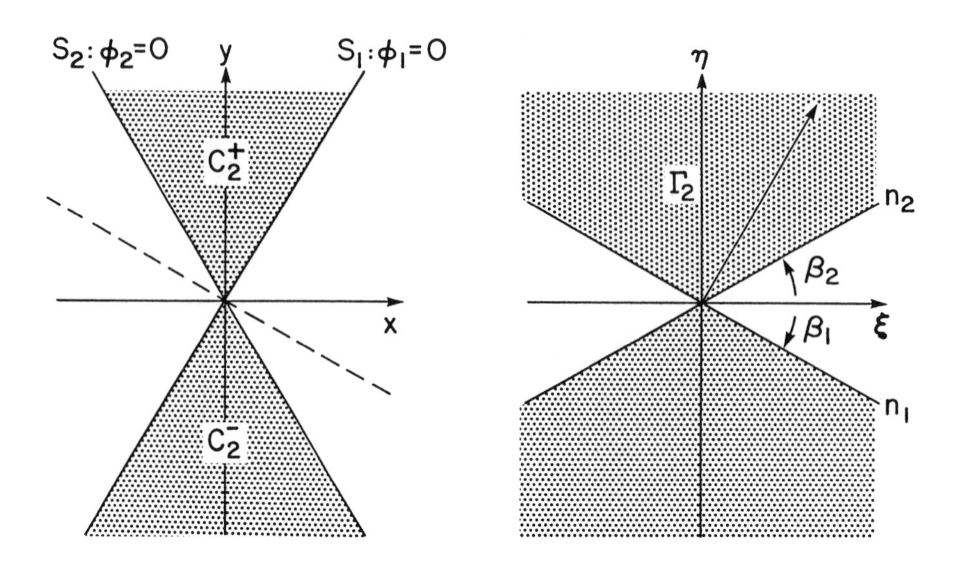

3.1 (a) Characteristic surfaces (b) Characteristic conormals

Now, for a constant $w_0 \in \mathcal{H}$, the linearized system corresponding to (2.1), $P(w_0\partial)w$, is a constant-coefficient hyperbolic operator; it admits a fundamental solution with support in any one of the four quadrants separated by $\mathcal{S}_1$ and $\mathcal{S}_2$ in Figure 3.1 (a). Selecting one of these quadrants as the *forward light cone* (for example, C_2^+ in Figure 3.1 (a)), determines a complementary set of *spacelike curves* with the property that their conormals lie in the open cone Γ_2 of *timelike conormals* in the dual space. Near the sonic line, β_1 is near β_2 and the two possible conormal cones are qualitatively different. We define

$$
\begin{aligned}
\Gamma_1 &= \{\zeta = \rho(\cos\theta, \sin\theta), \beta_1 < \theta < \beta_2, \rho \in \mathbf{R}\backslash\{0\}\} \\
\Gamma_2 &= \{\zeta = \rho(\cos\theta, \sin\theta), -\frac{\pi}{2} < \theta < \beta_1 \text{ or } \beta_2 < \theta \leq \frac{\pi}{2}, \rho \in \mathbf{R}\backslash\{0\}\};
\end{aligned}
$$

then Γ_1 is squeezed to a narrow cone for w near $\mathcal{B}$, and is empty if $w \in \mathcal{B}$, while Γ_2 contains many directions if w is near $\mathcal{B}$ and consists of two open half-planes if $w \in \mathcal{B}$. Corresponding to Γ_1, the light cone, $C_1 = C_1^+ \cup C_1^-$, is, by contrast, wide, while C_2, corresponding to Γ_2, is narrow. (To include the case $\beta_0 = \frac{\pi}{2}$, or to give a global smooth definition of Γ, we can extend this definition in the obvious way.)

We make a couple of remarks here. First, the fact that a system like (2.1) does not have a uniquely defined light cone is an anomaly occuring only in the case of two independent variables: for systems in three or more variables, there is only one cone - C_1 or C_2 - supporting a fundamental solution, and there is a corresponding choice for the timelike conormals. Near the sonic line for a system which changes type, there will still be only two qualitatively different types of behavior: squeezing or expansion of the conormal cone and the opposite behavior for supports. For an abstract equation like (2.1), either choice is a priori possible: we shall see, soon, that Γ_1 is the appropriate choice for transonic flow and Γ_2 for the unsteady model equations we have studied, of the type (1.3). However, there is no a priori reason that steady equations could not have a conormal cone like Γ_2, and there are possibly examples, in unsteady granular flow models for instance, where Γ_1 is appropriate for an unsteady problem. Here we are interested in both choices, and we will consider both.

A second comment concerns the choice of a forward light cone: C always has two components, one of which is designated C^+, and the choice of Γ_1 or Γ_2 does not suffice to determine C^+. This ambiguity is present also for more general systems: it expresses the fact that (2.1) remains hyperbolic under various changes of variables, $x \longmapsto -x$ or $t \longmapsto -t$, for instance, for an evolution equation, or $x \longmapsto -x$ and $u \longmapsto -u$ for the transonic equations. As is well-known, classical solutions of (2.1) are also invariant under these coordinate reversals, but admissible weak solutions are not, and the choice of C^+, from the two connected components of a given C, is bound up with other criteria for admissibility. For example, when (1.4) is strictly hyperbolic, the standard admissibility criteria for a weak solution (1.2) containing a single shock can be simply expressed in these terms. We define the shock, $\mathcal{S} = \{x \cos\beta + y \sin\beta = 0\}$, to be *spacelike with respect to a state* w_i if it forms a spacelike curve with respect to the conormal cone Γ determined by the state w_i. We say further that w_i is an *upstream state* for a spacelike shock $\mathcal{S}$ if the forward light cone C^+, from any point (x, y) where $w = w_i$, intersects $\mathcal{S}$, and the backward cone does not. Classical shocks (1.2) are admissible when precisely one of w_0 and w_1 is an upstream state. It is easy to verify that for unsteady systems of the form (1.3) this is equivalent to the Lax condition if the choice of C^+ is the classical forward cone: $\{t > 0, \lambda_1 t < x < \lambda_2 t\}$. For the steady transonic flow or transonic small disturbance equations, C^+ is the connected component of the complement of $\mathcal{S}_1 \cup \mathcal{S}_2$ which contains the flow vector $w = (u, v)$. This motivatives the following definition:

DEFINITION 3.1. Suppose given a determination $C^+(w)$ for all $w \in \mathbb{R}^2$. A shock $\{w_1, w_0, \beta\}$ *satisfies the C^+ criterion* if there is precisely one upstream state.

We have already commented on the following classical observations.

PROPOSITION 3.2. *Let $C^+(w) = \{t > 0, \lambda_1(w)t < x < \lambda_2(w)t\}$ for a strictly hyperbolic genuinely nonlinear system (1.3). Then a shock satisfies the C^+ criterion if and only if it satisfies the Lax geometric entropy criterion.*

PROPOSITION 3.3. *Let $C^+(w)$ be the component of $\mathbb{R}^2 \backslash \{\mathcal{S}_1 \cup \mathcal{S}_2\}$ containing $w = (u, v)$ if w is a supersonic state of (1.1), and let $C^+(w)$ be the open half-plane $C^+(w_0), w_0 \in \mathcal{B}$, otherwise. Then the shocks that satisfy the C^+ criterion are precisely the compressive shocks.*

(The extension of C^+ to $w \in \mathcal{E}$ is arbitrary: no state in $\mathcal{E}$ is ever upstream.)

We note that for hyperbolic but nonstrictly hyperbolic systems, an overcompressive shock may be upstream with respect to both states, while a noncompressive shock never satisfies the C^+ criterion. These observations, for a hyperbolic system, are not at all new. Definition 3.1 is an extremely special case of Majda's multidimensional shock stability condition [9], which he proves equivalent to Lax's geometric condition. In fact, for a system that is not genuinely nonlinear, where Lax's condition is not sufficient, the C^+-criterion is not either. This definition, however, allows an extension to systems without a physical time variable, such as (1.1), and the possibility of extension (as in Proposition 3.3) to systems that change type.

Definition 3.1 may be useful in a couple of ways: it throws a light on the contrast between transonic and unsteady change of type by showing that these two systems differ already in the strictly hyperbolic region in their choice of C^+. And it is closely connected with the construction of viscous profiles, as we will show in the next section. It is limited in some obvious ways: it does not admit a simple extension to systems of more than two equations except for extreme shocks. And it does not stand alone as a physical or mathematical admissibility criterion but merely suggests a method for proving a stability theorem.

Finally, the definition is not complete without some words on the selection of $C^+(w)$ for $w \in \mathcal{O}$: for smooth systems, the mapping $w \longmapsto C^+(w)$ is smooth in $\mathcal{H}$, and $w \longmapsto C_1^+(w)$ can be extended smoothly to $\mathcal{O}$, as indicated in Proposition 3.3. However, the only continuous extension of C_2^+ into $\mathcal{E}$ is $C_2^+(w) = \phi$ (the empty set) for $w \in \mathcal{E}$, and this gives $\Gamma_2 = \mathbb{R}^2 \backslash \{0\}$, according to which all shocks are spacelike. There is not a clear generalization of "upstream"; however, if we replace C^+ by $\overline{C}^+$ (the closure of C^+) then clearly the choice C_2^+ results, for $w_0 \in \mathcal{B}$, in a cone that degenerates to a line along which the fundamental solution is more singular. For $w_0 \in \mathcal{E}$, the support of the fundamental solution is $\mathbb{R}^2$, but one can choose a fundamental solution with certain growth properties. For the moment, fix any such specification continuously in $\mathcal{E}$. We have the following.

THEOREM 3.4. *Consider the states $\{w_1, w_0, \beta\}$ for (1.4), parameterized by β, in $\mathcal{O}$, with $w_0 \in \mathcal{H}$. Then*

a) with the specification of the conormal cone $\Gamma_1(w)$, the shock is spacelike with respect to w_0 for all pairs $\{w_1, w_0\}$ on the loop. In particular, there exists a smooth specification of $C_1^+(w)$ such that w_0 is the upstream state for all pairs.

b) for the conormal cone $\Gamma_2(w)$, the shock is spacelike with respect to w_1 for $\beta \in (\beta_1(w_0), \beta_1^*) \cup (\beta_2^*, \beta_2(w_0))$, and there is a locally smooth specification of $C_2^+(w)$ such that w_1 is the upstream state for all such pairs. However, there is no smooth extension of $C_2^+(w)$ to $w \in \mathcal{E}$ such that w_1 is the upstream state on the entire loop.

Proof. Recall that $\mathcal{S}$ is spacelike for a state w_i and cone Γ_j if its normal $\zeta = (\cos\beta, \sin\beta)$ is in $\Gamma_j(w_i)$. For $\Gamma_1(w_0), \mathcal{S}$ is clearly spacelike since $\beta_1(w_0) < \beta < \beta_2(w_0)$ along the entire loop. From the definition, w_0 is either an upstream or a downstream state, and there is a choice of C_1^+ that makes w_0 the upstream state. Since $\mathcal{S}$ is never characteristic for w_0, this determination of C_1^+ is smooth. Furthermore, $\mathcal{S}$ is not spacelike with respect to w_1 for $w_1 \in \mathcal{H}$, and under the natural choice of Γ_1 in $\mathcal{E}(\Gamma_1 = \phi), \mathcal{S}$ is not spacelike for $w_1 \in \mathcal{E}$ either.

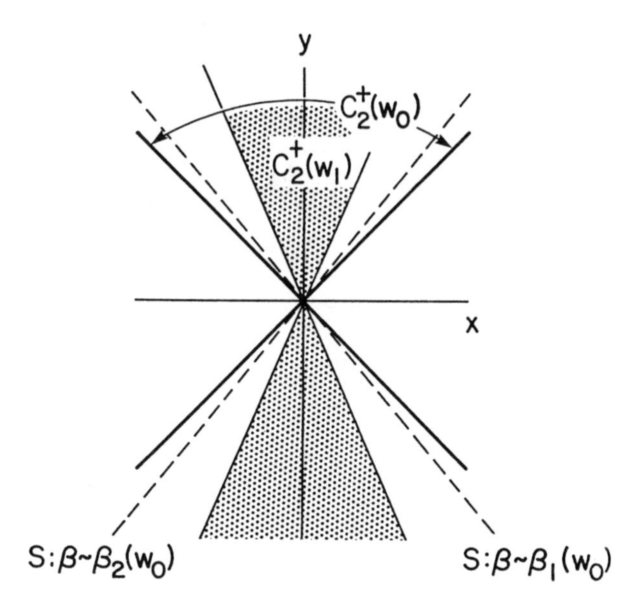

3.2 The geometry of the light cones

(b) For Γ_2, we have, immediately, that $\mathcal{S}$ is never spacelike with respect to w_0. On $(\beta_1(w_0), \beta_1^*)$ and $(\beta_2^*, \beta_2(w_0))$, $\mathcal{S}$ is spacelike with respect to $\Gamma_2(w_1)$. To see that there is no smooth determination of $C_2^+(w)$ for $w \in \mathcal{E}$ that makes w_2 an upstream state everywhere, we make a geometric/topological argument (See Figure 3.2). For $w_1 \in \mathcal{H}, C_2^+(w_1) \subset C_2^+(w_0)$, because of the monotonicity of characteristic speeds. Now $\mathcal{S}$ always lies in one or other component of $C_2(w_0)\backslash C_2(w_1)$, by the first statement after (b). However, for β near $\beta_1(w_0), \mathcal{S}$ is almost characteristic and is near the $\mathcal{S}_1$ characteristic; near $\beta_2(w_0), \mathcal{S}$ is near the $\mathcal{S}_2$ characteristic. Thus $\mathcal{S}$ cannot be exterior to $C_2(w_1)$ for all w_1 on the loop for any determination of C_2. This argument does not even depend on the choice of which component is $+$, but says $\mathcal{S}$ will actually fail to be spacelike. This has important implications, some of which we discuss in the next section.

4. The C^+ criterion, viscous profiles and vectorfield dynamics. In this section we will outline the relation between a definition of admissibility based on the C^+ criterion and the construction of travelling wave solutions, or viscous profiles, for an associated perturbed system.

A perturbation of (1.4) by higher-derivative terms would take the form

$$f(w)_x + g(w)_y = \varepsilon\{\partial_x(Dw_x + Ew_y) + \partial_y(Fw_x + Gw_y)\} \tag{4.1}$$

where D, E, F, G are 2×2 matrices which, typically, might depend on w, and ε measures the strength of the perturbation. In specific applications, introduction of such terms might be motivated by considerations of viscosity or other dissipative or dispersive mechanisms. The idea is well known for unsteady systems where, if y represents time, the perturbation is $(Dw_x)_x$ and (4.1) should have a uniformly parabolic character [10]. For a discussion of the physically motivated viscosity terms in the transonic case, see [7]. We will consider the general case for (4.1) elsewhere, observing here only that a parabolic character is what is desired, in order that solutions of (4.1) will converge to admissible weak solutions: the directions of rapid decay can be related to the light cone and conormal cone for (2.1). Here we will motivate the general case and illustrate the C^+ criterion by considering self-similar travelling wave solutions of (4.1) with similarity parameter

$$\chi = (x \cos\beta + y \sin\beta)/\varepsilon. \tag{4.2}$$

Solutions approaching the shock (1.2) for $\varepsilon \to 0$ satisfy

$$w(\chi) \to w_0 \text{ as } \chi \to -\infty, w \to w_1 \text{ as } \chi \to \infty, w'(\pm\infty) = 0. \tag{4.3}$$

The solution whose limit is the "mirror image" shock (2.20) satisfy

$$w \to w_1 \text{ as } \chi \to -\infty, w \to w_0 \text{ as } \chi \to +\infty. \tag{4.4}$$

Substituting $w(\chi)$ in (4.1), integrating once and using (4.3) or (4.4) results in

$$\begin{aligned} &(D\cos^2\beta + (E+F)\cos\beta\sin\beta + G\sin^2\beta)w' \\ &\qquad = (f(w) - f(w_0))\cos\beta + (g(w) - g(w_0))\sin\beta. \end{aligned} \tag{4.5}$$

We abbreviate this as

$$M(w,\beta)w' = V(w, w_0, \beta), \tag{4.6}$$

where M is a 2×2 matrix that measures the effect of the higher order terms and V is the mapping introduced and studied in section two.

Remark. We have already introduced the notion of the shock $\mathcal{S}$ as a spacelike surface, at least with respect to one state w_i. Now χ, defined in (4.2), which measures orthogonal distance from this surface, might be thought of as representing a "stream coordinate" in physical space; in that case $\chi \to -\infty$ will represent "upstream" and $\chi \to +\infty$ "downstream" limits, and these intuitive notions may be helpful. Since we are using the convention that w_0 is the state that is always in $\mathcal{H}$, we need different boundary conditions, (4.3) or (4.4), to cover the possibilities - both of which arise - that it be the upstream or downstream state. However, χ does not necessarily represent "time". For example, in unsteady systems, (1.3), the characteristic speed is $s = -\tan\beta$ and $\chi = \frac{\cos\beta}{\varepsilon}(x - st)$; since $\cos\beta > 0$ in our normalization, χ increases with increasing t only if $s < 0$. This emphasizes the fact that "timelike" is properly defined in the cotangent space.

We now consider separately the cases TS and US. We have noted that the state w_0 satisfies the C_1^+ criterion for all w_1 on its Hugoniot loop. That is, w_0 is an upstream state. Furthermore, for states $w_1 \in \mathcal{O}$ beyond the loop, w_1 is the upstream state. That is, if there is a solution to (4.6) satisfying (4.3) or (4.4) according as w_1 is in the loop or not, then (4.6) will be consistent with a well-known physical admissibility criterion, namely, admissibility of compressive shocks. We remark that the scalar dynamical system

$$\dot{x} = x(x - \mu^2 + a) \tag{4.7}$$

completely describes these dynamics; here $h = x(x - \mu^2 + a)$ is the function obtained in Theorem 2.2 and both the cases $a > 0 (w_0 \in \mathcal{H})$ and $a < 0 (w_0 \in \mathcal{E})$ are covered. See Figure 4.1. Now the full dynamics of (4.6) will be either one- or two-dimensional. (In [7] we showed that the addition of physical viscosity results in a one-dimensional system.) We may speak of M, or of the perturbation (4.1), as *consistent with TS dynamics in* $\mathcal{O}$ if (4.6) is vectorfield equivalent to (4.7). In [7] we showed that a system obtained by adding physical viscosity to (1.1) was consistent with TS dynamics. The general result is

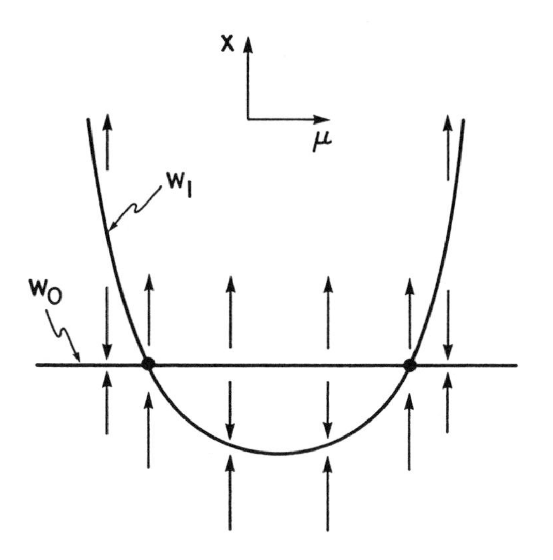

4.1 The reduced vectorfield dynamics in the TS case.

THEOREM 4.1. *Suppose M in (4.6) is either of rank 1 or of rank 2 uniformly in $\mathcal{O}$. Then in either case the reduction in Theorem 2.2 can be extended to (4.6). If M has rank 2, this will be a centermanifold reduction of*

$$w' = M^{-1}V(w)$$

and will reduce to (4.7) if the eigenvalues of $d(M^{-1}V)$ at w_0^* have the appropriate signs. If M has rank 1, then the dynamics of (4.6) are already one dimensional and will coincide with (4.7) for an open set of rank 1 matrices.

A proof is sketched in [7].

By contrast, when the unsteady case, (1.3), is perturbed by addition of a viscosity type term, one does not anticipate a simple addition of one-dimensional dynamics to the bifurcation equations (2.16). We give some heuristic arguments before summarizing a theorem that describes a special case.

First, we have noted that in the US case, where w_1 is the upstream direction in the loop near w_0, we cannot consistently choose w_1 to be upstream over the whole loop. Thus we do not expect to be able to reduce the dynamics to $-\dot{x} = x(x-\mu^2+a)$ (which is Figure 4.1 with the arrows reversed) for any M which would be consistent with the parabolic nature of the perturbed system. In fact, could we do so, w_1 would be a source in the 1-dimensional dynamics and w_0 a sink ; for a nonsingular M, then, w_1 would be a saddle in the 2-dimensional dynamics (before the center manifold reduction) and w_0 a sink. However, for equation (1.3), where one flux function is the identity mapping, it can be checked that the eigenvalues of dV are complex in the region $\mathcal{E}$ and that it is impossible for $M^{-1}V$ to be a saddle for w_1 on the entire loop; if M is singular, it is still impossible to do this smoothly. Finally, we quote a theorem from [6] which describes one possible situation in this case.

THEOREM 4.2. *Consider*

$$u_t + f(u)_x = \varepsilon M u_{xx} \tag{4.8}$$

at $(w_0^*, \beta_0) \in \mathcal{B}$. If (2.13) holds:

$$a = l_0 \cdot d^2 f \cdot r_0 r_0 \neq 0$$

and also the condition

$$b = l_0 \cdot d^2 f \cdot r_0 l_0 + r_0 \cdot d^2 f \cdot r_0 r_0 \neq 0$$

then (4.6) is equivalent, at $w_0 = w_1 = w_0^*$ and $M = I$, to the codimension two Takens-Bogdanov vectorfield

$$\begin{cases} \dot{x} = y \\ \dot{y} = ax^2 - bxy. \end{cases} \tag{4.9}$$

Furthermore, if $w_0 \in \mathcal{H}$ and $\{w_1, w_0, \beta\}$ satisfies the Hugoniot relation in $\mathcal{O}$, and M is near I, a complete unfolding of (4.9), up to t-equivalence, is obtained:

$$
\begin{aligned}
\dot{x} &= y \\
\dot{y} &= ax^2 - bxy + \mu_1 x + \mu_2 y.
\end{aligned}
\tag{4.10}
$$

The unfolding of the singularity is described in [8], though not in normal form; unfolding of the normal form under equivalence (not t-equivalence) is given in [4], and applied to (4.8) in [6]. Neither of the treatments emphasizes the distinguished parameter, β. In the unfolding, we may consider $(\mu_1(\beta), \mu_2(\beta))$ in (4.10) as a path through the unfolding space: the path begins and ends on the μ_2 axis, where the critical points coincide, corresponding to the ends of the loop, and $\mu_1 < 0$ gives a path with $w_0 \in \mathcal{H}$ and $\beta \in (\beta_1(w_0), \beta_2(w_0))$ such as we have been considering. The sketch in Figure 4.2 shows representative vectorfields in the three qualitatively different segments of the loop. Comparison of characteristic and shock speeds here shows that w_1 is the upstream state in the two intervals where connecting orbits exist (and the similarity parameter χ in (4.2) need not approach the upstream state at $-\infty$). This is further evidence of the impossibility of any admissibility condition that includes all states in this case. However, we note that the points where the curve $(\mu_1(\beta), \mu_2(\beta))$ meets B_h (locus of Hopf bifurcation) and B_{sc} (homoclinic bifurcation) are in the interior of $\mathcal{E}$ for M near I. Thus there are some unsteady shocks joining points in regions of different types that are admissible under the viscous profile criterion.

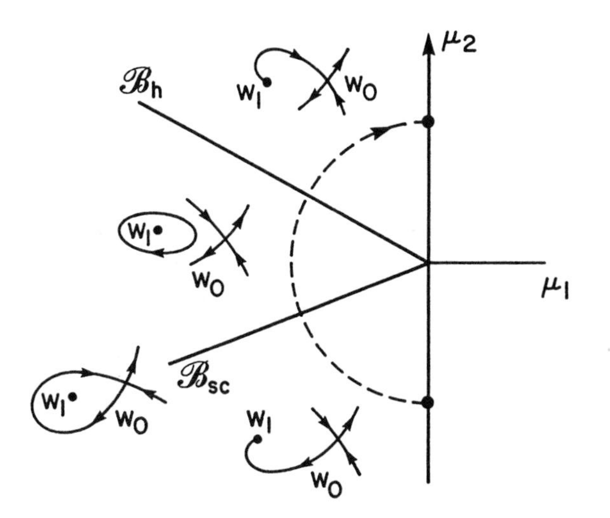

4.2 Vectorfield dynamics in the US case.

In this section, we have not given a complete discussion of the relation between the C^+ criterion and viscous perturbation, but several points have emerged:

1. The question of "suitability" of viscous perturbations is related to stability of the associated linearized parabolic system, as observed by Majda and Pego [10]; since that stability question is studied by examining the forward light cone and comparing it to evolution of the parabolic system, we note that here, too, the choice of "suitable" viscosities is related to a "suitable" definition of $C^+(w)$. In particular, the perturbation must make the system "parabolic" and not "elliptic".

2. Travelling wave solutions of the parabolic system, because they are self-similar, reduce the study of the PDE to an ODE in the similarity variable. Locally we are interested in rest points of the vectorfield; these are given, up to some notion of equivalence, by the development in section 2: the steady state theory identities the rest points (w_0 and w_1), and the C^+-criterion specifies something about their type (repellors for upstream limits, attractors for downstream).

3. The two prototype systems we have studied, TS and US, exemplify two qualitatively different kinds of behavior: in the TS case, there is *no* vectorfield bifurcation along the loop, and a one-dimensional theory is adequate. In the US case, it is necessary to consider a fully two-dimensional vectorfield unfolding.

4. These observations give a framework in which it may be possible to perform rigorous stability analysis of TS and US shocks.

REFERENCES

[1] M.B. Allen III, G.A. Behie and J.A. Trangenstein, Multiphase Flow in Porous Media, Springer, New York, 1988.

[2] R. Courant and K.O. Friedrichs, Supersonic Flow and Shock Waves, Wiley Interscience, New York, 1948.

[3] M. Golubitsky and D.G. Schaeffer, Singularities and Groups in Bifurcation Theory, Springer, New York, 1985.

[4] J. Guckenheimer and P. Holmes, Nonlinear Oscillations, Dynamical Systems, and Bifurcation of Vector Fields, Springer, New York, 1983.

[5] B.L. Keyfitz, " A criterion for certain wave structures in systems that change type ", to appear in Current Progress in Hyperbolic Systems, Contemp. Math., AMS, Providence, 1989.

[6] B.L. Keyfitz, "Admissibility conditions for shocks in conservation laws that change type", to appear in Proceedings of GAMM International Conference on Problems Involving Change of Type, ed K. Kirchgassner.

[7] B.L. Keyfitz and G.G. Warnecke, "The existence of viscous profiles and admissibility for transonic shocks", preprint.

[8] N. Kopell and L.N. Howard, "Bifurcations and trajectories connecting critical points", Adv. Math., 18 (1975), pp. 306-358.

[9] A. Majda, "The stability of multi-dimensional shock fronts", AMS Memoirs, 275 (1983).

[10] A. Majda and R.L. Pego, "Stable viscosity matrices for systems of conservation laws ", Jour. Diff. Eqns, 56 (1985), pp. 229-262.

[11] M. Renardy, W.J. Hrusa and J.A. Nohel, Mathematical Problems in Viscoelasticity, Longman, New York, 1987.

[12] D.G. Schaeffer, "Instability in the evolution equations describing incompressible granular flow", Jour. Diff. Eqns, 66 (1987), pp. 19-50.

[13] H.B. Stewart and B. Wendroff, "Two-phase flow: models and methods", Jour. Comp. Phys., 56 (1984), pp. 363-409.

A STRICTLY HYPERBOLIC SYSTEM OF CONSERVATION LAWS ADMITTING SINGULAR SHOCKS

HERBERT C. KRANZER† AND BARBARA LEE KEYFITZ‡

1. Introduction. The system

$$\begin{cases} u_t + (u^2 - v)_x = 0 \\ v_t + (\frac{1}{3}u^3 - u)_x = 0 \end{cases} \tag{1}$$

is an example of a strictly hyperbolic, genuinely nonlinear system of conservation laws. Usually the Riemann problem for such a system is well-posed: centered weak solutions consisting of combinations of simple waves and admissible jump discontinuities (shocks) exist and are unique for each set of values of the Riemann data [1-3]. The characteristic speeds λ_1 and λ_2 of system (1), however, do not conform to the usual pattern for strictly hyperbolic, genuinely nonlinear systems: although locally separated, they overlap globally (cf. Keyfitz [4] for a more general discussion of the significance of overlapping characteristic speeds). In particular, $\lambda_1 = u - 1$ and $\lambda_2 = u + 1$ are real and unequal at any particular point $U = (u, v)$ of state space (as strict hyperbolicity requires), and $\lambda_2 - \lambda_1 = 2$ is even bounded away from zero globally, but λ_1 at one point U_1 may be equal to λ_2 at a different point U_2. The corresponding right eigenvectors $\mathbf{r}_1 = (1, u+1)$ and $\mathbf{r}_2 = (1, u-1)$ of the gradient matrix for (1) display genuine nonlinearity, since $\mathbf{r}_i \cdot \nabla\lambda_i > 0$ for $i = 1, 2$ but the two eigenvalues vary in the same direction: $\mathbf{r}_i \cdot \nabla\lambda_j > 0$ for $i \neq j$, rather than the usual "opposite variation" $\mathbf{r}_i \cdot \nabla\lambda_j < 0$ familiar from (say) gas dynamics. As a result, classical global existence and uniqueness theorems [3,5] no longer apply.

In Section 2, we investigate the Riemann problem for system (1). We find that the rarefaction curves cover the u, v-plane smoothly, but that the Hugoniot loci are compact curves. As a result, for each fixed left-hand state U_L there is a large region of the plane which cannot be reached from U_L by any combination of rarefaction waves and admissible shocks, even if we were to admit as shocks jump discontinuities which violate the Lax entropy condition.

In Section 3, we introduce a new type of solution to (1), called a *singular shock*, which might be used to connect the left state U_L to right states U_R in this inaccessible region. We discuss how singular shocks may appear as limits of solutions to the Dafermos-DiPerna viscosity approximation

$$\begin{cases} u_t + (u^2 - v)_x = \epsilon t u_{xx} \\ v_t + (\frac{1}{3}u^3 - u)_x = \epsilon t v_{xx} \end{cases} \tag{2}$$

---

†Department of Mathematics and Computer Science, Adelphi University, Garden City, New York 11530.

‡Department of Mathematics, University of Houston, Houston, Texas 77204-3476. Research of BLK supported by AFOSR, under Grant Number 86-0088.

Solutions to (2) do not always remain uniformly bounded as $\epsilon \to 0^+$, but may instead approach singular distributions similar to modified Dirac δ-functions. (Singular solutions of this type were first found by Korchinski [6] for a nonstrictly hyperbolic system.) We investigate the asymptotic behavior of these solutions for small ϵ and discuss how their limits may be regarded as shocks with internal structure. We derive a generalized form of the Rankine-Hugoniot condition for these singular shocks and introduce two additional admissibility conditions for them. These conditions allow us to prove our principal result (Theorem 2), which asserts that the Riemann problem for (1) becomes well-posed for all Riemann data when the category of solutions is enlarged to include admissible singular shocks.

Finally, in Section 4 we attempt to show that the admissible singular shocks which we have defined are actually limits of solutions to the Dafermos-DiPerna approximation (2). We present some analytic results and some numerical calculations as supporting evidence for this conjecture, and we describe what would be needed to convert the conjecture into a theorem.

2. The classical solution and its limitations. Since system (1) is strictly hyperbolic and genuinely nonlinear, the Riemann problem

$$U(x,0) = \begin{pmatrix} u \\ v \end{pmatrix}(x,0) = \begin{cases} U_L, & x < 0 \\ U_R, & x > 0 \end{cases} \tag{3}$$

for (1) has a classical solution when U_R is sufficiently close to U_L. To describe this classical solution, we rewrite (1) as

$$U_t + F_x \equiv U_t + AU_x = 0 \tag{4}$$

with

$$F = \begin{pmatrix} u^2 - v \\ \frac{1}{3}u^3 - u \end{pmatrix}$$

and

$$A = \frac{\partial F}{\partial U} = \begin{pmatrix} 2u & -1 \\ u^2 - 1 & 0 \end{pmatrix}.$$

The eigenvalues of A are the characteristic speeds $\lambda_1 = u - 1$ and $\lambda_2 = u + 1$ with corresponding right eigenvectors

$$\mathbf{r}_1 = \begin{pmatrix} 1 \\ u + 1 \end{pmatrix}, \qquad \mathbf{r}_2 = \begin{pmatrix} 1 \\ u - 1 \end{pmatrix}.$$

The rarefaction curves R_i are the integral curves of the $\mathbf{r}_i$, namely

$$R_1 : v = \frac{1}{2}u^2 + u + c_1$$

and

$$R_2 : v = \frac{1}{2}u^2 - u + c_2.$$

The Rankine-Hugoniot condition $s[U] = [F]$ defines the *Hugoniot locus* $H(U_0)$, the set of U-points which can be connected across a jump discontinuity to U_0, as

$$[v] = [u]\left(\frac{u+u_0}{2} \quad \pm \quad \sqrt{1-[u]^2/12}\right). \tag{5}$$

where $[u] = u - u_0, [v] = v - v_0$; the Rankine-Hugoniot condition also determines the propagation speed s of the discontinuity as

$$s = u_0 + [u]/2 \mp \sqrt{1-[u]^2/12}. \tag{6}$$

We note that the Hugoniot locus is restricted to the strip $|u - u_0| \leq \sqrt{12}$ and consists of four branches in the neighborhood of (u_0, v_0) which join to form a figure eight (see Figure 2.1). In particular, the locus is compact, and its compactness places a finite upper bound on the strength of any discontinuity which can occur as part of a weak solution of (4).

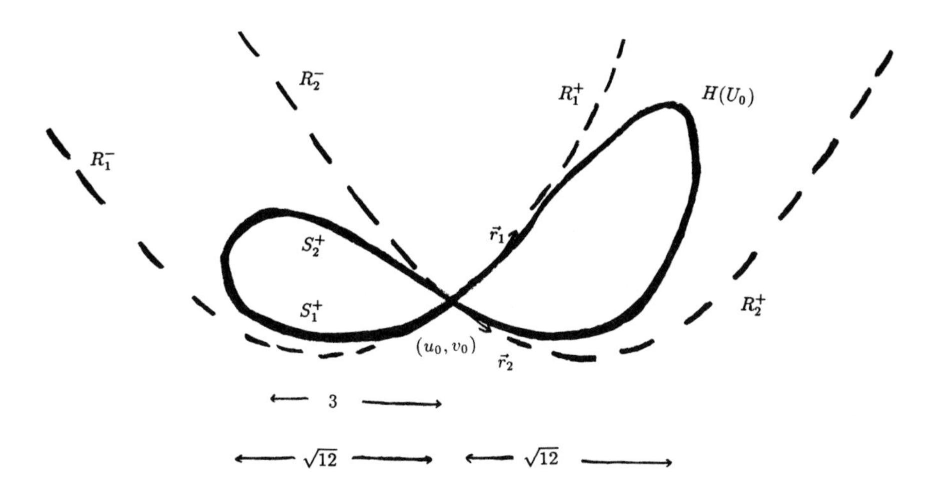

Figure 2.1

To solve the Riemann problem, we need to identify the admissible (+) portions of H and R_i with respect to any state U_0, considered to be on the left. For the rarefaction curves, we have

$$\begin{aligned} R_1^+(U_0) &: v = v_0 + u^2/2 + u - u_0^2/2 - u_0, u_0 \leq u; \\ R_2^+(U_0) &: v = v_0 + u^2/2 - u - u_0^2/2 + u_0, u_0 \leq u. \end{aligned} \tag{7}$$

The shock curves $S_i^+(U_0)$ are classically defined as those portions of $H(U_0)$ which satisfy the Lax entropy condition for discontinuities of the i-th family. A straightforward calculation identifies the S_i^+ as those portions of the curves (5) which lie within the narrower strip $u_0 - 3 \le u \le u_0$; S_1^+ has the upper sign in (5) and S_2^+ the lower sign. For a classical Riemann solution to exist with $U_L = U_0$, the right-hand state U_R must lie on either $R_2^+(U_m)$ or $S_2^+(U_m)$ for some intermediate state U_m which, in turn, lies on either $R_1^+(U_0)$ or $S_1^+(U_0)$, leading to the standard partition of the U-plane into four regions. However, in the present case the four "standard" regions do not fill the entire U-plane. Instead, they cover only the domain $Q = Q(U_0)$ described by

$$v_-(u) \le v \le v_+(u), \qquad u_0 - \sqrt{12} \le u \le +\infty,$$

with

$$(8) \qquad v_\pm(u) = \begin{cases} v_0 + [u](\frac{1}{2}(u+u_0) \mp \sqrt{1-[u]^2/12}), & u_0 - \sqrt{12} \le u \le u_0 - 3; \\ v_0 + u^2/2 - u_0^2/2 \pm [u] \pm \frac{9}{2}, & u_0 - 3 \le u < +\infty \end{cases}$$

(See Figure 2.2). When U_R lies in the complement of $Q(U_L)$ (which in the curvilinear $R_1 - R_2$ coordinate system occupies approximately three fourths of the plane), no classical solution to the Riemann problem (1), (3) exists.

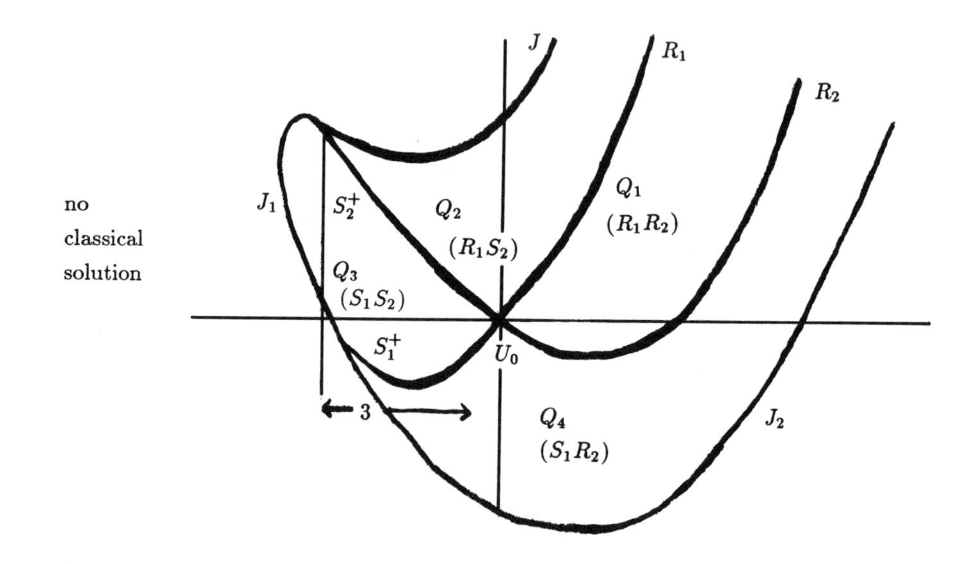

Figure 2.2

3. Singular shocks. Following the ideas of Dafermos [7] and Dafermos and DiPerna [8], we consider the viscosity approximation (2), i.e.

$$U_t + F_x = \epsilon t U_{xx}, \tag{9}$$

which reduces to (4) when $\epsilon = 0$. This approximation was designed for looking at solutions to the Riemann problem because it has centered solutions of the form $U = U(x/t)$ which satisfy the system of ordinary differential equations

$$\epsilon \ddot{U} = (A(U) - \xi)\dot{U}, \tag{10}$$

where $\xi = x/t$ and $\dot{} = d/d\xi$. In contrast to the viscosity approximation generally used in studying classical shocks, system (10) is not autonomous; it depends explicitly on ξ (and also on ϵ). The approximate Riemann problem for (9) provides boundary conditions

$$U(-\infty) = U_L, \qquad U(+\infty) = U_R \tag{11}$$

for (10).

The classical solutions to (10), (11) can be divided into *rarefactions*, for which U and $\dot{U}$ remain uniformly bounded as $\epsilon \to 0$, and *shocks*, for which U remains bounded but $\dot{U}$ approaches infinity. As we saw in Section 2, we should not expect a classical solution unless $U_R \in Q(U_L)$. We consider the possibility that *singular solutions* of (10) exist, in which U itself becomes unbounded as $\epsilon \to 0$ for ξ in the vicinity of some value s, which represents the speed of propagation of the singularity. Following the route which we previously took in [9], we may try the substitution

$$U(\xi) = \begin{pmatrix} \frac{1}{\epsilon^p}\tilde{u}(\frac{\xi - s}{\epsilon^q}) \\ \frac{1}{\epsilon^r}\tilde{v}(\frac{\xi - s}{\epsilon^q}) \end{pmatrix} \tag{12}$$

with undetermined positive constants p, q and r. For nontrivial solutions to exist we must balance at least two terms in each equation (after expansion); this leads to the relations

$$q = 1 + p, \quad r = 2p. \tag{13}$$

If we then expand $\tilde{u}$ and $\tilde{v}$ as series in ϵ, their lowest-order terms $\tilde{u}_0$ and $\tilde{v}_0$ can be shown (cf. [9]) to satisfy the autonomous system

$$\begin{aligned} \tilde{u}_0' &= \tilde{u}_0^2 - \tilde{v}_0 \\ \tilde{v}_0' &= \tfrac{1}{3}\tilde{u}_0^3 \end{aligned} \tag{14}$$

where $' = d/d\eta$, $\eta = (\xi - s)/\epsilon^q$. This system was found in [9] to have a one-parameter family of closed trajectories beginning and ending at $(0,0)$ (cf. Figure 3.1).

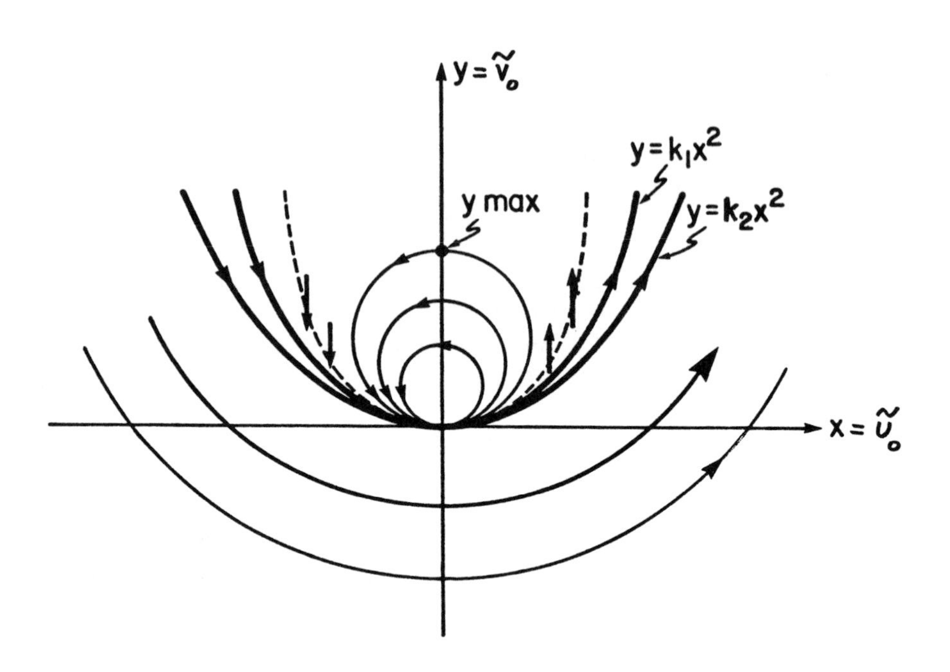

Figure 3.1

These trajectories represent functions whose essential support lies in a layer of width $|\xi - s| = O(\epsilon^q) = o(\epsilon)$ by (13). Thus they do not by themselves solve Riemann problems, since necessarily $\tilde{U}(\pm\infty) = 0$. However, their singularities lie in a zone narrower than a conventional shock profile, which has width $O(\epsilon)$, so that we are naturally led to the idea of embedding a singular solution within a shock profile of conventional type (Figure 3.2 and Figure 3.3). We shall call such a combination, if it exists, a *singular shock*.

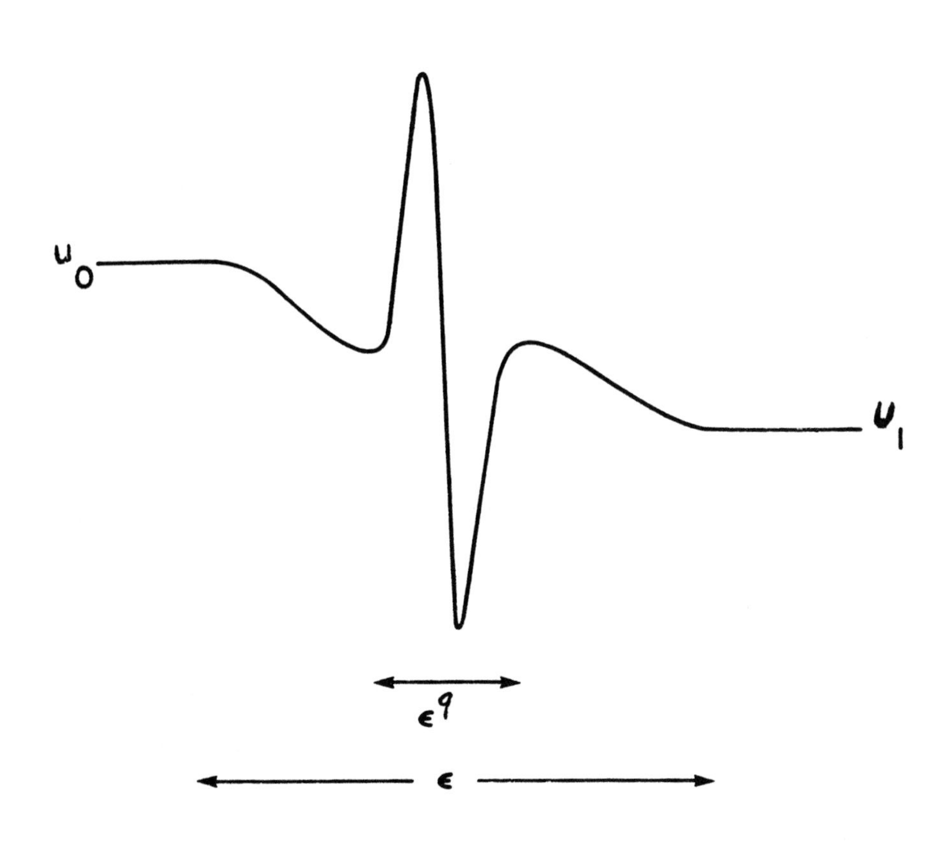

Figure 3.2

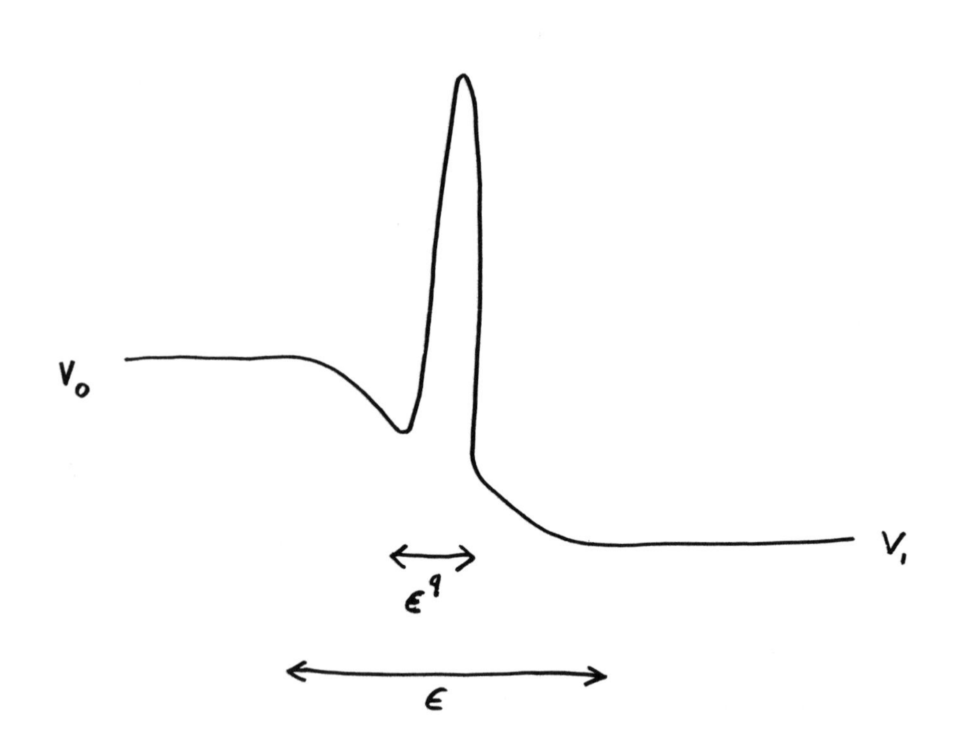

Figure 3.3

The following theorem helps us to determine the precise degree p of the singularity as well as the propagation speed s.

THEOREM 1. *Let $W = W(\xi)$ be a two-vector defined by*

$$W(\xi) = \begin{cases} U_L, & \xi < s \\ U_R, & \xi > s \end{cases} \tag{15}$$

Let $U = U(\xi)$ be any solution of (10) and (11). Then

$$\int_{-\infty}^{\infty} (U - W)d\xi = s(U_R - U_L) - (F(U_R) - F(U_L)).$$

Proof. Integrating (10) from $-\infty$ to ∞, we find

$$\int_{-\infty}^{\infty} A(U)\dot{U}\,d\xi - \int_{-\infty}^{\infty} \xi\dot{U}\,d\xi = \varepsilon \int_{-\infty}^{\infty} \ddot{U}\,d\xi = \varepsilon\dot{U}\Big|_{-\infty}^{\infty} = 0.$$

Rewriting $\xi = s + (\xi - s)$ in the second integral yields

$$\int_{-\infty}^{\infty} (\dot{F} - s\dot{U})d\xi = \int_{-\infty}^{\infty} (\xi - s)\dot{U}\,d\xi$$

or (since $\dot{W} = 0$ except at $\xi = s$)

$$\begin{aligned}(F(U(\xi)) - sU(\xi))\Big|_{-\infty}^{\infty} &= \int_{-\infty}^{\infty} (\xi - s)(\dot{U} - \dot{W})d\xi \\ &= (\xi - s)(U - W)\Big|_{-\infty}^{s} + (\xi - s)(U - W)\Big|_{s}^{\infty} - \int_{-\infty}^{\infty} (U - W)\frac{d}{d\xi}(\xi - s)d\xi \\ &= 0 + 0 - \int_{-\infty}^{\infty} (U - W)d\xi.\end{aligned}$$

Interchanging right and left hand sides then yields the theorem. □

Suppose now that a singular shock directly connects two states U_L and U_R, where $U_R \notin Q(U_L)$. Theorem 1 then tells us that

$$\int_{\infty}^{\infty} (U - W)d\xi = \begin{pmatrix} c_1 \\ c_2 \end{pmatrix} = C = s[U] - [F],$$

a constant two-vector independent of ε. Now, $U_R \notin H(U_L) \subseteq Q(U_L)$, so $C \neq 0$. However, by letting $\varepsilon \to 0$ we find from (12) and (13) that

$$c_1 = \lim_{\varepsilon \to 0} \frac{1}{\varepsilon^p} \int_{-\infty}^{\infty} \tilde{u}\left(\frac{\xi - s}{\varepsilon^q}\right) d\xi = \lim_{\varepsilon \to 0} \varepsilon^{q-p} \int_{-\infty}^{\infty} \tilde{u}(\eta)d\eta = 0, \tag{16}$$

so that the speed of a singular shock must satisfy the *generalized Rankine-Hugoniot condition*

$$s(u_R - u_L) = f(U_R) - f(U_L), \tag{17}$$

where $f(u,v) = u^2 - v$ is the first component of the flux function $F(U)$. A similar calculation applied to the second component shows that

$$c_2 = \lim_{\varepsilon \to 0} \varepsilon^{q-r} \int_{-\infty}^{\infty} \tilde{v}(\eta) d\eta.$$

But c_2 must be a finite nonzero constant (since $C \neq 0$ and $c_1 = 0$), which implies that $q = r$, and therefore from (13) we get

$$p = 1, \quad q = r = 2. \tag{18}$$

How large a family of right states can be connected through a singular shock to a given left state U_0? For conventional shocks, satisfying the full vector Rankine-Hugoniot condition, U_0 together with the shock speed s determines U_R, so the standard theory yields a one-parameter family of right states, those on the Hugoniot locus $H(U_0)$. For singular shocks, the "singular shock strength"

$$c_2 = \int_{-\infty}^{\infty} \tilde{v}(\eta) d\eta = \int_{-\infty}^{\infty} (v - W_2) d\xi, \tag{19}$$

which measures the amount by which the full Rankine-Hugoniot condition fails to hold, becomes a second, independent parameter. Hence the states connectible to U_0 by singular shocks form a two-parameter family, which can fill a two-dimensional subset (region) of the U-plane.

We next attempt to identify this "singular region" more precisely. In doing so, we temporarily turn our attention away from the internal structure of singular shocks and introduce a definition of admissibility that depends only on the two states which are connected. We shall return to the internal structure in Section 4.

DEFINITION. A jump discontinuity propagating with speed s in the x,t-plane is called an *admissible singular shock* if its right and left states

$$U_{\pm} = \lim_{x \to st \pm 0} U(x,t)$$

satisfy the following three conditions:

1) The generalized Rankine-Hugoniot condition

$$s(u_+ - u_-) = (u_+^2 - u_-^2) - (v_+ - v_-).$$

2) The singular shock strength condition

$$c_2 \doteq s(v_+ - v_-) - \frac{1}{3}(u_+^3 - u_-^3) + (u_+ - u_-) > 0.$$

3) The characteristic speeds condition

$$\lambda_2(U_-) > \lambda_1(U_-) \geq s \geq \lambda_2(U_+) > \lambda_1(U_+).$$

This definition can be motivated as follows. Condition 1 is just (17). Condition 2 is suggested by the first equality in (19) and the asymptotic behavior of $\tilde{v}$ as $\varepsilon \to 0$: the closed trajectories for the leading term $\tilde{v}_0$ all lie in the half-plane $\tilde{v}_0 \geq 0$ (see Figure 3.1). Condition 3 might be thought of as a requirement of "over-compression" (cf. Shearer [10]). It states that four characteristic curves enter the discontinuity and none leave. Since a conventional (internally bounded) shock has three incoming characteristic curves and one outgoing, it seems plausible that an additional incoming characteristic should be needed to provide "energy" (integrated $|\dot{U}|$) to maintain a singularity.

Let us denote by $Q_S(U_0)$ the region of the u, v-plane consisting of those right states U_+ which can be joined to the left state $U_- = U_0$ by an admissible singular shock. An easy calculation shows that $Q_S(U_0)$ is disjoint from $Q(U_0)$, shares with $Q(U_0)$ as common boundary the portion J of $H(U_0)$ lying in the strip $u_0 - \sqrt{12} \leq u \leq u_0 - 3$, and has as its other two boundaries the ray

$$E : v = v_0 + (u_0 - 1)(u - u_0), \quad -\infty < u \leq u_0 - 3$$

and the parabolic segment

$$D : v = v_0 + u^2 + (1 - u_0)u - u_0, \quad -\infty < u \leq u_0 - 3$$

(see Figure 3.4). The remainder of the plane, not contained in either Q_S or Q, falls into two regions, $Q'(U_0)$ lying below E and $Q''(U_0)$ lying above and to the right of D.

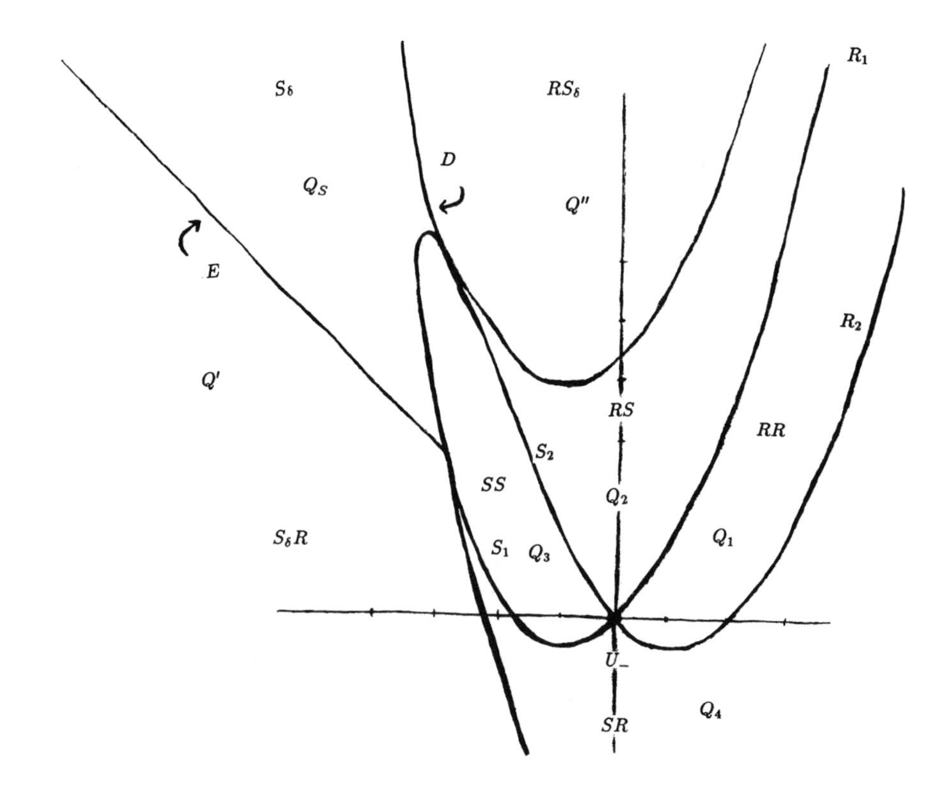

S_δ
RS_δ
R_1
D
Q_S
Q''
E
R_2
Q'
RS
RR
S_2
SS
Q_2
$S_\delta R$
S_1
Q_3
Q_1
U_-
Q_4
SR

Figure 3.4

For a singular shock which connects U_0 to a state in the interior of Q_S, both inequalities in Condition 3 are strict. Such a shock is therefore not continuable on either the right or the left by a conventional wave (shock or rarefaction) of either family. However, a singular shock connecting to a point (say U_m) on E has $s = \lambda_2(U_+)$ and can be continued on the right by a rarefaction wave along $R_2^+(U_m)$. Note that the curves $R_2^+(U_m), U_m \in E$, exactly fill Q', so that all points in Q' can be reached from U_0 by a singular shock followed by a rarefaction of the second family. Similarly, a singular shock connecting U_0 to a point on D has $s = \lambda_1(U_-)$ and can be preceded on the left by a rarefaction wave of the first family; solutions of this type exactly fill Q''. [More precisely, such solutions go from U_0 to a point $U_m \in R_1^+(U_0)$, then to $U_R \in Q_S(U_m)$.]

We have accordingly obtained the following result.

THEOREM 2. *A unique solution to the Riemann problem (1), (3), composed of constant states, centered rarefaction waves, Lax entropy shocks and admissible singular shocks, exists for every pair U_L, U_R of Riemann data.*

4. Admissible singular shocks as Dafermos-DiPerna limits. In this section, we investigate more closely the connection between admissible singular shocks and solutions of the Dafermos-DiPerna viscosity approximation (10), (11).

Conjecture. If $U_R \in Q_S(U_L)$, then for sufficiently small positive ε equations (10), (11) possess solutions

$$U = U_\varepsilon(\xi) = \begin{pmatrix} u_\varepsilon \\ v_\varepsilon \end{pmatrix}$$

whose asymptotic behavior as $\varepsilon \to 0$ is described by

$$\left\{ \begin{array}{l} u_\varepsilon(\xi) \approx W_1(\xi) + \frac{1}{\varepsilon}\tilde{u}_0\left(\frac{\xi-s}{\varepsilon^2}\right) \\ v_\varepsilon(\xi) \approx W_2(\xi) + \frac{1}{\varepsilon^2}\tilde{v}_0\left(\frac{\xi-s}{\varepsilon^2}\right) \end{array} \right. \tag{20}$$

here $(\tilde{u}_0, \tilde{v}_0)$ is a closed-trajectory solution of (14) and s is determined by the generalized Rankine-Hugoniot relation (17), while W_1 and W_2 are the components of the Heaviside function $W(\xi)$ defined by (15).

We present three types of evidence in support of this conjecture, and we indicate what is still needed to convert the conjecture into a theorem.

First of all, we observe that the conjecture would follow from Theorem 3.1 of Dafermos [7] provided we could verify the hypotheses of that theorem. To do that, we must embed (10), (11) in a two-parameter family of boundary-value problems

$$\left\{ \begin{array}{l} \varepsilon\ddot{U} = (aA(U) - \xi)\dot{U}, \\ U(-b) = aU_L, \ \ U(b) = aU_R, \end{array} \right. \tag{21}$$

where $0 \le a \le 1$ and $b \ge 1$. Dafermos shows that, if all solutions of (21) satisfy an a priori bound

$$|U(\xi)| < M \text{ for } -b \le \xi \le b, \tag{22}$$

where M may depend on U_L, U_R, ε and A but not on a or b, then (10), (11) does have a solution for each $\varepsilon > 0$ which satisfies the same bound. Once we knew that solutions to (10), (11) existed, we could apply the asymptotic analysis described in section 3 to obtain (20). Thus an a priori estimate on the maximum norm of U_ε for each fixed ε would be sufficient to establish the conjecture. While we do not yet have a maximum-norm estimate, it is suggestive in this regard to note that our Theorem 1 does provide an L^1 estimate, which is even uniform in ε.

The second bit of supporting evidence arises when we look at the actual maxima (and minima) of solutions of (21). The Riemann invariants $\pi = v - \frac{1}{2}u^2 + u$ and $\rho = v - \frac{1}{2}u^2 - u$ corresponding to these solutions satisfy the equations

$$\begin{aligned} \varepsilon\ddot{\pi} &= [a(u-1) - \xi]\dot{\pi} - \varepsilon\dot{u}^2, \\ \varepsilon\ddot{\rho} &= [a(u+1) - \xi]\dot{\rho} - \varepsilon\dot{u}^2. \end{aligned} \tag{23}$$

Hence, for nonconstant solutions of (21), $\dot{\pi} = 0$ implies $\ddot{\pi} < 0$ and $\dot{\rho} = 0$ implies $\ddot{\rho} < 0$. Thus each Riemann invariant has no (relative) minimum and at most one maximum on any solution trajectory. From this it can be shown (see [11] for details) that u has at most one maximum point $\xi = \xi_1$ and one minimum point $\xi = \xi_2$, while v has only a maximum at $\xi = \xi_3$, and that $\xi_1 < s < \xi_3 < \xi_2$. Thus the trajectory of a solution to the Dafermos-DiPerna viscosity approximation in the u, v-plane should look like Figure 4.1. Observe that Figure 4.1 is related to the asymptotic picture of Figure 3.1 in exactly the same manner as (20) is related to (12), namely by addition of the Heaviside function $W(\xi)$.

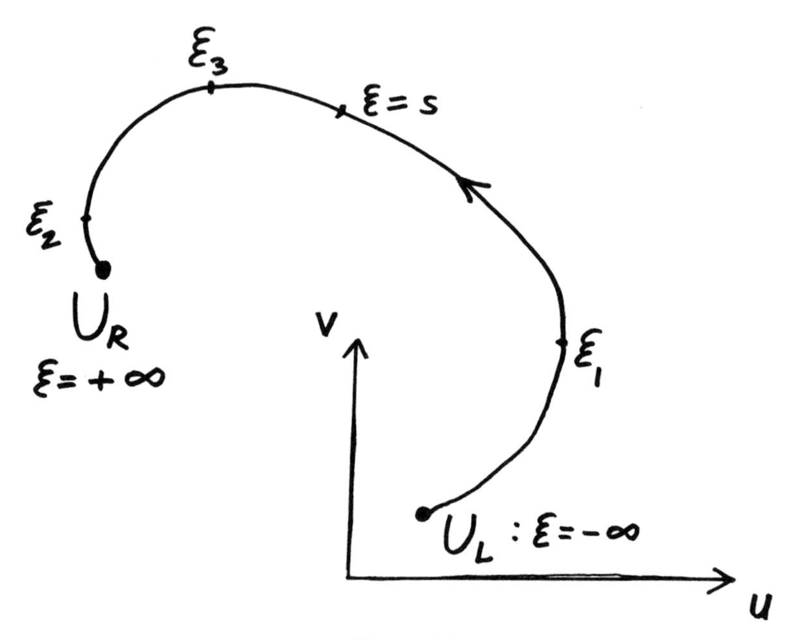

Figure 4.1

As our third piece of evidence, we present the results of a numerical solution

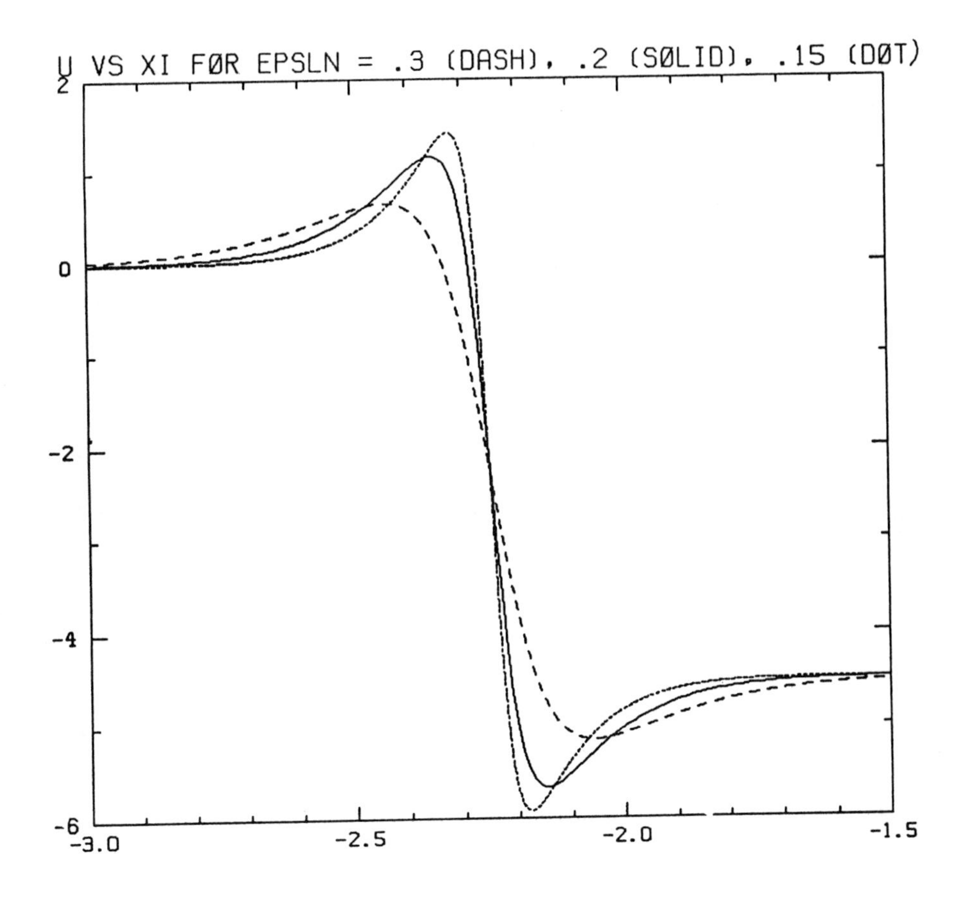

Figure 4.2

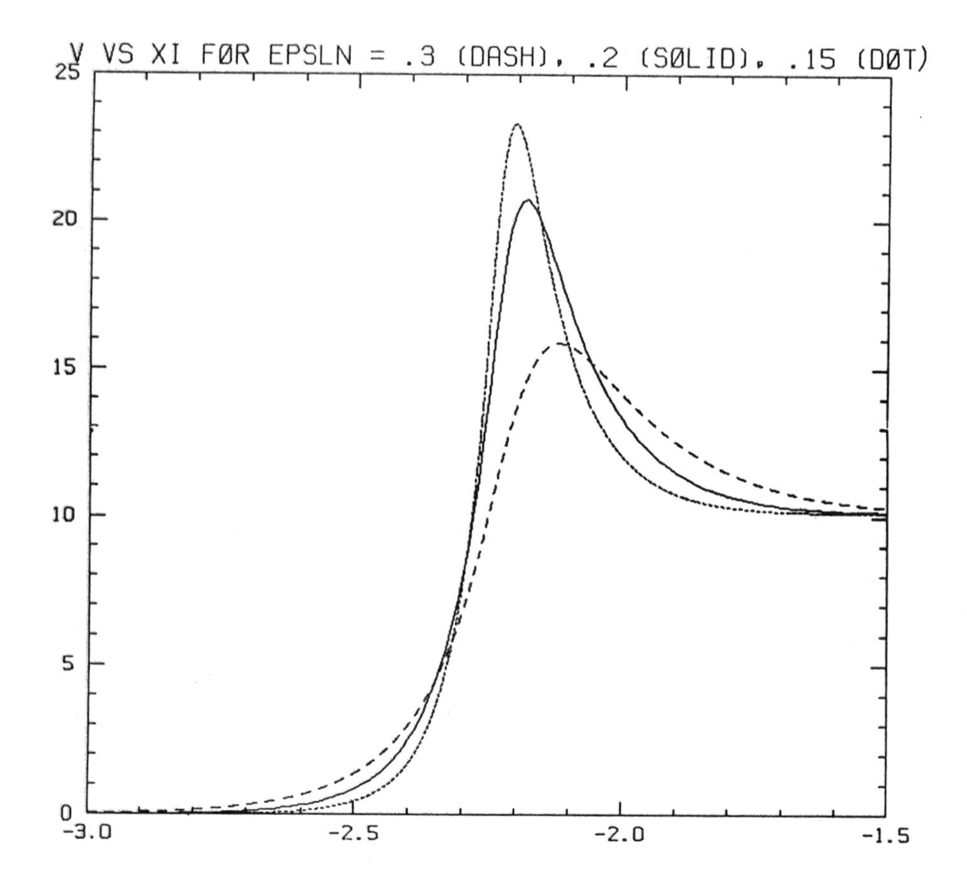

Figure 4.3

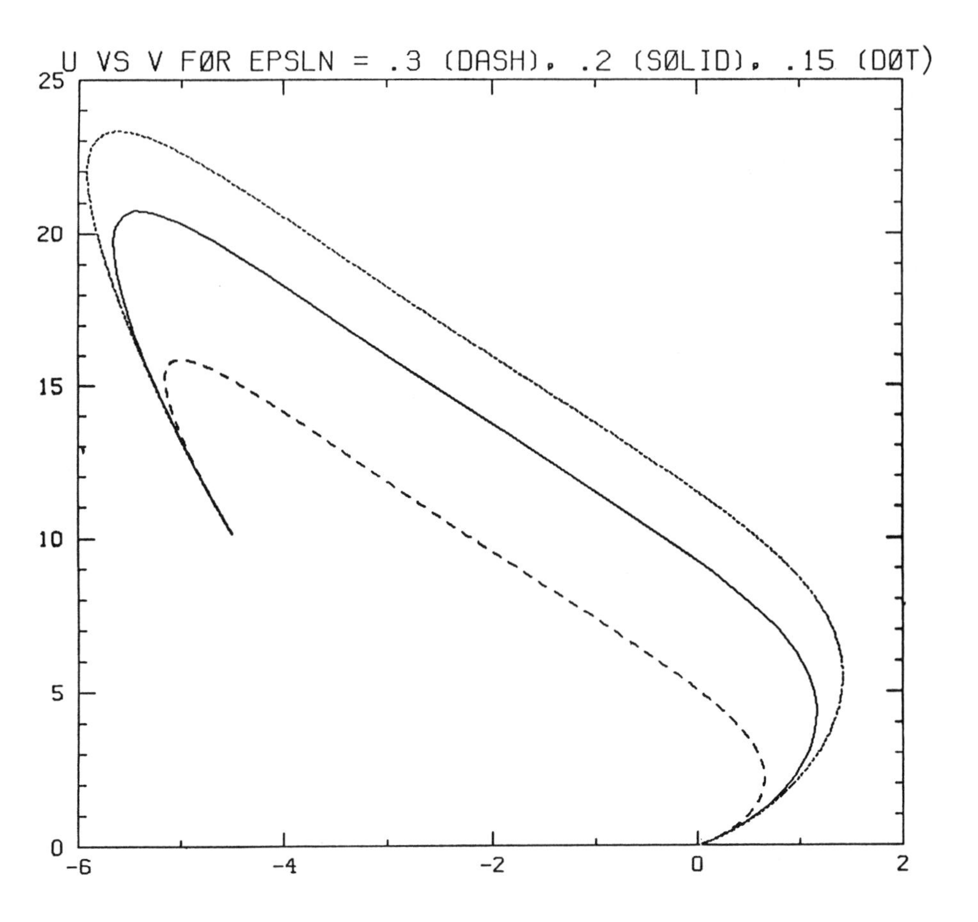

Figure 4.4

of a two-point boundary-value problem similar to (21). We chose $U_L = (0,0)$ and $U_R = (-4.5, 10.125)$, with the Dafermos parameter $a = 1$ and boundaries symmetrically situated (at $\xi = -4$ and $\xi = -0.5$) with respect to the theoretical singular shock speed $s = -2.25$. With a grid size $\Delta\xi = .0025$, we approximated $\dot{U}$ and $\ddot{U}$ by centered differences. We solved the resulting non-linear system of difference equations by an iteration procedure. At each step we determined the solution of the linear system which results when $A(U)$ is computed from the U found at the previous step. Approximately 80 iterations were required for convergence, and the results did not change significantly when the grid size was halved, nor when the length of the ξ-interval was increased or decreased by 50%.

Computations were carried out for three values of ε, namely $\varepsilon = .3, \varepsilon = .2$ and $\varepsilon = .15$, and the numerical solutions are compared in the accompanying figures. Figure 4.2 shows the first component $u(\xi)$, Figure 4.3 shows the second component $v(\xi)$, and Figure 4.4 shows the solution trajectory in the u,v-plane. Note the degree of resemblance between the computed Figure 4.4 and the theoretically derived Figure 4.1. Note further that the upward and downward "bumps" in u (Figure 4.2) do seem to roughly double in height and reduce fourfold in width when ε is halved from .3 to .15, while the height of the single bump in v (Figure 4.3) does seem to more than double (though not quite quadruple). This at least approximately confirms the $1/\varepsilon$ and $1/\varepsilon^2$ scalings of (20). Finally, the theoretical value of the singular shock strength parameter c_2 is 3.09375, while the value computed from the numerical solution for $\varepsilon = .2$, using the trapezoidal rule for the second integral in (19), was $c_2 \approx 3.11$.

On all these points, there seems to be sufficient agreement between numerical and theoretical solutions to support the conjecture, at least for the particular values of U_L and $U_R \in Q_s(U_L)$ used in the computation. Computations involving other values of the Riemann data, including cases where U_R is in $Q'(U_L)$ or $Q''(U_L)$, are consistent with this specimen result.

Acknowledgement. Much of the research described in this article was carried out while the authors were visitors at the Institute for Mathematics and its Applications at the University of Minnesota. It was influenced generally by the stimulating scientific atmosphere at the Institute, and in particular by fruitful conversations with Mitchell Luskin, Bruce Pitman, Bradley Plohr and Michael Shearer.

REFERENCES

[1] Lax, P.D., *Hyperbolic systems of conservation laws II*, Comm. Pure Appl. Math., 10 (1957), pp. 537-566.

[2] Smoller, J.A., and Johnson, J.L., *Global solutions for an extended class of hyperbolic systems of conservation laws*, Arch. Rat. Mech. Anal., 32 (1969), pp. 169-189.

[3] Keyfitz, B.L., and Kranzer, H.C., *Existence and uniqueness of entropy solutions to the Riemann problem for hyperbolic systems of two nonlinear conservation laws*, Jour. Diff. Eqns., 27 (1978), pp. 444-476.

[4] Keyfitz, B.L., *Some elementary connections among nonstrictly hyperbolic conservation laws*, Contemporary Mathematics, 60 (1987), pp. 67-77.

[5] Borovikov, V.A., *On the decomposition of a discontinuity for a system of two quasilinear equations*, Trans. Moscow Math. Soc., 27 (1972), pp. 53-94.

[6] Korchinski, D., *Solution of the Riemann problem for a* 2×2 *system of conservation laws possessing no classical weak solution*, Thesis, Adelphi University, 1977.

[7] Dafermos, C.M., *Solution of the Riemann problem for a class of hyperbolic systems of conservation laws by the viscosity method*, Arch. Rat. Mech. Anal., 52 (1973), pp. 1-9.

[8] Dafermos, C.M., and DiPerna, R.J., *The Riemann problem for certain classes of hyperbolic systems of conservation laws*, Jour. Diff. Eqns., 20 (1976), pp. 90-114.

[9] Keyfitz, B.L., and Kranzer, H.C., *A viscosity approximation to a system of conservation laws with no classical Riemann solution*, Proceedings of International Conference on Hyperbolic Problems, Bordeaux, 1988.

[10] Shearer, M., *Riemann problems for systems of nonstrictly hyperbolic conservation laws*, This volume.

[11] Keyfitz, B.L., and Kranzer, H.C., *A system of conservation laws with no classical Riemann solution*, preprint, University of Houston (1989).

AN EXISTENCE AND UNIQUENESS RESULT FOR TWO NONSTRICTLY HYPERBOLIC SYSTEMS*

PHILIPPE LE FLOCH†

Abstract. We prove a result of existence and uniqueness of entropy weak solutions for two nonstrictly hyperbolic systems, both a *nonconservative* system of two equations

$$\partial_t u + \partial_x f(u) = 0, \quad \partial_t w + a(u)\partial_x w = 0,$$

and a *conservative* system of two equations

$$\partial_t u + \partial_x f(u) = 0, \quad \partial_t v + \partial_x (a(u)v) = 0,$$

where $f : \mathbf{R} \to \mathbf{R}$ is a given strictly convex function and $a = \frac{d}{du} f$. We use the Volpert's product ([19], see also Dal Maso - Le Floch - Murat [1]) and find entropy weak solutions u and w which have bounded variation while the solutions v are *Borel measures.* The equations for w and v can be viewed as linear hyperbolic equations with discontinuous coefficients.

1. Introduction. The theory of nonlinear hyperbolic systems assumes usually the systems to be strictly hyperbolic with genuinely nonlinear or linearly degenerate characteristic fields and to be in conservation form. Moreover, general results of existence of entropy weak solutions to these systems are known only for initial data with small total variation (see Glimm [6] and Lax [11]). But, most of the hyperbolic systems used by physicists have not genuinely nonlinear or linearly degenerate fields, are not strictly hyperbolic and sometimes are even not written in conservation form. This is for instance the case of the systems used in the modeling of great deformations of elastoplastic materials (Trangenstein - Colella [18]) and the modeling of two-phase flows as mixtures of liquid and vapor (Stewart - Wendroff [17]).

It is recognized that the most part of the physical systems does not fit into the standard theory of conservation laws. The aim of the present paper is precisely to consider two examples of *nonstrictly hyperbolic* nonlinear systems, both a conservative and a nonconservative one, for which new concepts of weak solutions must be introduced. Namely, for these systems there is no existence and no uniqueness of (entropy weak) solutions in the usual context, say L^∞ solutions in the sense of distributions. But it is possible to extend the notion of weak solution and derive a formulation of a *well - posed* Cauchy problem for these systems. The main ingredient we shall use in this paper is the Volpert's product [19] for functions of bounded variation (BV). This work follows also the general theory of weak solutions to nonconservative nonlinear hyperbolic systems by Dal Maso - Le Floch - Murat [1,2] and Le Floch ([12] - [15]).

*This research was supported in part by the Institute for Mathematics and its Applications with funds provided by the National Science Foundation.

†Centre de Mathématiques Appliquées, URAD0756, Ecole Polytechnique, 91128 Palaiseau Cedex (FRANCE), E-mail address: UMAP067 at FRORS12.BITNET.

We consider the Cauchy problem for both a nonconservative system of two equations

$$\partial_t u + \partial_x f(u) = 0, \quad \partial_t w + a(u)\partial_x w = 0, \tag{1.1}$$

and a system of two conservation laws

$$\partial_t u + \partial_x f(u) = 0, \quad \partial_t v + \partial_x(a(u)v) = 0, \tag{1.2}$$

where $f : \mathbf{R} \to \mathbf{R}$ is a given smooth convex function and $a = \frac{d}{du} f$. These systems have only one characteristic speed, $a(u)$, so that they are *not strictly hyperbolic.*

It is easy to show that the Riemann problem for the conservative system (1.2) does not possess weak L^∞-solution, except for some particular initial data, even if the initial data is assumed to be small (in L^∞ and BV norm). This fact contrasts with the standard results of existence of weak solutions to strictly hyperbolic systems (Glimm [6] and Lax [11]). On the other hand, we know that the usual notion of weak L^∞ solution in the sense of distributions does not make sense for the nonconservative system (1.1).

In this paper, we use the Volpert's product (Volpert [19] and Le Floch [12]) which yields a notion of entropy weak solutions to nonconservative systems in the frame of functions of bounded variation. We prove that the Cauchy problem for the nonconservative system (1.1) is well-posed *in the space BV*, provided that the initial data for u is a BV function with only nonincreasing jumps (this means that $u(0, \cdot)$ is entropy) and the initial data for w is a Lipschitz continuous function. The solutions w we find are thus in BV and generally speaking are indeed discontinuous. A corresponding existence and uniqueness result of entropy weak solutions to the *conservative* system (1.2) follows by derivation, with respect to x, of the equation satisfied by the function w (set $v = \partial_x w$). We emphasize that, with our definition below, the solutions v in system (1.2) are *not functions* but *Borel measures* on $\mathbf{R}_+ \times \mathbf{R}$. The initial data for v is only assumed to be in $L^\infty(\mathbf{R})$, so that our results apply in particular to the Riemann problem for the conservative system (1.2).

We remark that, in the systems (1.1) and (1.2), the equation for u is uncoupled from the second one and can be solved first, so that the equations for w and u,

$$\partial_t w + a(u)\partial_x w = 0$$

and

$$\partial_t v + \partial_x(a(u)v) = 0$$

can be viewed as linear hyperbolic equations with *discontinuous* coefficients. Such equations have been recently investigated by Di Perna - Lions [4], but their results do not apply to the above equations. We mention also that the ideas of this paper could be also applied to other systems which does not possess L^∞ solutions (Korchinski [8], Kranzer [9]). See also Keyfitz [7].

The plan of the paper is as follows. In Section 2, we recall the definition of weak solution to systems in nonconservative form. Section 3 contains the existence and uniqueness result for the nonconservative system. In Section 4, we deduce a result for the conservative system and we solve the Riemann problem.

2. Nonconservative products in BV space. In this section, we recall the definition of weak BV solutions to nonlinear hyperbolic systems in nonconservative form which is based on a notion of averaged superposition due to Volpert [19] and investigated in Le Floch ([12] - [14]).

Let Ω be an open subset of $\mathbf{R}^m$ and $p \geq 1$ be given. The space $BV(\Omega, \mathbf{R}^p)$ of *functions of bounded variation* consists of all integrable functions $u : \Omega \to \mathbf{R}^p$ each of whose first order partial derivatives $\frac{\partial u}{\partial y_i}(i = 1, 2, \cdots, m)$ is represented by a finite Borel measure. The total variation $TV(u)$ is by definition the sum of the total masses of these Borel measures. We recall that results of regularity concerning BV functions are proved by Volpert [19]. For an element u in $BV(\Omega, \mathbf{R}^p)$, it occurs that with the exception of a set with zero $(m-1)$-dimensional Hausdorff measure, each point y of Ω is *regular*, i.e. either a point of approximate continuity (Lebesque point) or a point of approximate jump. At a point y_0 of approximate jump, there exists a unit normal $\nu(y_0) \in \mathbf{R}^m$, a left value $u_-(y_0)$ and a right value $u_+(y_0)$. Moreover these jump points form an at most countable set of curves. We refer to Volpert [19] for the precise definition and results.

An interesting concept proposed in [19] is the notion of averaged superposition.

DEFINITION 2.1. Let g be in $\mathcal{C}^1(\mathbf{R}^p, \mathbf{R}^p)$ and u be in $L^\infty(\Omega, \mathbf{R}^p) \cap BV(\Omega, \mathbf{R}^p)$. The *averaged superposition* of the function u by the function g is defined by
(2.1)

$$\hat{g}(u)(y) = \begin{cases} g(u(y)), & \text{if } y \text{ is a Lebesque point of } u, \\ \int_0^1 g((1-\alpha)u_-(y) + \alpha u_+(y))d\alpha, & \text{if } y \text{ is a jump point of } u, \end{cases}$$

for H_{m-1}-almost every y in Ω (H_{m-1} denotes the $(m-1)$ dimensional Hausdorff measure).

We emphasize that the basic definition of BV functions considers functions which are defined almost everywhere in the sense of the *Lebesque* measure, while Definition 2.1 treats functions which are defined a.e. with respect to the Hausdorff measure H_{m-1}.

Using the notion of averaged superposition, Volpert proves:

THEOREM 2.1. *(Volpert [19]) Let u and v be in $L^\infty(\Omega, \mathbf{R}^p) \cap BV(\Omega, \mathbf{R}^p)$ and g be in $\mathcal{C}^1(\mathbf{R}^p, \mathbf{R}^p)$. Then, for each $i = 1, 2, \cdots, m$, the function $\hat{g}(u)$ given by Definition 2.1 is measurable and integrable with respect to the Borel measure $\frac{\partial v}{\partial y_i}$, so that the nonconservative product*

$$\hat{g}(u)\frac{\partial v}{\partial y_i}$$

makes sense as a finite Borel measure.

This result is very useful even in the context of conservation laws (Di Perna [3], Di Perna - Majda [5]). It leads also to a definition of weak solution to systems in nonconservative form, i.e. hyperbolic systems

$$\partial_t u + A(u)\partial_x u = 0, \quad u(t,x) \in \mathbf{R}^p, \tag{2.2}$$

where A is a given smooth function of u, *not* assumed to be a Jacobian matrix.

Definition 2.2. A function u in $L^\infty(\mathbf{R}_+ \times \mathbf{R}, \mathbf{R}^p) \cap BV(\mathbf{R}_+ \times \mathbf{R}, \mathbf{R}^p)$ is said to be a weak BV-solution to the system in nonconservative form (2.2) if it satisfies

$$\partial_t u + \hat{A}(u)\partial_x u = 0 \tag{2.3}$$

as Borel measures on $\mathbf{R}_+ \times \mathbf{R}$.

This notion of weak solutions has been investigated in [12]. In particular, if the system is strictly hyperbolic with genuinely nonlinear or linearly degenerate characteristic fields, the Riemann problem for (2.2) can be solved in the class of piecewise smooth solutions. In other words, one has a generalization of Lax's theorem for the Riemann problem.

In fact, Dal Maso - Le Floch - Murat [1,2] (see also Le Floch [15]) have recently shown that in general, Definition 2.2 is not sufficient and have proposed an extension of the Volpert's product. However, we restrict ourselves in this paper to Definitions 2.1 and 2.2 since it will be sufficient for our purpose.

3. A nonstrictly hyperbolic and nonconservative 2×2 system. We consider the following *nonconservative* system of two equations:

$$\partial_t u + \partial_x f(u) = 0, \tag{3.1a}$$

$$\partial_t w + a(u)\partial_x w = 0, \tag{3.1b}$$

where $f : \mathbf{R} \to \mathbf{R}$ is a given strictly convex function and $a = \frac{d}{du} f$. Equation (3.1a) can also be written in the form

$$\partial_t u + a(u)\partial_x u = 0, \tag{3.2}$$

so that system (3.1) is of the general form

$$\partial_t U + A(U)\partial_x U = 0, \tag{3.3}$$

with here

$$U = \begin{pmatrix} u \\ w \end{pmatrix} \quad \text{and} \quad A(U) = \begin{pmatrix} a(u) & 0 \\ 0 & a(u) \end{pmatrix}. \tag{3.4}$$

The matrix $A(U)$ is diagonal, thus (3.3) is hyperbolic. But it is *not strictly hyperbolic* since $a(u)$ is a *double* eigenvalue of $A(U)$. Furthermore, $A(U)$ admits a genuinely nonlinear characteristic field and a linearly degenerate one.

To apply the general theory of systems in nonconservative form (see Section 2) to solve system (3.1), we set:

Definition 3.1. A function (u, v) in $L^\infty(\mathbf{R}_+, BV(\mathbf{R}, \mathbf{R}^2))$ is said to be an entropy weak BV-solution to the *nonconservative* system (3.1) if

1) u is an entropy weak solution to the scalar conservation law (3.1a) in the usual sense of distributions (see Kruskov [10]);

2) w is a weak solution to (3.1b) in the sense of the Volpert's product (see Section 2 and [12]), i.e.

$$\partial_t w + \hat{a}(u)\partial_x w = 0. \tag{3.5}$$

With this definition, we are going to prove that, under suitable assumptions on the initial data, the Cauchy problem for system (3.1) has one and only one entropy weak BV-solution.

Our first result treats of existence of solutions when the initial data for w is a Lipschitz continuous function.

THEOREM 3.1. *(Existence) Let u_0 be in $BV(\mathbf{R})$ and w_0 in $W^{1,\infty}(\mathbf{R})$. Then system (3.1) has at least one entropy weak solution (u, w) in $L^\infty(\mathbf{R}_+, BV(\mathbf{R}, \mathbf{R}^2))$ satisfying the initial condition*

$$u(0,x) = u_0(x), \quad w(0,x) = w_0(x), \quad a.e.\ x \in \mathbf{R}. \tag{3.6}$$

In Theorem 3.1, the initial condition at $t = 0$ is understood in the usual strong L^1 sense, i.e. (for instance for u)

$$\lim_{\substack{t\to 0\\ t>0}} \frac{1}{t}\int_0^t \int_{\mathbf{R}} |u(s,x) - u_0(x)|dxds = 0.$$

Proof of Theorem 3.1. Equation (3.1a) is uncoupled from equation (3.1b) so that it can be solved first. Volpert has shown in [19] that it admits one (unique) entropy weak solution u in BV satisfying the initial data $u_0 \in BV$. Then, when u is known, we have to solve for w a linear hyperbolic equation with *discontinuous* coefficients.

To find a weak solution to (3.1b), we will use an explicit formula due to Lax [11] for the entropy weak solution to (3.1a). We recall this formula briefly. Let $G : \mathbf{R}_+ \times \mathbf{R}^2 \to \mathbf{R}$ be the function given by

$$G(t,x,y) = \int_{-\infty}^{y} u_0(z)dz + tg(\frac{x-y}{t}), \quad (t,x,y) \in \mathbf{R}_+ \times \mathbf{R}^2, \tag{3.7}$$

where g is the Legendre transform of the (convex) function f. Denote by $\xi : \mathbf{R}_+ \times \mathbf{R} \to \mathbf{R}$ the function characterized by the property

$$\min_{y\in\mathbf{R}} G(t,x,y) = G(t,x,\xi(x,t)), \quad a.e.(t,x) \in \mathbf{R}_+ \times \mathbf{R}. \tag{3.8}$$

Then, Lax shows that the solution u to (3.1a) is given explicitly by

$$u(t,x) = b\left(\frac{x-\xi(t,x)}{t}\right), \quad a.e.\ (t,x) \in \mathbf{R}_+ \times \mathbf{R}, \tag{3.9}$$

where we have set $b = a^{-1}$. We note in passing that by (3.9) the function ξ *belongs to* $L^\infty(\mathbf{R}_+, BV)$ as u does.

Then, we claim that the formula

$$w(t,x) = w_0(\xi(t,x)), \quad a.e. \ (t,x) \in \mathbf{R}_+ \times \mathbf{R}^2, \tag{3.10}$$

defines a weak BV-solution to equation (3.1b). Namely it is proved by Dal Maso - Le Floch - Murat [1] that $w(\xi)$ *is in* $L^\infty(\mathbf{R}_+, BV)$ when ξ is in $L^\infty(\mathbf{R}_+, BV)$ but w_0 is (only) in $W^{1,\infty}$. On the other hand, to prove that the function w given by (3.10) is a weak solution, we can set

$$\mathbf{R}_+ \times \mathbf{R} = \mathcal{C} \cup J \cup \mathcal{N}, \tag{3.11}$$

where $\mathcal{C}$ (respectively J) contains the points of approximate continuity (resp. jump) of u (and thus those of ξ) and $\mathcal{N}$ has zero one-dimensional Hausdorff measure. We know that (3.9) defines a weak solution to (3.1a), thus

$$\partial_t b\left(\frac{x-\xi(t,x)}{t}\right) + \partial_x f\left(b\left(\frac{x-\xi(t,x)}{t}\right)\right) = 0.$$

On the set of approximate continuity $\mathcal{C}$, we deduce that

$$-\frac{1}{t^2}(x-\xi) - \frac{1}{t}\partial_t\xi + a(u)\cdot\frac{1}{t}\cdot(1-\partial_x\xi) = 0, \ \text{on } \mathcal{C},$$

thus

$$\partial_t\xi + a(u)\partial_x\xi = 0, \ \text{on } \mathcal{C}. \tag{3.12}$$

But

$$\partial_t w + a(u)\partial_x w = w_0'(\xi)\{\partial_t\xi + a(u)\partial_x\xi\}, \ \text{on } \mathcal{C},$$

so we find by (3.12)

$$\partial_t w + a(u)\partial_x w = 0, \ \text{on } \mathcal{C}. \tag{3.13}$$

Then, we turn to the subset J where u and ξ are discontinuous. Let (t_0, x_0) be a point in J and u_-, u_+, σ denote the left-value, the right-value and the speed at the discontinuity point of u respectively. We denote also by w_- and w_+ the left and right value of the function w at (t_0, x_0) (these two values may be equal). Then, the value of the Borel measure

$$\mu \equiv \partial_t w + \hat{a}(u)\partial_x w$$

at the point (t_0, x_0) is

$$\mu(\{(t_0,x_0)\}) = -\sigma(w_+ - w_-) + \int_0^1 a(u_- + \alpha(u_+ - u_-))d\alpha \cdot (w_+ - w_-) \tag{3.14}$$

by an immediate application of the Volpert's definition of averaged superposition (see Def. 2.1 and 2.3 and Volpert [19]). But, we have

$$\int_0^1 a(u_- + \alpha(u_+ - u_-))d\alpha = \frac{f(u_+) - f(U_-)}{u_+ - u_-} = \sigma,$$

since u is a weak solution to equation (3.1a). Hence (3.14) becomes

$$\mu((t_0, x_0)) = \left(-\sigma + \frac{f(u_+) - f(u_-)}{\sigma}\right)(w_+ - w_-) = 0,$$

whatever the jump of w is. We conclude that

$$\partial_t w + \hat{a}(u)\partial_x w = 0, \text{ on } J.$$

Finally, (3.11), (3.13) and (3.14) prove that (3.10) defines a weak BV solution to equation (3.1b). □

Remark 3.1. The regularity on the initial data w_0 (i.e. $w_0 \in W^{1,\infty}$) is needed just to ensure that the function $w_0(\xi)$ has bounded variation (see (3.10)).

Recall that, since the flux-function f is assumed to be convex, the entropy condition for a solution u to the scalar conservation law (3.1a) is

$$u(x-, t) \geq u(x+, t), \qquad x \in \mathbf{R}, \quad a.e.\ t > 0. \tag{3.15}$$

We shall use a more precise statement obtained by Smoller [16]

$$\frac{\partial u}{\partial x} \leq \frac{k}{t}, \text{ in the sense of distributions,} \tag{3.16}$$

where k is a constant independent of x and t.

We prove now a result of uniqueness of weak BV-solutions for system (3.1) in the case that the initial data u_0 to (3.1a) does satisfy the entropy condition (3.15).

THEOREM 3.2. *(Uniqueness) Let the initial data u_0 and w_0 be in $BV(\mathbf{R})$ and u_0 be an entropy initial data, i.e. there exists a constant K_0 such that*

$$\frac{d}{dx}u_0 \leq K_0, \textit{ in the sense of distributions.} \tag{3.17}$$

Then, problem (3.1), (3.6) has at most one entropy weak solution (u, w) in L^∞ $(\mathbf{R}_+, BV(\mathbf{R}, \mathbf{R}^2))$.

Proof of Theorem 3.2. It is well-known that the entropy weak solution u to the scalar conservation law (3.1a) is unique, so we focus on equation (3.1b) for the function w. The proof below uses standard arguments for BV functions, so we only sketch it.

First at all, since u_0 satisfies (3.17) and the function f is strictly convex, then inequality (3.16) can be improved to

$$\frac{\partial u}{\partial x} \leq K, \text{ in the sense of distributions,} \tag{3.18}$$

where K is a constant *independent* of x and t.

On the other hand, if w is a solution to equation (3.1b), then for every smooth function $\eta : \mathbf{R} \to \mathbf{R}$, the function $(x,t) \to \eta(w(x,t))$ is also a weak solution to (3.1b), i.e.

$$\partial_t \eta(w) + \hat{a}(u)\partial_x \eta(w) = 0. \tag{3.19}$$

The proof is based on the decomposition (3.11) for the function u. We omit it. By standard arguments of regularization, we also have (3.19) for the Kruskov entropies, i.e.

$$\partial_t |w-k| + \hat{a}(u)\partial_x |w-k| = 0 \tag{3.20}$$

for every constant k.

Then, let w_1 and w_2 be two weak solutions to equation (3.1b). From the equation (3.20), it is classical to deduce (Kruskov [10], Volpert [19]) that

$$\partial_t |w_2 - w_1| + \hat{a}(u)\partial_x |w_2 - w_1| = 0. \tag{3.21}$$

Finally, we rewrite the *nonconservative* equation (3.21) in a *conservative* form plus a *measure* source-term

$$\partial_t |w_2 - w_1| + \partial_x (a(u)|w_2 - w_1|) = \mu \tag{3.22}$$

with

$$\mu = \partial_x (a(u)|w_2 - w_1|) - \hat{a}(u)\partial_x |w_2 - w_1|. \tag{3.23}$$

On the set of approximate continuity of the function u, we have

$$\mu = |w_2 - w_1| \partial_x a(u) \leq K|w_2 - w_1| \tag{3.24}$$

thanks to condition (3.18). On the other hand, at a point (x,t) of approximate jump of u, we have with the same notation as previously

$$\begin{aligned}
\mu(\{(x,t)\}) &= a(u_+)|w_2^+ - w_1^+| - a(u_-)|w_2^+ - w_1^-| \\
&\quad - \int_0^1 a(u_- + \alpha(u_+ - u_-))d\alpha \cdot (|w_2^+ - w_1^+| - |w_2^- - w_1^-|) \\
&= |w_2^+ - w_1^+| \left\{ a(u_+) - \int_0^1 a(u_- + \alpha(u_+ - u_-))d\alpha \right\} \\
&\quad + |w_2^- - w_1^-| \left\{ \int_0^1 a(u_- + \alpha(u_+ - u_-))d\alpha - a(u_-) \right\}.
\end{aligned}$$

But, since

$$\int_0^1 a(u_- + \alpha(u_+ - u_-))d\alpha = \frac{f(u_+) - f(u_-)}{u_+ - u_-},$$

we get

$$\mu(\{(x,t)\}) = |w_2^+ - w_1^+| \left\{ a(u_+) - \frac{f(u_+) - f(u_-)}{u_+ - u_-} \right\}$$
$$+ |w_2^- - w_2^+| \left\{ \frac{f(u_+) - f(u_-)}{u_+ - u_-} - a(u_-) \right\}.$$

Since $u_- > u_+$ and f is convex, we deduce that

$$\mu(\{(x,t)\}) \leq 0. \tag{3.25}$$

Finally, (3.24) (3.25) show that the measure μ given by (3.23) satisfies

$$\mu \leq K \cdot |w_2 - w_1|$$

so that (3.22) implies the inequality

$$\partial_t |w_2 - w_1| + \partial_x (a(u)|w_2 - w_1|) \leq K \cdot |w_2 - w_1|. \tag{3.26}$$

By integration in time and space and Gronwall's Lemma, we conclude that

$$\int_{\mathbf{R}} (w_2(t,x) - w_1(t,x)|dx \leq e^{K \cdot t} \int_{\mathbf{R}} |w_2(0,x) - w_1(0,x)|dx \tag{3.27}$$

which prove the uniqueness result concerning the equation (3.1b). □

Remark 3.2.

1) The same arguments prove also that equation (3.19) yields the following inequality

$$\partial_t \eta(w) + \partial_x (a(u)\eta(w)) \leq K\eta(w). \tag{3.28}$$

2) If u_0 does not satisfy (3.17), then inequality (3.16) holds *but* inequality (3.18) does not. Inequality (3.16) would be not sufficient to conclude with the arguments of the proof of Theorem 3.2. In fact, in that case, problem (3.1), (3.6) admits an *infinity* of entropy weak solutions. For instance, the function

$$u(t,x) = \begin{cases} u_L, & x \leq ta(u_L) \\ a^{-1}(\frac{x}{t}), & ta(u_L) \leq x \leq ta(u_R) \\ u_R, & x \geq ta(u_R) \end{cases}$$

is an entropy weak solution to the equation (3.1a) provided that

$$u_L < u_R.$$

And, *for any* w_* *in* $\mathbf{R}$, the function

$$w(t,x) = \begin{cases} w_L, & x \leq ta(u_L) \\ w_*, & ta(u_L) \leq x \leq ta(u_R) \\ w_R, & x \geq ta(u_R) \end{cases}$$

is an entropy weak solution to (3.1b) corresponding to the initial data

$$w(0,x) = \begin{cases} w_L, & x < 0 \\ w_R, & x > 0. \end{cases}$$

From Theorems 3.1 and 3.2, we conclude that:

COROLLARY 3.3. *(Existence and uniqueness): Let w_0 be in $W^{1,\infty}(\mathbf{R})$ and u_0 be in $BV(\mathbf{R})$ and satisfy*

$$\frac{d}{dx}u_0 \leq K,$$

where K is a constant independent of x. Then, problem (3.1) (3.6) has one and only one entropy weak BV-solution in $L^\infty(\mathbf{R}_+, BV(\mathbf{R}, \mathbf{R}^2))$.

4. A nonstrictly hyperbolic system of two conservation laws. We are now interested in the following *conservative* system of two equations

$$\partial_t u + \partial_x f(u) = 0, \tag{4.1a}$$

$$\partial_t v + \partial_x (a(u)v) = 0, \tag{4.1b}$$

where, as previously, f is strictly convex and $a = \frac{d}{du}f$. This system can be written as

$$\begin{cases} \partial_t u + a(u)\partial_x u = 0, \\ \partial_t v + va'(u)\partial_x u + a(u)\partial_x v = 0. \end{cases} \tag{4.2}$$

So, it is of the general form

$$\partial_t U + A(U)\partial_x U = 0 \tag{4.3}$$

by setting

$$U = \begin{pmatrix} u \\ v \end{pmatrix}, \quad A(U) = \begin{pmatrix} a(u) & 0 \\ va'(u) & a(u) \end{pmatrix}. \tag{4.4}$$

The matrix $A(U)$ has a double real eigenvalue, $a(u)$, and it does not admit a basis of eigenvectors (except at $v = 0$). Thus, system (4.1) is *not strictly hyperbolic.* Nevertheless, it makes up a well-posed system for small time when the initial data to (4.1) is smooth, as we can see by solving equation (4.1a) first, and then equation (4.1b).

It is easy to prove that, given a discontinuous initial data, system (4.1) has in general non existence of solution or non uniqueness, in the class of BV solutions (for instance). Thus, a more general notion of weak solution is needed.

Our aim here is to derive a theory of weak solutions (existence and uniqueness) for the *conservative* system (4.1) from the results of Section 3 where the *nonconservative* system (3.1) was studied. Our definition of weak solution below is motivated by the following remark: if (u, w) is a smooth solution to system (3.1), then the pair (u, v), where

$$v = \partial_x w, \tag{4.5}$$

is a solution to (4.1). We denote by $M_1(\mathbf{R})$ the space of all bounded Borel measures on $\mathbf{R}$.

DEFINITION 4.1. A pair (u, v), with

$$u \in L^\infty(\mathbf{R}_+, BV(\mathbf{R})), \quad v \in L^\infty(\mathbf{R}_+, M_1(\mathbf{R})), \tag{4.6}$$

is said to be an entropy weak solution to the *conservative system* (4.1) if the function (u, w), with w defined by

$$w(t,x) = v(t,]-\infty, x[), \quad \forall x \in \mathbf{R}, \quad a.e. \ t \in \mathbf{R}_+, \tag{4.7}$$

is an entropy weak solution to the (nonconservative) system (3.1).

Using this definition, we can solve the Cauchy problem for system (4.1) with discontinuous initial data. The following result treats the existence of solutions.

THEOREM 4.1. *Let u_0 be in $BV(\mathbf{R})$ and v_0 in $L^1(\mathbf{R})$. Then system (4.1) has at least one entropy weak solution (u, v) in $L^\infty(\mathbf{R}_+, BV(\mathbf{R})) \times L^\infty(\mathbf{R}_+, M_1(\mathbf{R}))$ which satisfies the initial condition*

$$u(0,x) = u_0(x), \quad v(0,x) = v_0(x), \quad a.e. x \in \mathbf{R} \tag{4.8}$$

In (4.8), the initial condition for μ is understood in the following sense

$$\lim_{t \to 0} \frac{1}{t} \int_0^t \int_{\mathbf{R}} |v(s,]-\infty, x[) - \int_{-\infty}^x v_0(y) dy| dx ds. \tag{4.9}$$

The proof of Theorem 4.1 is a corollary of Theorem 3.1; so we omit it.

Remark 4.1. By using the method of proof of Theorem 3.1, we could verify that a solution to (4.1b) is explicitly provided by the formula

$$v(t,x) = \frac{\partial}{\partial x} \int_{-\infty}^x v_0 \left(\frac{y - \xi(y,t)}{t} \right) dy \tag{4.10}$$

in the sense of distributions.

Moreover, as a corollary of Theorem 3.2, we get a result of uniqueness of weak solutions.

THEOREM 4.2. *Let u_0 be in $BV(\mathbf{R})$ and v_0 be in $M_1(\mathbf{R})$. Suppose that the initial data u_0 is entropy, i.e. there exists a constant K_0 such that*

$$\frac{du_0}{dx} \leq K_0, \text{ in the sense of distributions.} \tag{4.11}$$

Then, problem (4.1), (4.8) has at most one entropy weak solution (u, v) in L^∞ $(\mathbf{R}_+, BV(\mathbf{R})) \times L^\infty(\mathbf{R}_+, M_1(\mathbf{R}))$.

Finally, we get the following result of existence and uniqueness.

COROLLARY 4.3. *Let u_0 be in $BV(\mathbf{R})$ and v_0 in $L^1(\mathbf{R})$. Suppose that u_0 satisfies (4.11). Then problem (4.1) (4.8) has* ***one and only one*** *entropy weak solution in* $L^\infty(\mathbf{R}_+, BV(\mathbf{R})) \times L^\infty(\mathbf{R}_+, M_1(\mathbf{R}))$.

A simple but interesting Cauchy problem is the ***Riemann problem*** for which we have

$$u_0(x) = \begin{cases} u_L, & x < 0 \\ u_R, & x > 0 \end{cases}, \qquad v_0(x) = \begin{cases} v_L, & x < 0 \\ v_R, & x > 0 \end{cases}$$

with $u_L, u_R, v_L, v_R \in \mathbf{R}$. Our results above was stated for simplicity in the case of bounded Borel measures, but an extension to *locally* bounded Borel measures could be proved and thus would apply to the Riemann problem. We omit the details, and are contented here with its explicit solution.

When $u_L > u_R$, u is a shock wave

$$u(t,x) = \begin{cases} u_L, & \text{if } x < \sigma t, \\ u_R, & \text{if } x > \sigma t, \end{cases}$$

with

$$\sigma = \frac{f(u_R) - f(u_L)}{u_R - u_L}.$$

The measure v, solution to equation (4.1b), is given by

$$v = v_L + (v_R - v_L)H(x - \sigma t) + t\left\{(\sigma - a(u_R))v_R - (\sigma - a(u_L))v_L\right\}\delta_{x-\sigma t},$$

where $H(x - \sigma t)$ and $\delta_{x-\sigma t}$ are the Heaviside function and the Dirac mass at $\frac{x}{t} = \sigma$ respectively. The solution is *unique* in that case.

When $u_L < u_R$, u is a rarefaction wave

$$u(t,x) = \begin{cases} u_L, & \text{if } x \le a(u_L)t \\ a^{-1}(\frac{x}{t}), & \text{if } a(u_L)t \le x \le a(u_R)t \\ u_R, & \text{if } x \ge a(u_R)t. \end{cases}$$

In that case, (4.1b) has an *infinity* of solutions. In particular, for every w_* in $\mathbf{R}$, the measure v defined by

$$\text{(4.12)} \qquad \begin{cases} v(t,x) & = v_L(1 - H(x - ta(u_L))) - (x - ta(u_L))v_L\delta^*_{x-ta(u_L)} \\ & + v_R H(x - ta(u_R)) + (x - ta(u_R))v_R\delta^*_{x-ta(u_R)} \\ & + w_*(\delta_{x-ta(u_L)} - \delta_{x-ta(u_R)}) \end{cases}$$

is a weak solution to the conservative equation (4.1b).

Acknowledgements. I have appreciated very much the support and the hospitality of the IMA during my visit at the University of Minnesota. I wish to especially thank Barbara Keyfitz for her kindness.

REFERENCES

[1] G. Dal Maso, Ph. Le Floch, F.Murat, *Definition and weak stability of a nonconservative product*, Internal Report, Centre de Mathématiques Appliqueés, Ecole Polytechnique, Palaiseau (FRANCE) (to appear).

[2] G. Dal Maso, Ph Le Floch, F. Murat, *Definition et stabilité faible d'un produit nonconservatif*, Note C.R. Acad. Sc. Paris (to appear).

[3] R. DiPerna, *Uniqueness of solutions of hyperbolic conservation laws*, Ind. Univ. Math. J., 28 (1979), pp. 137-188.

[4] R.J. DiPerna, P.L. Lions, *Ordinary differential equation, transport theory and Sobolev spaces*, (to appear).

[5] R. Di Perna, A. Majda, *The validity of nonlinear geometric optics for weak solutions of conservation laws*, Comm. Math. Phys., 98 (1985), pp. 313-347.

[6] J. Glimm, *Solutions in the large for nonlinear systems of equations*, Comm. Pure Appl. Math., 18 (1965), pp. 697-715.

[7] B. Keyfitz, *A viscosity approximation to a system of conservation laws with no classical Riemann solution*, to appear in Proc. of Int. Conf. on Hyp. Problems, Bordeaux 1988.

[8] D.J. Korchinski, *Solution of a Riemann problem for a 2×2 system of conservation laws possessing no classical weak solution*, Ph. D. Thesis Adelphi University (1977).

[9] H. Kranzer, (to appear).

[10] N. Kruskov, *First order quasilinear equations in several independent variables*, Math. USSR Sb. 10 (1970), pp. 127-243.

[11] P.D. Lax, *Hyperbolic systems of conservation laws and the mathematical theory of shock waves*, CBMS monograph N^0 11, SIAM (1973).

[12] Ph. Le Floch, *Entropy weak solutions to nonlinear hyperbolic systems in nonconservative form*, Comm. Part. diff. Eq., 13 (6) (1988), pp. 669-727.

[13] Ph. Le Floch, *Solutions faibles entropiques des systèmes hyperboliques nonlinéaires sous forme nonconservative*, Note C.R. Acad. Sc. Paris, t. 306, Série 1 (1988), pp. 181-186.

[14] Ph. Le Floch, *Entropy weak solutions to nonlinear hyperbolic systems in nonconservative form*, in Proc. of the Second Int. Conf. on Nonlin. Hyp. Problems, (Aachen FRG) (1988), pp. 362-373.

[15] Ph. Le Floch, *Shock waves for nonlinear hyperbolic systems in nonconservative form*, IMA series, University of Minnesota, USA, (to appear).

[16] J. Smoller, *Shock waves and reaction-diffusion equations*, Springer-Verlag, New York (1983).

[17] H.B. Stewart, B. Wendroff, *Two-phase flow: models and methods*, J. Comp. Phys., 56 (1984), pp. 363-409.

[18] J. Trangenstein, P. Colella, *A higher-order godunov method for modeling finite deformations in elastic-plastic solids.*

[19] A.I. Volpert, *The space BV and quasilinear equations*, Math. USSR Sb. 2 (1967), pp. 225-267.

OVERCOMPRESSIVE SHOCK WAVES

TAI-PING LIU* AND ZHOUPING XIN†

Abstract. Overcompressive shock waves for nonstrictly hyperbolic conservation laws are stable in a sense different from that of the corresponding viscous traveling waves. We describe the difference and the reason for it.

Key words. Overcompressive shock, nonstrict hyperbolicity, nonlinear stability.

AMS(MOS) subject classifications. 35L65; 76L05

1. Introduction. Consider a system of conservation laws

$$\frac{\partial u}{\partial t} + \frac{\partial f(u)}{\partial x} = 0, \quad u \in \mathbb{R}^n . \tag{1.1}$$

The system is hyperbolic if $f'(u)$ has real eigenvalues $\lambda_i(\phi), i = 1, 2, \ldots, 4$. It is strictly hyperbolic if the eigenvalues are distinct:

$$\lambda_1(u) < \lambda_2(u) < \ldots < \lambda_n(u)$$

for all u under consideration. For such a system, nonlinear waves, shock waves, rarefaction waves and contact discontinuities, are nonlinearly stable, Liu [2].

In recent years much interest has centered on nonstrictly hyperbolic systems where $\lambda_i(u) - \lambda_{i+1}(u), 1 \leq i < n$, may change signs (see several other articles in this volume for such physical models). Two new types of shock waves, overcompressive and undercompressive (crossing) shock waves, arise for such systems. In this article we study the stability of overcompressive shock waves. The result is compared with the result of Liu-Xin [3] for the corresponding viscous conservation laws

$$\frac{\partial u}{\partial t} + \frac{\partial f(u)}{\partial x} = \frac{\partial}{\partial x}\left(B(u)\frac{\partial u}{\partial x}\right). \tag{1.2}$$

As in [3] we will illustrate the marked difference between (1.1) and (1.2) on the level of overcompressive shock waves by carrying out the analysis for the following simple models,

$$\begin{aligned} &\frac{\partial u}{\partial t} + \frac{\partial}{\partial x}\left(\frac{au^2}{2} + bv\right) = 0 \\ &\frac{\partial v}{\partial t} + \frac{\partial}{\partial x}\left(\frac{v^2}{2}\right) = 0, \quad (u, v) \in \mathbb{R}^2, \quad a > 0, \quad b > 0. \end{aligned} \tag{1.1'}$$

*Courant Institute of Mathematical Sciences, New York University, 251 Mercer Street, New York, NY 10012. Research supported in part by NSF grant DMS-847-03971, Army grant DAAL03-87-K-0063, and AFOSR-89-0203.

†Courant Institute of Mathematical Sciences, New York University, 251 Mercer Street, New York, NY 10012. Research supported in part by NSF grant DMS-88-06731.

$$
(1.2') \qquad \begin{aligned} &\frac{\partial u}{\partial t} + \frac{\partial}{\partial x}\left(\frac{au^2}{2} + bv\right) = \frac{\partial^2 u}{\partial x^2} \\ &\frac{\partial v}{\partial t} + \frac{\partial}{\partial x}\left(\frac{v^2}{2}\right) = \frac{\partial^2 v}{\partial x^2} \,. \end{aligned}
$$

Write $(1.1)'$ in matrix form

$$
\begin{pmatrix} u \\ v \end{pmatrix}_t + \begin{pmatrix} au & b \\ 0 & v \end{pmatrix} \begin{pmatrix} u \\ v \end{pmatrix}_x = 0
$$

The flux matrix is nondiagonalizable when its eigenvalues $\lambda_1 = au$ and $\lambda_2 = v$ are equal. A shock wave $(u_-, v_-; u_+, v_+)$ with speed σ satisfies the jump (Rankine–Hugoniot) conditions

$$
\begin{aligned} \sigma(u_+ - u_-) &= \frac{au_+^2}{2} + bv_+ - \frac{au_-^2}{2} - bv_-, \\ \sigma(v_+ - v_-) &= \frac{v_+^2}{2} - \frac{v_-^2}{2} \,. \end{aligned}
$$

For a classical shock wave three characteristic lines impinge on the shock and one leaves it, Lax [1]. A shock wave for two conservation laws is overcompressive if all characteristic lines impinge on the shock:

$$
\lambda_1(u_+) < \sigma < \lambda_1(u_-) \quad \text{and} \quad \lambda_2(u_+) < \sigma < \lambda_2(u_-).
$$

For hyperbolic conservation laws, $(1.1)'$, an overcompressive shock is a combination of two classical shock waves. In the next section we show that a perturbation of an overcompressive shock wave gives rise to two classical shock waves. The resulting waves and their location are determined by the two conservation laws

$$
(1.3) \qquad \int u \, dx = \text{const.}, \qquad \int v \, dx = \text{const.}
$$

For viscous conservation laws $(1.2)'$, (1.3) has different implications. In Section 3 we recall briefly the result for $(1.2)'$ in [3] and compare it to that of the previous section.

Although we carry out our analysis for the simple models $(1.1)'$, the phenomenon exhibited is universal of overcompressive shock waves. A plausible argument is presented in Section 4.

2. Inviscid Overcompressive Shock Waves. For simplicity, and without loss of generality, we only consider shock waves $(u_-, v_-; u_+, v_+)$ of $(1.1)'$ with zero speed. The jump condition becomes

$$
\begin{aligned} &\frac{au_+^2}{2} + bv_+ = \frac{au_-^2}{2} + bv_- \\ &\frac{v_+^2}{2} = \frac{v_-^2}{2} \,. \end{aligned}
$$

There are two possibilities. When $v_+ = v_-$ we adopt the usual entropy condition $u_+ < u_-$ for $u_t + (u^2/2)_x = 0$. Thus the above jump condition becomes

$$-u_+ = u_- > 0, \quad v_+ = v_-. \tag{2.1}$$

When $v_+ \neq v_-$ we impose the entropy condition $v_+ < v_-$ for the second equation in (1.1)′ $v_t + (v^2/2)_x = 0$ and obtain

$$\begin{aligned} -v_+ &= v_- \equiv v_0 > 0, \\ au_+^2 &= au_-^2 + 4bv_0. \end{aligned} \tag{2.2}$$

For each fixed u_- there are two values $\pm u_+, u_+, < 0$, satisfying (2.2). When $au_- > v_0$ then $(u_-, v_-; u_+, v_+)$ is overcompressive and $(u_-, v_-; -u_+, v_+)$ is a classical shock wave. $(u_-, v_-; u_+, v_+)$ can be viewed either as the combination of the classical waves $(u_-, v_-; -u_-, v_-)$ and $(-u_-, v_-; u_+, v_+)$ or as the combination of classical waves $(u_-, v_-; -u_+, v_+)$ and $(-u_+, v_+; u_+, v_+)$. (See Figure 2.1)

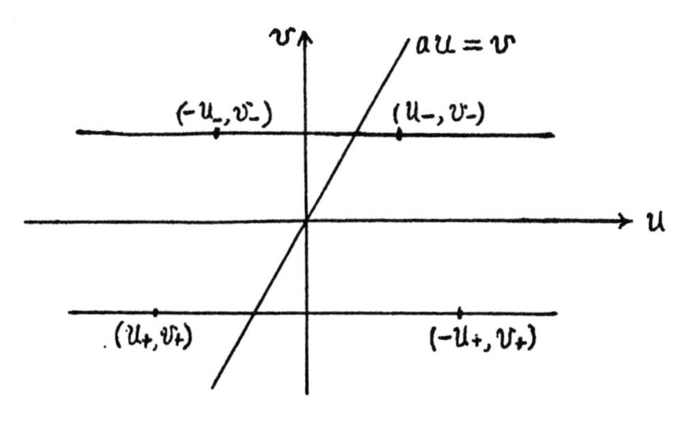

Figure 2.1

Consider the overcompressive shock wave

$$(\overline{u}, \overline{v})(x,t) = \begin{cases} (u_-, v_-) & \text{for} \quad x < 0 \\ (u_+, v_+) & \text{for} \quad x > 0 \end{cases}$$

Let $(u, v)(x, t)$ be a solution of (1.1)′ which is a perturbation of $(\overline{u}, \overline{v})(x, t)$:

$$\begin{aligned} &(u, v)(x,t) = (\overline{u}, \overline{v})(x,t) + (\tilde{u}, \tilde{v})(x,t) \\ &(\tilde{u}, \tilde{v})(x,t) = 0 \quad \text{for} \quad |x| > M \end{aligned}$$

To construct the solution $(u, v)(x, t)$, we first solve the second equation $v_t + (v^2/2)_x = 0$ with given initial data $v(x, 0)$. With $v(x, t)$ thus determined we can define $u_1(x, t)$ so that $(u_1(x,t),\ v(x,t))$ consists of waves pertaining to the eigenvalue $\lambda_2 = v$. That is, for each rarefaction wave in $v(x, t)$ we set $u_1(x, t)$ so that

$(u_1(x,t), v(x,t))$ is a rarefaction wave corresponding to λ_2; and the same for shock waves. This can be done explicitly for approximate solutions of $v(x,t)$ by shock and rarefaction waves through the random choice method or other characteristic methods. We set $u(x,t) = u_1(x,t) + u_2(x,t)$, $\quad u_2(x,0) = u(x,0) - u_1(x,0)$. Here $u_2(x,t)$ consists of waves pertaining to λ_1 only. Thus u_2 solves $u_t + (au^2/2) = 0$ almost always. Waves in u_2 are either those generated by the initial data $u_2(x,0)$ or through interaction of shock waves in (u_1, v). Since v solves the scalar equation $v(x,t)$ tends to a single shock wave (v_-, v_+) after finite time $t = T_1$. Since (u_1, v) consists of waves pertaining to λ_2, after time T_1 (u_1, v) is a single 2-shock. After time T_1 there is no interaction in $(u_1, v)(x,t)$ and so u_2 solves the scalar equation $u_t + (au^2/2)_x = 0$. Thus u_2 becomes a single 1-shock wave after finite time T_2. We thus conclude that after finite time the solution (u, v) consists of two shock waves connecting (u_-, v_-) on the left and (u_+, v_+) on the right. It is easy to see that the two waves are either $(u_-, v_-; -u_-, v_-)$ followed by $(-u_-, v_-; u_+, v_+)$ or $(u_-, v_-; -u_+, v_+)$ followed by $(-u_+, v_+; u_+, v_+)$.

To carry out the above analysis the initial perturbation $(\tilde{u}, \tilde{v})$ needs only to be bounded. Since v satisfies a scalar equation with convex flux $v^2/2$, $v(\cdot, t)$ is of bounded total variation for any $t > 0$. The interactions of shock waves in (u_1, v) give rise to 1-waves in u_2 which is also of bounded variation. Thus $u_2(\cdot, t)$ is of bounded total variation for any $t > 0$.

The location of the two shock waves in the asymptotic state can be determined through conservation law (1.3). The 2-shock coincides with the shock (v_-, v_+) for $v(x,t)$ and is determined by

$$v(x,t) = \begin{cases} v_- & \text{for} \quad x < x_0, \quad t \geq T_1 \\ v_+ & \text{for} \quad x > x_0, \quad t \geq T_1 \end{cases}$$

$$x_0 = (v_- - v_+)^{-1} \int_{-M}^{M} \tilde{v}(y)\, dy \ .$$

Having determined the location of the 2-shock, the location of the 1-shock is given by the first conservation law in (1.3) as follows:

When the 1-shock is located to the left of the 2-shock, $x_1 < x_0$, then the asymptotic state is the 1-shock $(u_-, v_-; -u_-, v_-)$ followed by the 2-shock $(-u_-, v_0; u_+, v_+)$. The integral in x of the asymptotic state minus $(\overline{u}, \overline{v})$ can easily be calculated to be $2(x_1 - x_0)u_- + x_0(u_- - u_+)$. When $x_1 > x_0$ and the asymptotic state is the 2-shock $(u_-, v_-; -u_+, v_+)$ followed by the 1-shock $(-u_+, v_+; u_+, v_+)$ the integral of its difference with $(\overline{u}, \overline{v})$ is $-2(x_1 - x_0)u_+ + x_0(u_- - u_+)$. Since the integral equals the integral of $\tilde{u}$, we have

Case 1. When $p > 0$ the asymptotic state is a 2-shock $(u_-, v_-; -u_+, v_+)$ followed by a 1-shock $(-u_+, v_+; u_+, v_+)$ located at $x = x_0$ and $x = x_1$, respectively; and x_1

is given by

$$x_1 = x_0 - p/2u_+;$$

$$p \equiv \int_{-M}^{M} \tilde{u}(x,0)dx + x_0(u_+ - u_-).$$

Case 2. When $p < 0$ the asymptotic state is a 1-shock $(u_-, v_-; -u_-, v_-)$ followed by a 2-shock $(-u_-, v_-; u_+, v_+)$ located at $x = x_1$ and $x = x_0$ respectively; and x_1 is given by

$$x_1 = x_0 + p/2u_- ,$$

$$p \equiv \int_{-M}^{M} \tilde{u}(x,0)dx + x_0(u_+ - u_-).$$

3. Viscous Overcompressive Shock Waves. For viscous conservation laws (1.2)$'$ shock waves are smooth traveling waves $(u,v)(x,t) = (\phi,\psi)(x - st)$. As before for simplicity we study overcompressive shock waves $(u_-, v_-; u_+, v_+)$ with speed $s = 0$. From (1.2)$'$ we have

$$\begin{aligned} &\phi' = a\phi^2/2 + b\psi + A \\ &\psi' = \psi^2/2 + B \\ &A \equiv -au_+^2/2 + bv_+ = -au_-^2/2 + bv_- , \\ &B \equiv v_+^2/2 = v_-^2/2 . \end{aligned} \tag{3.1}$$

For the overcompressive shock we have

$$au_+ < 0 < au_- , \quad v_+ < 0 < v_-,$$

and so the critical point (u_-, v_-) (or (u_+, v_+)) is an unstable (stable) node for (3.1). Consequently there are infinitely many connecting orbits for $(u_-, v_-; u_+, v_+)$. (cf. Figure 3.1.)

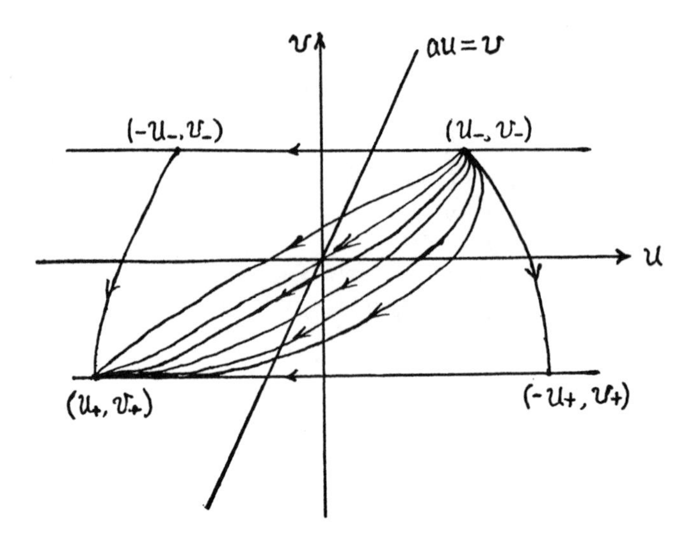

Figure 3.1

There exists a unique connecting orbit for each classical shock wave, $(u_-, v_-; -u_-, v_-)$, $(u_-, v_-; -u_+, v_+)$, $(-u_+, v_+; u_+, v_+)$, $(-u_-, v_-; u_+, v_+)$. The region between these orbits is filled with orbits for the overcompressive shock $(u_-, v_-; u_+, v_+)$. When a traveling wave $(\phi_1, \psi)(x)$ is perturbed the solution converges to another traveling wave $(\phi_2, \psi)(x + x_0)$ of Figure 3.1, properly translated. The new orbit ϕ_2 as well as the translation x_0 are determined by the conservation laws (1.3). For details see Liu–Xin [3].

4. Concluding Remarks. For two conservation laws an overcompressive shock wave absorbs all the characteristic curves around it. As a consequence a perturbation does not create diffusion waves propagating away from the shock wave, cf. Liu [4]. On the other hand, there are two time-invariants, (1.3). Thus a perturbation cannot just create a translation of the shock wave. As we have seen in the previous two sections, for the inviscid shock, it splits into two shocks; and for the viscous shock, it jumps to another viscous shock. Thus the two limits of viscosity tending to zero and time going to infinity do not commute. Although this phenomenon holds for diffusion waves of strictly hyperbolic systems, that it holds here for strongly nonlinear shock waves is significant. Although we carry out our analysis for simple models (1.1)′ and (1.2)′ the above phenomena would hold for general systems. For general n conservation laws, $n > 2$, an overcompressive shock wave with speed s satisfying

$$\lambda_i(u_+) < s < \lambda_i(u_-), \quad \lambda_{i+1}(u_+) < s < \lambda_{i+1}(u_-)$$

for some i, $1 \leq i < n$, would have the following properties: A perturbation of the inviscid shock wave would create N-waves for each genuinely nonlinear field λ_j, and a linear traveling wave for each linear degenerate field $\lambda_j, j \neq i,\ j \neq i+1$, and the splitting of the overcompressive shock into classical i and $i+1$ shocks. For the corresponding viscous shock wave, a perturbation would give rise to diffusion waves pertaining to $\lambda_j, j \neq i, j \neq i+1$, fields and the jumping of the viscous profile to another profile. The equations of magneto-hydrodynamics are such an important physical model.

Acknowledgment. The authors would like to thank IMA for the opportunity to visit the Institute during the Spring of 1989. Part of the research done in the present paper was initiated during the visit to the Institute.

REFERENCES

[1] LAX, P.D., *Hyperbolic systems of conservation laws, II*, Comm. Pure Appl. Math. 10 (1957), pp. 537-566.

[2] LIU, T.-P., *Admissible solutions of hyperbolic conservation laws*, Memoirs, Amer. Math. Soc., No. 240 (1981).

[3] LIU, T.-P. AND XIN, Z., *Stability of viscous shock waves associated with a non strictly hyperbolic conservation law*, (to appear).

[4] LIU, T.-P., *Nonlinear stability of shock waves for viscous conservation laws*, Memoirs, Amer. Math. Soc., No 328 (1985).

QUADRATIC DYNAMICAL SYSTEMS DESCRIBING SHEAR FLOW OF NON-NEWTONIAN FLUIDS*

D. S. MALKUS†1, J. A. NOHEL†2, AND B. J. PLOHR†3

Abstract. Phase-plane techniques are used to analyze a quadratic system of ordinary differential equations that approximates a single relaxation-time system of partial differential equations used to model transient behavior of highly elastic non-Newtonian liquids in shear flow through slit dies. The latter one-dimensional model is derived from three-dimensional balance laws coupled with differential constitutive relations well-known by rheologists. The resulting initial-boundary-value problem is globally well-posed and possesses the key feature: the steady shear stress is a non-monotone function of the strain rate. Results of the global analysis of the quadratic system of ode's lead to the same qualitative features as those obtained recently by numerical simulation of the governing pde's for realistic data for polymer melts used in rheological experiments. The analytical results provide an explanation of the experimentally observed phenomenon called spurt; they also predict new phenomena discovered in the numerical simulation; these phenomena should also be observable in experiments.

1. Introduction. The purpose of this paper is to analyze novel phenomena in dynamic shearing flows of non-Newtonian fluids that are important in polymer processing [17]. One striking phenomenon, called "spurt," was apparently first observed by Vinogradov *et al.* [19] in experiments concerning quasi static flow of monodispersive polyisoprenes through capillaries or equivalently through slit dies. They found that the volumetric flow rate increased dramatically at a critical stress that was independent of molecular weight. Until recently, spurt has been associated with the failure of the flowing polymer to adhere to the wall [5]. The focus of our current research is to offer an alternate explanation of spurt and related phenomena.

Understanding these phenomena has proved to be of significant physical, mathematical, and computational interest. In our recent work [12], we found that satisfactory explanation and modeling of the spurt phenomenon requires studying the full dynamics of the equations of motion and constitutive equations. The common and key feature of constitutive models that exhibit spurt and related phenomena is a non-monotonic relation between the steady shear stress and strain rate. This allows jumps in the steady strain rate to form when the driving pressure gradient exceeds a critical value; such jumps correspond to the sudden increase in volumetric flow rate observed in the experiments of Vinogradov *et al.* The governing systems used to model such one-dimensional flows are analyzed in [12] by numerical techniques and simulation, and in the present work by analytical methods. The systems derive from fully three-dimensional differential constitutive relations with m-relaxation times (based on work of Johnson and Segalman [8] and Oldroyd [16]).

*Supported by the U. S. Army Research Office under Grant DAAL03-87-K-0036, the National Science Foundation under Grants DMS-8712058 and DMS-8620303, and the Air Force Office of Scientific Research under Grants AFOSR-87-0191 and AFOSR-85-0141.

†Center for the Mathematical Sciences University of Wisconsin–Madison Madison, WI 53705

1Also Department of Engineering Mechanics.

2Also Department of Mathematics.

3Also Computer Sciences Department.

They are evolutionary, globally well posed in a sense described below, and they possess discontinuous steady states of the type mentioned above that lead to an explanation of spurt. The governing systems for shear flows through slit-dies are formulated from balance laws in Sec. 2.

Specifically, we model these flows by decomposing the total shear stress into a polymer contribution, evolving in accordance with a differential constitutive relation with a single relaxation time and a Newtonian viscosity contribution (see system (JSO) in Sec. 2.). The flows can also be modelled by a system based on a differential constitutive law with two widely spaced relaxation times (see system (JSO_2) in [13].) but no Newtonian viscosity contribution. Numerical simulation [9, 12] of transient flows at high Weissenberg (Deborah) number and very low Reynolds number using the model (JSO) exhibited spurt, shape memory, and hysteresis; furthermore, it predicted other effects, such as latency, normal stress oscillations, and molecular weight dependence of hysteresis, that should be analysed further and tested in rheological experiment.

In earlier work, Hunter and Slemrod [7] used techniques of conservation laws to study the qualitative behavior of discontinuous steady states in a simple one-dimensional viscoelastic model of rate type with viscous damping. They predicted shape memory and hysteresis effects related to spurt. A salient feature of their model is linear instability and loss of evolutionarity in a certain region of state space.

The objective of the present paper is to develop analytical techniques, the results of which verify these rather dramatic implications of numerical simulation. Based on scaling introduced in [12], appropriate for the highly elastic and very viscous polyisoprenes used in the spurt-experiment, we are led to study the following pair of quadratic autonomous ordinary differential equations that approximates the governing system (JSO) in the relevant range of physical parameters for each fixed position in the channel:

$$\begin{aligned} \dot{\sigma} &= (Z+1)\left(\frac{\overline{T}-\sigma}{\varepsilon}\right) - \sigma \,, \\ \dot{Z} &= -\sigma\left(\frac{\overline{T}-\sigma}{\varepsilon}\right) - Z \,. \end{aligned} \tag{1.1}$$

Here the dot denotes the derivative d/dt, $\overline{T}$ is a parameter that depends on the driving pressure gradient as well as position x in the channel, and $\varepsilon > 0$ is a ratio of viscosities. System (1.1) is obtained by setting $\alpha = 0$ in the momentum equation in system (JSO); this approximation is reasonable because α is at least several orders of magnitude smaller than ε. We show that steady states of system (JSO), some of which are discontinuous for non-monotone constitutive relations, correspond to to critical points of the quadratic system. We deduce the local characters of the critical points, and we prove that system (1.1) has no periodic orbits or closed separatrix cycles. Moreover, this system is endowed with a natural Lyapunov-like function with the aid of which we are able to determine the global dynamics of the approximating quadratic system completely and thus identify its globally asymptotically stable crical points (i.e. steady states) for each position x. This analysis is carried out in Sec. 3 When α, the ratio of Reynolds to Deborah numbers, is strictly positive, the stability of discontinuous steady states of system (JSO) remains to be settled. Recently, Nohel, Pego and Tzavaras [15] established such a result for simple model in which the polymer contribution to the shear stress satisfies a single differential constitutive relation; for a particular choice, their model and system (JSO) with $\alpha > 0$ have the same behavior in steady shear. Their asymptotic stability result, combined with numerical experiments and research in progress, suggest that the same result holds for the full system (JSO), at least when α is sufficiently small.

In Sec. 4.,the analysis of Sec. 3. is applied to each point x in the channel, allowing us to explain spurt, shape memory, hysteresis, and other effects originally observed in the numerical simulations in terms of a continuum of phase portraits. We discuss asymptotic expansions of solutions of systems (JSO) and (JSO_2) of Ref. [13] in powers of ε that enable us to explain latency (a pseudo-steady state that precedes spurt). The asymptotic analysis also permits a more quantitative comparison of the dynamics of the two models when ε is sufficiently small. In Sec. 5., we discuss physical implications of the analysis, particularly those that suggest new experiments. In Sec. 6., we draw certain conclusions. Although the analysis in this paper applies only to the special constitutive models we have studied, we expect that the qualitative features of our results appear in a broad class of non-Newtonian fluids. Indeed, numerical simulation by Kolkka and Ierley [10] using another model with a single relaxation time and Newtonian viscosity exhibits very similar character.

2. A Johnson-Segalman-Oldroyd Model for Shear Flow. The motion of a fluid under incompressible and isothermal conditions is governed by the balance of mass and linear momentum. The response characteristics of the fluid are embodied in the constitutive relation for the stress. For viscoelastic fluids with fading memory, these relations specify the stress as a functional of the deformation history of the fluid. Many sophisticated constitutive models have been devised; see Ref. [2] for a survey. Of particular interest is a class of differential models with m-relaxation times, derived in a three-dimensional setting in Refs. [12] and [13]; these models can be regarded as a special cases of the Johnson-Segalman model [8]when the memory function is a linear combination of m-decaying exponentials with positive coefficients or of the Oldroyd differential constitutive equation [16].

Essential properties of constitutive relations are exhibited in simple planar Poiseuille shear flow. We study shear flow of a non-Newtonian fluid between parallel plates, located at $x = \pm h/2$, with the flow aligned along the y-axis, symmetric about the center line, and driven by a constant pressure gradient $\overline{f}$. We restrict attention to the simplest model of a single relaxation-time differential model that possesses steady state solutions exhibiting a non-monotone relation between the total steady shear stress and strain rate, and thereby reproduces spurt and related phenomena discussed below. The total shear stress T is decomposed into a polymer contribution and a Newtonian viscosity contribution. When restricted to one space dimension the initial-boundary value problem, in non-dimensional units with distance scaled by h, governing the flow can be written in the form (see Refs. [9, 12]):

$$\text{(JSO)}\qquad \begin{aligned} \alpha v_t - \sigma_x &= \varepsilon v_{xx} + \overline{f} \ , \\ \sigma_t - (Z+1)v_x &= -\sigma \ , \\ Z_t + \sigma v_x &= -Z \end{aligned}$$

on the interval $[-1/2, 0]$, with boundary conditions

$$\text{((BC))}\qquad v(-1/2,t) = 0 \quad \text{and} \quad v_x(0,t) = 0$$

and initial conditions
((IC))

$$v(x,0) = v_0(x) \ , \quad \sigma(x,0) = \sigma_0(x) \ , \quad \text{and} \quad Z(x,0) = Z_0(x) \ , on -1/2 \leq x \leq 0;$$

symmetry of the flow and compatibility with the boundary conditions requires that $v_0(-1/2) = 0$, $v_0'(0) = 0$ and $\sigma_0(0) = 0$.

The evolution of σ, the polymer contribution to the shear stress, and of Z, a quantity proportional to the normal stress difference, are governed by the second and third equations in system (JSO). As a result of scaling motivated by numerical simulation and introduced in Ref. [12], there are only three essential parameters: α is a ratio of Reynolds number to Deborah number, ε is a ratio of viscosities, and f is the constant pressure gradient.

When $\varepsilon = 0$, and $Z + 1 \geq 0$, system (JSO) is hyperbolic, with characteristics speeds $\pm[(Z+1)/\alpha]^{1/2}$ and 0. Moreover, for smooth intial data in the hyperbolic region and compatible with the boundary conditions, techniques in [18] can be used to establish global well-posedness (in terms of classical solutions) if the data are small, and finite-time blow-up of classical solutions if the data are large. If $\varepsilon > 0$, system (JSO) for any smooth or piece-wise smooth data; indeed, general theory developed in [15] (see Sec. 3 and particularly Appendix A) yields global existence of classical solutions for smooth initial data of arbitrary size, and also existence of almost classical, strong solutions with discontinuities in the initial velocity gradient and in stress components; the latter result allows one to prescribe discontinuous initial data of the same type as the discontinuous steady states studied in this paper.

The steady-state solutions of system (JSO) play an important role in our discussion. Such a solution, denoted by $\overline{v}$, $\overline{\sigma}$, and $\overline{Z}$, can be described as follows. The stress components $\overline{\sigma}$ and $\overline{Z}$ are related to the strain rate $\overline{v}_x$ through the relations

$$(2.1)\qquad \overline{\sigma} = \frac{\overline{v}_x}{1+\overline{v}_x^2} \ , \quad \overline{Z} + 1 = \frac{1}{1+\overline{v}_x^2} \ .$$

Therefore, the steady total shear stress $\overline{T} := \overline{\sigma} + \varepsilon \overline{v}_x$ is given by $\overline{T} = w(\overline{v}_x)$, where

$$(2.2)\qquad w(s) := \frac{s}{1+s^2} + \varepsilon s \ .$$

The properties of w, the steady-state relation between shear stress and shear strain rate, are crucial to the behavior of the flow. By symmetry, it suffices to consider $s \geq 0$. For all $\varepsilon > 0$, the function w has inflection points at $s = 0$ and $s = \sqrt{3}$. When $\varepsilon > 1/8$, the function w is strictly increasing, but when $\varepsilon < 1/8$, the function w is not monotone. Lack of monotonicity is the fundamental cause of the non-Newtonian behavior studied in this paper; hereafter we assume that $\varepsilon < 1/8$.

The graph of w is shown in Fig. 1. Specifically, w has a maximum at $s = s_M$ and a minimum at $s = s_m$, where it takes the values $\overline{T}_M := w(s_M)$ and $\overline{T}_m := w(s_m)$ respectively. As $\varepsilon \to 1/8$, the two critical points coalesce at $s = \sqrt{3}$.

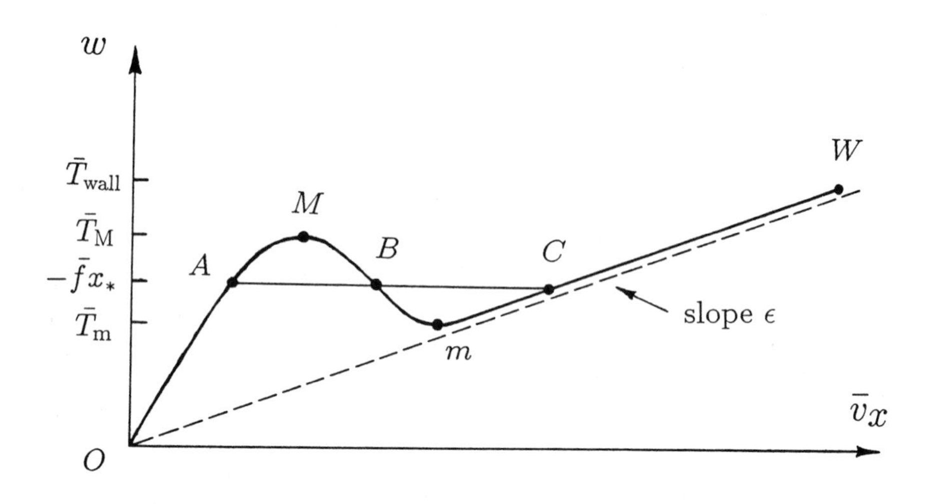

Fig. 1: Total steady shear stress $\overline{T}$ *vs.* shear strain rate $\overline{v}_x$ for steady flow. The case of three critical points is illustrated; other possibilities are discussed in Sec. 3.

The momentum equation, together with the boundary condition at the centerline, implies that the steady total shear stress satisfies $\overline{T} = -\overline{f}x$ for every $x \in [-\frac{1}{2}, 0]$. Therefore, the steady velocity gradient can be determined as a function of x by solving

$$w(\overline{v}_x) = -\overline{f}x \; . \tag{2.3}$$

Equivalently, a steady state solution $\overline{v}_x$ satisfies the cubic equation $P(\overline{v}_x) = 0$, where

$$P(s) := \varepsilon\, s^3 - \overline{T}\, s^2 + (1+\varepsilon)s - \overline{T} \; . \tag{2.4}$$

The steady velocity profile in Fig. 2 is obtained by integrating $\overline{v}_x$ and using the boundary condition at the wall. However, because the function w is not monotone, there might be up to three distinct values of $\overline{v}_x$ that satisfy Eq. (2.3) for any particular x on the interval $[-1/2, 0]$. Consequently, $\overline{v}_x$ can suffer jump discontinuities, resulting in kinks in the velocity profile (as at the point x_* in Fig. 2). Indeed, a steady solution must contain such a jump if the total stress $\overline{T}_{\text{wall}} = \overline{f}/2$ at the wall exceeds the total stress $\overline{T}_M$ at the local maximum M in Fig. 1.

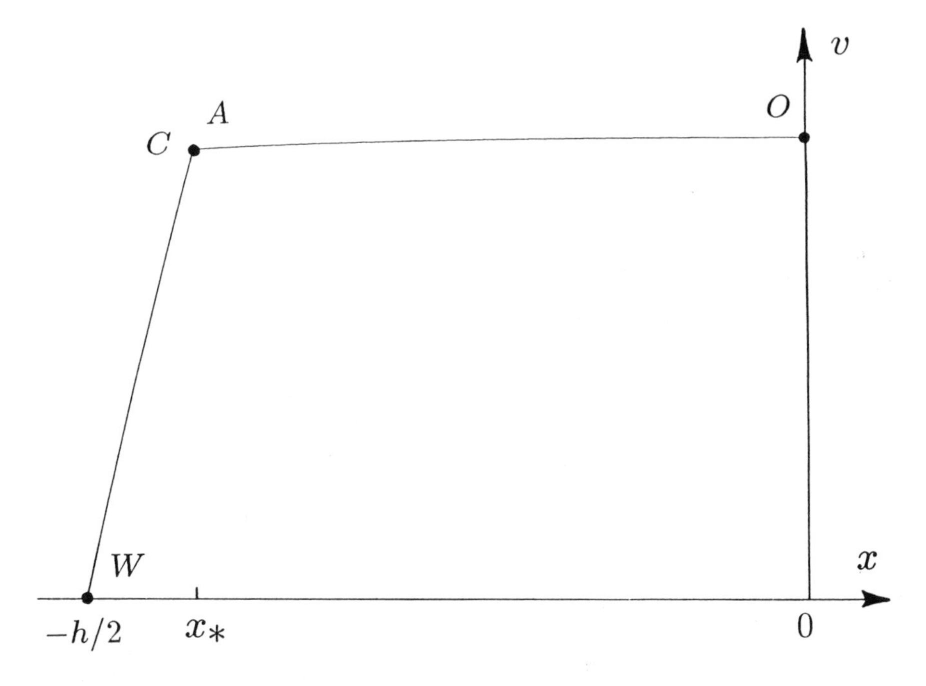

Fig. 2: Velocity profile for steady flow.

Finally, we remark that the flow problem discussed here can also be modelled by a system based on a differential constitutive law with two widely spaced relaxation times but no Newtonian viscosity contribution (see system (JSO_2) in Sec. 2. of [13]); with an appropriate choice of relevant parameters, the resulting problem exhibits the same steady states and the same characteristics as (JSO).

3. Phase Plane Analysis for System (JSO) When $\alpha = 0$. When α is not zero, numerical simulation developed in [9, 11, 12] discovered striking phenomena in shear flow and suggested the analysis that follows. A great deal of information about the structure of solutions of system (JSO) can be garnered by studying a quadratic system of ordinary differential equations that approximates it in a certain parameter range, the dynamics of which is determined completely. Motivation for this approximation comes from the following observation: in experiments of Vinogradov *et al.* [19], α is of the order 10^{-12}; thus the term αv_t in the momentum equation of system (JSO) is negligible even when v_t is moderately large. This led us to the approximation to system (JSO) obtained when $\alpha = 0$.

When $\alpha = 0$, the momentum equation in system (JSO) can be integrated to show that the total shear stress $T := \sigma + \varepsilon v_x$ coincides with the steady value $\overline{T}(x) = -\overline{f}x$. Thus $T = \overline{T}(x)$ is a function of x only, even though σ and v_x are functions of both x and t. The remaining equations of system (JSO) yield, for each fixed x, the autonomous, quadratic, planar system of ordinary differential equations

$$\begin{aligned} \dot{\sigma} &= (Z+1)\left(\frac{\overline{T}-\sigma}{\varepsilon}\right) - \sigma \ , \\ \dot{Z} &= -\sigma\left(\frac{\overline{T}-\sigma}{\varepsilon}\right) - Z \ . \end{aligned} \tag{3.1}$$

Here the dot denotes the derivative d/dt. We emphasize that for each $\overline{f}$, a different dynamical system is obtained at each x on the interval $[-1/2, 0]$ in the channel because $\overline{T} = -\overline{f}x$. By symmetry, we may focus attention on the case $\overline{T} > 0$; also recall from Sec. 2 that $\varepsilon < 1/8$; these are assumed throughout. The dynamical system (3.1) can be analyzed completely by a phase-plane analysis outlined below; the reader is referred to Sec. 3 in [13] for further details. Here we state the main results.

The critical points of system (3.1) satisfy the algebraic system

$$\begin{aligned} &(Z+1+\varepsilon)\left(\frac{\sigma}{\overline{T}} - 1\right) + \varepsilon = 0 \ , \\ &\frac{\overline{T}^2}{\varepsilon}\frac{\sigma}{\overline{T}}\left(\frac{\sigma}{\overline{T}} - 1\right) - Z = 0 \ . \end{aligned} \tag{3.2}$$

These equations define, respectively, a hyperbola and a parabola in the σ-Z plane; these curves are drawn in Fig. 3, which corresponds to the most comprehensive case of three critical points. The critical points are intersections of these curves. In particular, critical points lie in the strip $0 < \sigma < \overline{T}$.

Eliminating Z in these equations shows that the σ-coordinates of the critical points satisfy the cubic equation $Q(\sigma/\overline{T}) = 0$, where

$$Q(\xi) := \left[\frac{\overline{T}^2}{\varepsilon}\xi(\xi-1) + 1 + \varepsilon\right](\xi - 1) + \varepsilon \ . \tag{3.3}$$

A straightforward calculation using Eq. (2.4) shows that

$$P(\overline{v}_x) = P\left(\frac{\overline{T}-\sigma}{\varepsilon}\right) = -\frac{\overline{T}}{\varepsilon}Q(\sigma/\overline{T}) \ . \tag{3.4}$$

Thus each critical point of the system (3.1) defines a steady-state solution of system (JSO): such a solution corresponds to a point on the steady total-stress curve (see Fig. 1) at which the total stress is $\overline{T}(x)$. Consequently, we have:

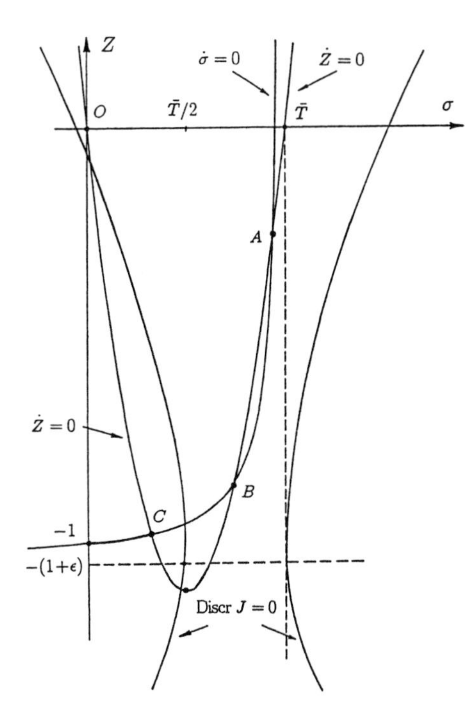

Fig. 3: The phase plane in the case of three critical points.

PROPOSITION 3.1:
For each position x in the channel and for each $\varepsilon > 0$, there are three possibilities:
(1) there is a single critical point A when $\overline{T} < \overline{T}_m$;
(2) there is also a single critical point C if $\overline{T} > \overline{T}_M$;
(3) there are three critical points A, B, and C when $\overline{T}_m < \overline{T} < \overline{T}_M$.

For simplicity, we ignore the degenerate cases, where $\overline{T} = \overline{T}_M$ or $\overline{T} = \overline{T}_m$, in which two critical points coalesce.

To determine the qualitative structure of the dynamical system (3.1), we first study the nature of the critical points. The behavior of orbits near a critical point depends on the linearization of system (3.1) at this point, *i.e.*, on the eigenvalues of the Jacobian matrix J associated with Eq. (3.1), evaluated at the critical point. To avoid solving the cubic equation $Q(\sigma/\overline{T}) = 0$, the character of the eigenvalues of $\mathbf{J}$ can be determined from the signs of the trace of $\mathbf{J}$ denoted by $\operatorname{Tr}\mathbf{J}$, the determinant of $\mathbf{J}$ denoted by $\operatorname{Det}\mathbf{J}$, and the discriminant of $\mathbf{J}$ denoted by $\operatorname{Discrm}\mathbf{J}$ at the critical points. We omit these tedious calculations, a result of which is a useful fact: at a critical point, $\varepsilon \operatorname{Det}\mathbf{J} = Q'(\sigma/\overline{T})$. This relation is important because Q' is positive at A and C and negative at B. To assist the reader, Fig. 3 shows the hyperbola on which $\dot{\sigma} = 0$, the parabola on which $\dot{Z} = 0$ [see Eqs. (3.2)], and the hyperbola on which $\operatorname{Discrm}\mathbf{J}$ vanishes. As a result of the analysis above, we draw the following conclusions:

(1) $\operatorname{Tr}\mathbf{J} < 0$ at all critical points;

(2) $\operatorname{Det}\mathbf{J} > 0$ at A and C, while $\operatorname{Det}\mathbf{J} < 0$ at B; and

(3) $\operatorname{Discrm}\mathbf{J} > 0$ at A and B, whereas $\operatorname{Discrm}\mathbf{J}$ can be of either sign at C. (For typical values of ε and $\overline{T}$, $\operatorname{Discrm}\mathbf{J} < 0$ at C; in particular, $\operatorname{Discrm}\mathbf{J} < 0$ if C is the only critical point. But it is possible for $\operatorname{Discrm}\mathbf{J}$ to be positive if $\overline{T}$ is sufficiently close to $\overline{T}_m$.)

Standard theory of nonlinear planar dynamical systems (see, e.g., Ref. [3, Chap. 15]) now establishes the local characters of the critical points A, B, C in Proposition 3.1:

PROPOSITION 3.2:

(1) A is an attracting node (called the classical attractor);

(2) B is a saddle point;

(3) C is either an attracting spiral point or an attracting node (called the spurt attractor).

The next task is to determine the global structure of the orbits of system (3.1). In this direction, we modify an argument suggested by A. Coppel [4] and establish the crucial result, the proof of which involves a change in the time scale and an application of Bendixson's theorem:

PROPOSITION 3.3:.

System (3.1) has neither periodic orbits nor separatrix cycles.

To understand the global qualitative behavior of orbits, we construct suitable invariant sets. In this regard, a crucial tool is that system (3.1) is endowed with the identity (3.1)

$$\frac{d}{dt}\left\{\sigma^2 + (Z+1)^2\right\} = -2\left[\sigma^2 + (Z+\tfrac{1}{2})^2 - \tfrac{1}{4}\right] . \tag{3.5}$$

Thus the function $V(\sigma, Z) := \sigma^2 + (Z+1)^2$ serves as a Lyapunov function for the dynamical system. Notice that identity (3.5) is independent of $\overline{T}$ and ε.

Let Γ denote the circle on which the right side of Eq. (3.5) vanishes, and let C_r denote the circle of radius r centered at $\sigma = 0$ and $Z = -1$, i.e. $C_r := \{(\sigma, Z) : V(\sigma, Z) = r, r > 0\}$; each C_r is a level set of V. The circles Γ and C_1 are shown in Fig. 4, which corresponds to the case of a single critical point, the spiral point C. Eq. (3.5) also implies the critical points of system (3.1) lie on Γ. If $r > 1$, Γ lies strictly inside C_r. Consequently, Eq. (3.5) shows that the dynamical system (3.1) flows inward at points along C_r. Thus the interior of C_r is a positively invariant set for each $r > 1$. Furthermore, the closed disk bounded by C_1, which is the intersection of these sets, is also positively invariant. Therefore the above argument establishes:

PROPOSITION 3.4: *Each closed disk bounded by the circle $C_r, r \geq 1$ is a positively invariant set for the system (3.1).*

The above results combined with identification of suitable invariant sets were used to determine the global structure of the orbits of system (3.1) in the cases of one and three critical points, and to analyze the stable and unstable manifolds of the saddle point at B. These results are shown in Figs. 5 and 6 and summerized in the following result.

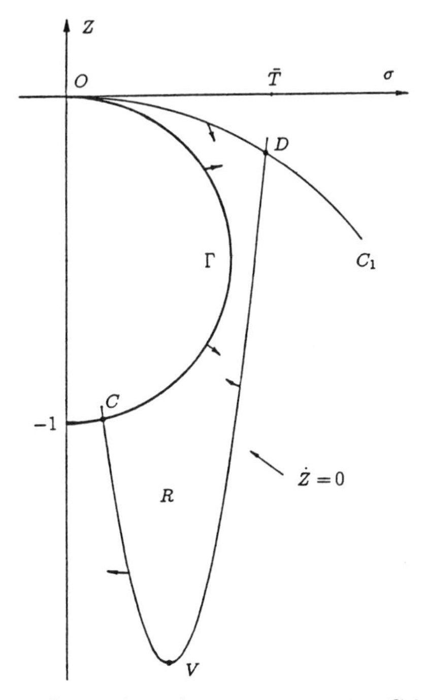

Fig. 4: The phase plane when the spurt attractor C is the only critical point.

PROPOSITION 3.5:

The basin of attraction of A, i.e., the set of points that flow toward A as $t \to \infty$, comprises those points on the same side of the stable manifold of B as is A; points on the other side are in the basin of attraction of C. Moreover, the arc of the circle Γ through the origin, between B and its reflection B' is contained in the basin of attraction of A. In particular, the stable manifold for B cannot cross its boundary, so that it cannot cross Γ between B and B'.

All qualitative features of the dynamics of system (3.1) (except possibly whether C is a node or a focus) carry over to one that approximates the system (JSO_2) in the case of two widely separated relaxation times (see system (4.3) in [13]).

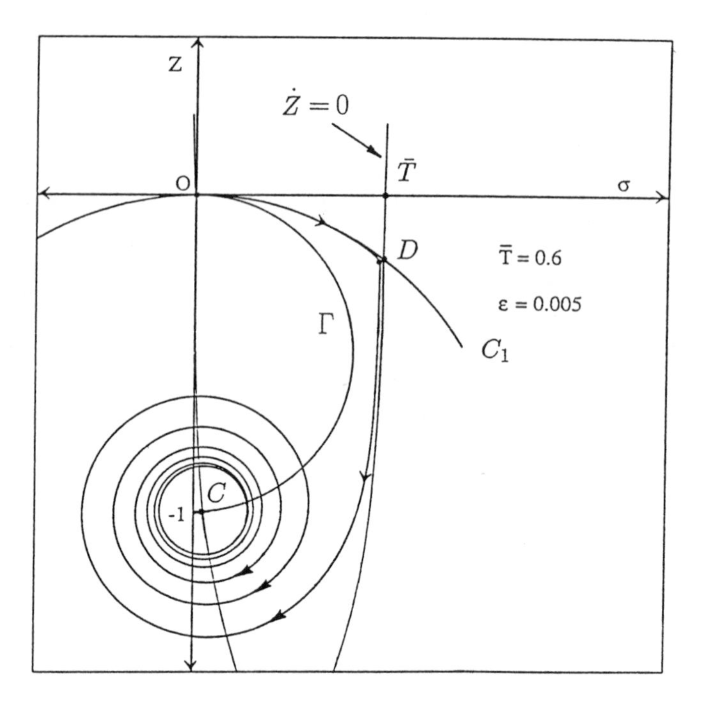

Fig. 5: The orbit through origin when the spurt attractor C is the only critical point.

4. Qualitative Features of (JSO) Based on Phase Plane Analysis. The discussion that follows sketches an explanation of recent numerical simulations of (JSO) described in Refs. [9, 12]. These exhibited several effects related to spurt: latency, shape memory, and hysteresis. Fig. 7 shows the result of simulating a "quasi-static" loading sequence in which the pressure gradient $\overline{f}$ is increased in small steps, allowing sufficient time between steps to achieve steady flow [9]. The loading sequence is followed by a similar quasi-static unloading sequence, in which the driving pressure gradient is decreased in steps. The initial step used zero initial data, and succeeding steps used the results of the previous step as initial data. The resulting hysteresis loop includes the shape memory predicted by Hunter and Slemrod [7] for a simpler model by a different approach. The width of the hysteresis loop at the bottom can be related directly to the molecular weight of the sample [9].

We explain spurt, shape memory, hysteresis and latency. We consider experiments of the following type: the flow is initially in a steady state corresponding to a forcing $\overline{f}_0$, and the forcing is suddenly changed to $\overline{f} = \overline{f}_0 + \Delta\overline{f}$. We call this process "loading" (resp. "unloading") if $\Delta\overline{f}$ has the same (resp. opposite) sign as $\overline{f}_0$. The initial flow can be described by specifying, for each channel position x, whether the flow is at a classical attractor A (x is a "classical point") or a spurt attractor C (x is a "spurt point") for the system (3.1) with $\overline{T} = -\overline{f}_0 x$. We shall say that any point lying on the same side of the stable manifold of B as is A lies on

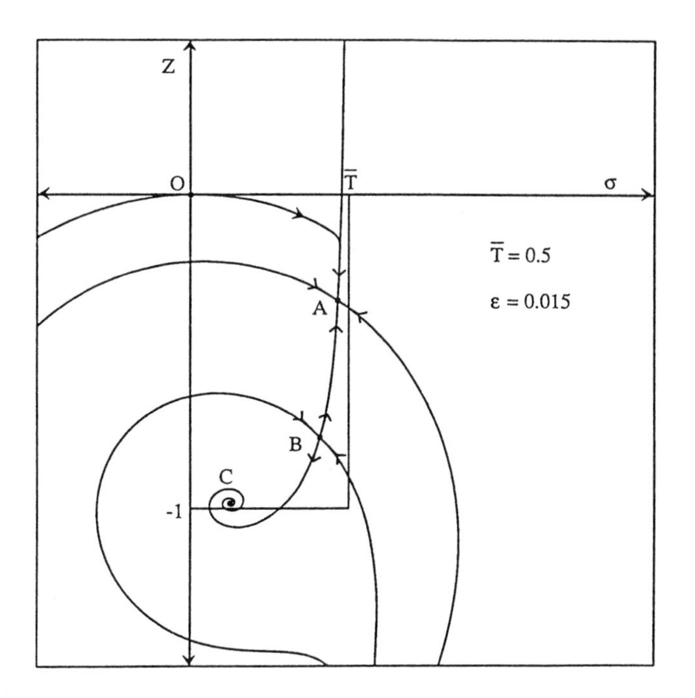

Fig. 6: Phase portrait in the case of three critical points, with C being a spiral.

the "classical side"; points lying on the other side are said to be on the "spurt side." The outcome of the experiment depends on the character of the phase portrait with $\overline{T} = -\overline{f}x$. To determine this outcome, we need only decide when a classical point becomes a spurt point or *vice versa*.

The principle mathematical properties of the dynamical system (3.1) that determine the outcome of loading and unloading experiments are embodied in the following consequence of the phase plane analysis.

PROPOSITION 4.1:

(1) A classical point A_0 for the initial forcing $\overline{f}_0$ lies in the domain of attraction of the classical attractor A for $\overline{f}$, provided that A exists (i.e., $|\overline{f}x| < \overline{T}_M$);

(2) A spurt point C_0 for the initial forcing $\overline{f}_0$ lies in the domain of attraction of the spurt attractor C for $\overline{f}$ unless (a) *C does not exist (i.e., $|\overline{f}x| < \overline{T}_m$);* or (b) *$C$ lies on the classical side of the stable manifold of the saddle point B for $\overline{f}$.*

Consider starting with $\overline{f}_0 = 0$ and loading to $\overline{f} > 0$. Thus the initial state for each x lies at the origin $\sigma = 0$, $Z = 0$. Then according to 4.1(1) above, each $x \in [-1/2, 0]$ such that $\overline{f}|x| < \overline{T}_M$ is a classical point, while the x for which $\overline{f}|x| > \overline{T}_M$ are spurt points (because there is no classical attractor). Consequently, we draw two conclusions:

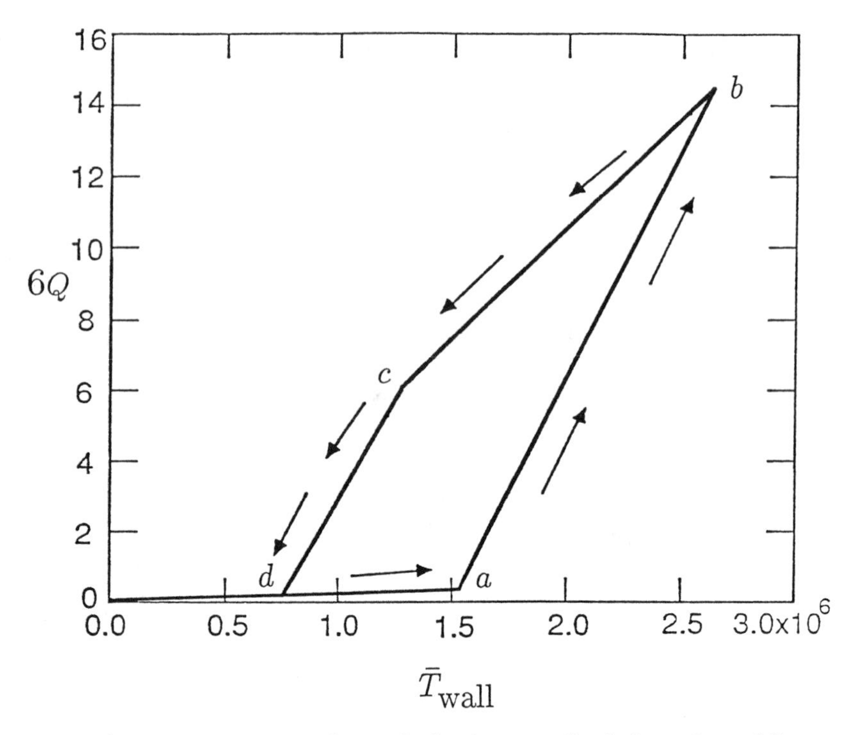

Fig. 7: Hysteresis under cyclic load: normalized throughput $6Q$ vs. wall shear stress $\overline{T}_{\text{wall}}$ [9].

PROPOSITION 4.2:

(a) *If the forcing is subcritical (i.e., $\overline{f} < \overline{f}_{\text{crit}} := 2\overline{T}_M$), the asymptotic steady flow is entirely classical.*

(b) *If the forcing is supercritical ($\overline{f} > \overline{f}_{\text{crit}}$), there is a single kink in the velocity profile (see Fig. 2), located at $x_* = -\overline{T}_M/\overline{f}$; those $x \in [-1/2, x_*)$, near the wall, are spurt points, whereas $x \in (x_*, 0]$, near the centerline, are classical.*

The solution in case (b) can be described as "top jumping" because the stress $\overline{T}_* = \overline{T}_M$ at the kink is as large as possible, and the the kink is located as close as possible to the wall.

Next, consider increasing the load from $\overline{f}_0 > 0$ to $\overline{f} > \overline{f}_0$. A point x that is classical for $\overline{f}_0$ remains classical for $\overline{f}$ unless there is no classical attractor for $\overline{T} = -\overline{f}x$, i.e., $\overline{f}|x| > \overline{T}_M$. A spurt point x for $\overline{f}_0$, on the other hand, is always a spurt point for $\overline{f}$. As a result, a point in x in the channel can change only from a classical attractor to a spurt attractor, and then only if $\overline{f}|x|$ exceeds $\overline{T}_M$. When $\overline{f}$ is chosen to be supercritical, loading causes the position x_* of the kink in Fig. 2 to move away from the wall, but only to the extent that it must: a single jump in strain rate occurs at $x_* = -\overline{T}_M/\overline{f}$, where the total stress is $\overline{T}_* = \overline{T}_M$. These conclusions are valid, in particular, for a quasi-static process of gradually increasing the load from $\overline{f}_0 = 0$ to $\overline{f} > \overline{f}_{\text{crit}}$.

Now consider unloading from $\overline{f}_0 > 0$ to $\overline{f} < \overline{f}_0$; assume, for the moment, that $\overline{f}$ is positive. Here, the initial steady solution need not correspond to top jumping. For this type of unloading, a point x that is classical for $\overline{f}_0$ always remains classical for $\overline{f}$: the classical attractor for $\overline{f}$ exists because $\overline{f}|x| < \overline{f}_0|x|$. By contrast, a spurt point x for $\overline{f}_0$ can become classical at $\overline{f}$. This occurs if: (a) the total stress $\overline{T} = -\overline{f}x$ falls below $\overline{T}_m$; or (b) the spurt attractor C_0 for $\overline{T} = -\overline{f}_0 x$ lies on the classical side of the stable manifold of the saddle point B for $\overline{T} = -\overline{f}x$ (see Proposition 4.1(2b)).

Combining the analysis of loading and unloading leads to the following summary of quasi-static cycles and the resulting flow hysteresis.

Kinks move away from the wall under top jumping loading; they move toward the wall under bottom jumping unloading; otherwise they remain fixed. The hysteresis loop opens from the point at which unloading commences; no part of the unloading path retraces the loading path until point d in Fig. 7.

To explain the latency effect that occurs during loading, assume that ε is small. It is readily seen that the total stress $\overline{T}_M$ at the the local maximum M is $1/2+O(\varepsilon)$, while the local minimum m corresponds to a total stress $\overline{T}_m$ of $2\sqrt{\epsilon}\,[1 + O(\epsilon)]$. Furthermore, for x such that $\overline{T}(x) = O(1)$, $\sigma = \overline{T} + O(\varepsilon)$ at an attracting node at A, while $\sigma = O(\epsilon)$ at a spurt attractor C (which is a spiral). Consider a point along the channel for which $\overline{T}(x) > \overline{T}_M$, so that the only critical point of the system (3.1) is C, and suppose that that $\overline{T} < 1$. Then the evolution of the system exhibits three distinct phases, as indicated in Fig. 6: an initial "Newtonian" phase (O to N); an intermediate "latency" phase (N to S); and a final "spurt" phase (S to C).

The Newtonian phase occurs on a time scale of order ε, during which the system approximately follows an arc of a circle centered at $\sigma = 0$ and $Z = -1$. Having assumed that $\overline{T} < 1$, Z approaches

$$Z_N = (1 - \overline{T}^2)^{\frac{1}{2}} - 1 \tag{4.1}$$

as σ rises to the value $\overline{T}$. (If, on the other hand, $\overline{T} \geq 1$, the circular arc does not extend as far as $\overline{T}$, and σ never attains the value $\overline{T}$; rather, the system slowly spirals toward the spurt attractor. Thus the dynamical behavior does not exhibit distinct phases.)

The latency phase is characterized by having $\sigma = \overline{T} + O(\epsilon)$, so that σ is nearly constant and Z evolves approximately according to the differential equation

$$\dot{Z} = -\frac{\overline{T}^2}{Z+1} - Z\,. \tag{4.2}$$

Therefore, the shear stress and velocity profiles closely resemble those for a steady solution with no spurt, but the solution is not truly steady because the normal stress difference Z still changes. Integrating Eq. (4.2) from $Z = Z_N$ to $Z = -1$ determines the latency period. This period becomes indefinitely long when the forcing decreases to its critical value; thus the persistence of the near-steady solution with no spurt can be very dramatic. The solution remains longest near point L where $Z = -1+\overline{T}$. This point may be regarded as the remnant of the attracting node A and the saddle point B. Eventually the solution enters the spurt phase and tends to the critical point C. Because C is an attracting spiral, the stress oscillates between the shear and normal components while it approaches the steady state.

Asymptotic analysis carried out in Sec. 6 of [13] shows that when ε is sufficiently small, system (JSO_2) of [13] has the same asymptotic properties as system (JSO). Thus system (JSO) approximates (JSO_2) quantitatively as well as qualitatively.

5. Physical Implications. One of the widely accepted explanations of spurt and similar observations is that the presence of the wall affects the dynamics of the polymer system near the wall. Conceivably, there could be a variety of "wall effects," the most obvious is the loss of chemical bond between wall and fluid, or wall slip [5]. Perhaps the most distinguishing feature of our alternative approach is: it predicts that spurt stems from a material property of the polymer and is not related to any external interaction. The spurt layer forms at the wall in situations such as top jumping because the stresses are higher there; for the same reason, of course, is chemical bonds would break at the wall;however, our approach predicts that the layer of spurt points spreads into the interior of the channel on continued loading. Layer thickness is predicted to grow continuously in loading to a thickness that should be observable, provided secondary (two-dimensional) instabilities do not develop.

Our analysis suggests other ways in which experiments might be devised to verify the dependence of spurt on material properties: (i) produce multiple kinks with spurt layer separated from the wall, (ii) produce hysteresis in flow reversal (Fig. 9). Our model predicts circumstances under which a different path can be followed in sudden reversal of the flow than would be followed by a sequence of solutions in which the pressure gradient is reduced to zero and reloaded again (with the opposite sign) to a value of somewhat smaller magnitude. Such behavior does not seem likely to be explainable by a wall effect.

The most important and perhaps the easiest experiment to perform to verify our theory is to produce latency. Our analysis predicts long latency times for data corresponding to realistic material data; no sophisticated timing device would be required, nor would the onset of the instability be hard to identify. The increase in throughput is predicted to be so dramatic that simple visual inspection of the exit flow would probably be sufficient.

6. Conclusions. Although our analysis applies only to the special constitutive models we have studied, we expect that the qualitative features of our results appear in a broad class of non-Newtonian fluids. Our analysis has identified certain universal mathematical features in the shear flow of viscoelastic fluids described by differential constitutive relations that give rise to spurt and related phenomena. The key feature is that there are three widely separated time scales, each associated with an important non-dimensional number (α, ε, and 1, respectively), when scaled by the dominant relaxation time, λ^{-1}. Each of these time scales can be associated with a particular equation in system (JSO) [13]. The key to understanding the dynamics of such systems is fixing the location of the discontinuity in the strain rate induced by the non-monotone character of the steady shear stress vs. strain rate.

Acknowledgments. We thank Professor A. Coppel for suggesting an elegant argument that rules out the existence of periodic and separatrix cycles for the systems (3.1). We also acknowledge helpful discussions with D. Aronson, M. Denn, G. Sell, M. Slemrod and A. Tzavaras, and M. Yao.

References

1. A. Andronov and C. Chaikin, *Theory of Oscillations*, Princeton Univ. Press, Princeton, 1949.

2. R. Bird, R. Armstrong, and O. Hassager, *Dynamics of Polymeric Liquids*, John Wiley and Sons, New York, 1987.

3. E. Coddington and N. Levinson, *Theory of Ordinary Differential Equations*, McGraw-Hill, New York, 1955.

4. A. Coppel, , 1989. private communication.

5. M. Denn, "Issues in Viscoelastic Fluid Dynamics," *Annual Reviews of Fluid Mechanics*, 1989. to appear.

6. M. Doi and S. Edwards, "Dynamics of Concentrated Polymer Systems," *J. Chem. Soc. Faraday* **74** (1978), pp. 1789–1832.

7. J. Hunter and M. Slemrod, "Viscoelastic Fluid Flow Exhibiting Hysteretic Phase Changes," *Phys. Fluids* **26** (1983), pp. 2345–2351.

8. M. Johnson and D. Segalman, "A Model for Viscoelastic Fluid Behavior which Allows Non-Affine Deformation," *J. Non-Newtonian Fluid Mech.* **2** (1977), pp. 255–270.

9. R. Kolkka, D. Malkus, M. Hansen, G. Ierley, and R. Worthing, "Spurt Phenomena of the Johnson-Segalman Fluid and Related Models," *J. Non-Newtonian Fluid Mech.* **29** (1988), pp. 303–325.

10. R. Kolkka and G. Ierley, "Spurt Phenomena for the Giesekus Viscoelastic Liquid Model," *J. Non-Newtonian Fluid Mech.*, 1989. To appear.

11. D. Malkus, J. Nohel, and B. Plohr, "Time-Dependent Shear Flow Of A Non-Newtonian Fluid," in *Conference on Current Problems in Hyberbolic Problems: RiemannProblems and Computations (Bowdoin,1988)*, ed. B. Lindquist, Amer. Math. Soc., Providence, 1989. Contemporary Mathematics, to appear.

12. D. Malkus, J. Nohel, and B. Plohr, "Dynamics of Shear Flow of a Non-Newtonian Fluid," *J. Comput. Phys.*, 1989. To appear.

13. D. Malkus, J. Nohel, and B. Plohr, "Analysis of New Phenomena In Shear Flow of Non-Newtonian Fluids," *SIAM J. Appl. Math.*, 1989. Submitted.

14. T. McLeish and R. Ball, "A Molecular Approach to the Spurt Effect in Polymer Melt Flow," *J. Polymer Sci.* **24** (1986), pp. 1735–1745.

15. J. Nohel, R. Pego, and A. Tzavaras, "Stability of Discontinuous Steady States in Shearing Motions of Non-Newtonian Fluids," *Proc. Roy. Soc. Edinburgh, Series A*, 1989. submitted.

16. J. Oldroyd, "Non-Newtonian Effects in Steady Motion of Some Idealized Elastico-Viscous Liquids," *Proc. Roy. Soc. London* **A 245** (1958), pp. 278–297.

17. J. Pearson, *Mechanics of Polymer Processing*, Elsevier Applied Science, London, 1985.

18. M. Renardy, W. Hrusa, and J. Nohel, *Mathematical Problems in Viscoelasticity*, Pitman Monographs and Surveys in Pure and Applied Mathematics, Vol. 35, Longman Scientific & Technical, Essex, England, 1987.

19. G. Vinogradov, A. Malkin, Yu. Yanovskii, E. Borisenkova, B. Yarlykov, and G. Berezhnaya, "Viscoelastic Properties and Flow of Narrow Distribution Polybutadienes and Polyisoprenes," *J. Polymer Sci., Part A-2* **10** (1972), pp. 1061–1084.

20. M. Yao and D. Malkus, "Analytical Solutions of Plane Poiseuille Flow of a Johnson-Segalman Fluid," in preparation, 1989.

DYNAMIC PHASE TRANSITIONS: A CONNECTION MATRIX APPROACH*

KONSTANTIN MISCHAIKOW†

Abstract. Travelling wave solutions between liquid and vapor phases in a van der Waals fluid are shown to exist. The emphasis, however, is on the techniques, namely the Conley index and connection matrix. In particular it is suggested that these methods provide a natural approach for solving problems of this nature.

1. Introduction. In this paper we shall discuss the existence of travelling wave solutions between liquid and vapor phases in a van der Waals fluid. The particular equations to be studied were derived by Marshall Slemrod [Sl1], and are meant to model an elastic fluid with pressure given by the van der Waals equations of state and possessing a higher order correction term given by Korteweg's theory of capillarity. A complete description of the analysis of shocks and phase transitions in such a fluid would have to begin with a discussion of a set of partial differential equations of mixed type. However, since we are only interested in the travelling wave solutions, we shall begin with the reduced problem, namely, a system of ordinary differential equations. The interested readers should consult [Sl1], [Sl2], [HS], [G1], [G2], and [G3] for a discussion of the full problem.

It must be emphasized that most of the results which we shall describe were obtained by Slemrod [Sl1] and M. Grinfeld [G1], [G3]. The purpose for reproducing these results is to demonstrate the power of recent developments in the Conley index. Both Slemrod and Grinfeld made extensive use of this index. However, the techniques which we shall use, most prominently, the connection matrix, were not available when their work was being done. It is our contention that these new techniques simplify the computations sufficiently to warrant this presentation. Furthermore, this problem is typical of a wide variety of phase transition problems and these techniques should prove equally useful in these other applications.

Before considering the specific equations let us consider four important characteristics of this problem. First, the system of ordinary differential equations is n-dimensional (in this case $n = 3$), and phase plane techniques are difficult if not impossible to apply in $n \geq 3$. Second, the number of critical points is large (i.e. greater than or equal to 4). Third, the goal is to find heteroclinic orbits, i.e. solutions which asymptote to different critical points in forward and backward time. These three characteristics make the problem "challenging". Finally, there exists a Lyapunov function. This of course simplifies the problem.

We hope to convince the reader that for any problem with the above characteristics, the Conley index theory is a natural approach to adopt. This approach can be summarized in three steps:

1. Identify the critical points and their Conley indices.

*Partially supported by an AURIG from Michigan State University.

†Department of Mathematics, Michigan State University, East Lansing, MI 48824. Current address: School of Mathematics, Georgia Institute of Technology, Atlanta, GA 30332.

2. Show that the set of bounded solutions is compact and compute its index.
3. Apply the connection matrix theory.

This last step is, for the most part, just a matter of simple matrix computations.

The equations we shall study are

$$\mathbf{P}_U \qquad \begin{aligned} w' &= v \\ Av' &= -Uv - U^2(w - w_-) - p(w,\theta) + p_- \\ \theta' &= \frac{u}{\alpha} U \left\{ -(\varepsilon(w,\theta) - \varepsilon_-) - (w - w_-)p_- + \frac{U^2}{2}(w - w_-)^2 + \frac{Av^2}{2} \right\} \end{aligned}$$

where:

1. The pressure p is given by the van der Waals equation of state

$$p(w,\theta) = \frac{R\theta}{w-b} - \frac{a}{w^2} \qquad \text{for } 0 < b < w < \infty$$

where a, b, and R are all positive constants, w is the specific volume, and θ is the absolute temperature.

2. The specific internal energy function is given by

$$\varepsilon(w,\theta) = -\frac{a}{w^2} + F(\theta) - \theta F'(\theta)$$

where F is any function for which $F''(\theta) < 0$ and $\lim_{\theta\to\infty} \frac{F(\theta)}{\theta} = -\infty$. The most natural choice of F is $F^*(\theta) = -c_v\theta\ell n\theta +$ constant . We shall study the problem with the more general internal energy function only because it demonstrates some of the power of the index theory. In particular, for $F = F^*$ there are at most 4 critical points, two of which are of particular importance since they correspond to the liquid and vapor phases of the fluid. For a general F, there may be more critical points though again there are only two physically relevant ones. Using the index theory we shall show that there is a natural way to ignore the irrelevant critical points. In particular, under appropriate conditions (see the end of §3) treating the general F is no more difficult than studying $F = F^*$.

3. U is the wave speed of the travelling wave and as indicated above will be used as a parameter. One only expects to find traveling wave solutions for a small set of wave speeds.

4. α and μ are positive constants corresponding to the coefficient of thermal conductivity and the viscosity coefficient, respectively.

5. w_- and θ_- are the values of the specific volume and absolute temperature at the critical point corresponding to the liquid phase, $p_- = p(w_0, \theta_-)$ and $\varepsilon_- = \varepsilon(w_-, \theta_-)$.

Since we are studying travelling waves, we are interested in those solutions to $\mathbf{P}_U$ which satisfy certain boundary conditions, the explanation of which requires the following digression concerning the function p. In Figure 1.1, two θ isoclines of p have been plotted (clearly, fixing θ defines p as a function only of w). Choose θ_-

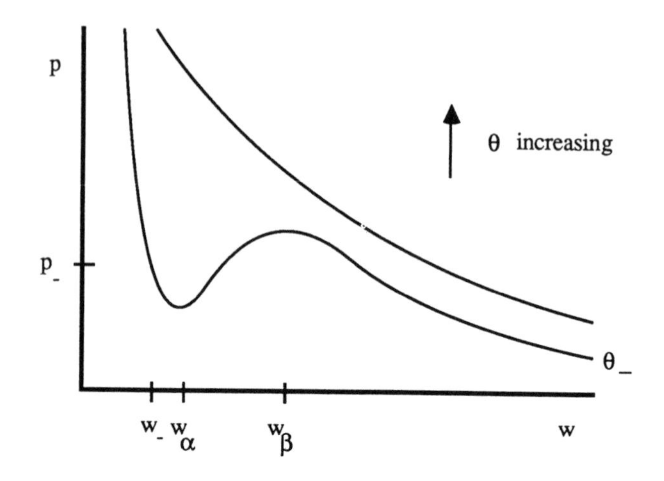

Figure 1.1

small enough that the function $p(w, \theta_-)$ has a local minimum and a local maximum. In the diagram the location of this minimum and maximum are labelled by w_α and w_β. Let $w_- \in (b, w_\alpha)$. This of course defines p_- and ε_-. One can now easily check that $(w, v, \theta) = (w_-, 0, \theta_-)$ is a critical point of $\mathbf{P}_U$.

We can now formally state the problem which needs to be solved. Prove the existence of wave speeds U for which there exist solutions $(w(t), v(t), \theta(t))$ to $\mathbf{P}_U$ satisfying the boundary conditions

$$\lim_{t \to -\infty} (w(t), v(t), \theta(t)) = (w_-, 0, \theta_-)$$

and

$$\lim_{t \to \infty} (w(t), v(t), \theta(t)) = (w_+, 0, \theta_+)$$

where $w_+ > w_\beta$. Solutions of this form shall be denoted by $\mathbf{w}_- \to \mathbf{w}_+$. If one defines $\mathfrak{U} = \{U > 0 | \mathbf{w}_- \to \mathbf{w}_+ \text{ exists }\}$ then the problem can be restated as: show that $\mathfrak{U} \neq \emptyset$.

The Lyapunov function for the system $\mathbf{P}_U$ is given by

$$V(w, v, \theta) = Rln(w-b) - F'(\theta) + \frac{1}{2}\Big\{-(\varepsilon(w,\theta) - \varepsilon_-) - (w - w_-)p_- + \frac{U^2}{2}(w - w_-)^2 + \frac{Av^2}{2}\Big\}$$

It is a simple computation to check that for the system $\mathbf{P}_U$,

$$\frac{dV}{dt} = -\frac{U}{\theta}w^2 - \frac{\mu U}{\alpha\theta^2}\Big\{-(\varepsilon(w,\theta) - \varepsilon_-) - (w - w_-)p_- + \frac{U^2}{2}(w - w_-)^2\Big\}^2 \leq 0.$$

The rest of the paper is organized as follows. We begin in §2 with a brief review of the relevant aspects of the Conley index theory, with a special emphasis on the connection matrix and transition matrix techniques. In §3 we analyze the existence, and compute the index, of the critical points. (Notice that it is not clear from the above discussion that the critical point $(w_+, 0, \theta_+)$ exists.) Our analysis is quite similar to that of Grinfeld [G1]. §4 is concerned with the set of bounded solutions. In §5 we compute the necessary connection matrices. And, finally in §6 we state and prove theorems guaranteeing the existence of the desired travelling waves.

A final comment is in order. As was indicated earlier the main goal of this paper is to show how the connection matrix theory can be applied to dynamic phase transition problems. Thus we are emphasizing the technique rather than the results. This is most apparent in §3 where certain necessary assumptions are made, and in §6 where the results are stated. In the first case we state the assumptions in the most convenient form for the connection matrix theory. We justify this approach by acknowledging that the work required to verify the assumptions can be found in [G3]. With regard to §6, we state three theorems, only two of which we prove. As the reader will see the proofs are trivial (given the connection matrix machinery). The theorem we do not prove is due to Grinfeld and requires a more subtle approach. We state it for two reasons: (one) to acknowledge that the connection matrix cannot do everything, and (two) because we feel that it would be desirable, and perhaps even possible, to extend Conley index techniques to handle such problems directly.

Acknowledgements. I would like to thank Marshall Slemrod and Michael Grinfeld for calling this problem to my attention.

2. Conley Index Theory. This section presents a brief review of the Conley index, connection matrices, and transition matrices. While it is assumed that the reader is familiar with the Conley index (see [C], [S], [Sm]), the basic properties of the connection matrices and transition matrices are described in sufficient detail to allow the reader to mechanically apply the theory. For a complete presentation of this material see [F], [Mc], [M1], [M2], and [R].

Throughout this paper S will denote an isolated invariant set and $\mathbf{h}(S)$, the Conley index of S. Recall that $\mathbf{h}(S)$ is the homotopy type of a pointed topological space, and therefore one can define a homology version of the Conley index by

$$CH_*(S; \mathbf{R}) \equiv H_*(\mathbf{h}(S); \mathbf{R})$$

where the latter expression denotes the singular homology of the topological space $\mathbf{h}(S)$ relative the special point with coefficients in a ring $\mathbf{R}$. The reason for using the homology index is that in general homology theory is more computable than homotopy theory. To simplify the computations even further we shall choose $\mathbf{R} = \mathbb{Z}_2$. Of course this implies that $CH_*(S) = CH_*(S; \mathbf{R})$ is a graded vector space.

PROPOSITION 2.1. *If S is a hyperbolic critical point with an unstable manifold of dimension k, then*

$$CH_n(S) \approx \begin{cases} \mathbb{Z}_2 & \text{if } n = k \\ 0 & \text{otherwise} . \end{cases}$$

If $S = \emptyset$, then $CH_*(S) \approx (0, 0, 0, \dots)$.

DEFINITION 2.2. Given two isolated invariant sets, S and S', the set of *connections* from S to S' is denoted by $C(S, S') = \{x | \omega(x) \subset S' \text{ and } \omega^*(x) \subset S\}$ where $\omega(x)$ and $\omega^*(x)$ denote the omega and alpha limit sets of x, respectively.

Using this language, our problem can be restated as; show that $C(w_-, w_+) \neq \emptyset$. Before considering how one decomposes isolated invariant sets we need to introduce some notation.

A partially order set,$(P, >)$, consists of a finite set P along with a strict partial order relation, $>$, which satisfies:

(i) $i > i$ never holds for $i \in P$, and

(ii) if $i > j$ and $j > k$, then $i > k$ for all $i, j, k \in P$.

An interval in $(P, >)$ is a subset, $I \subset P$, such that given $i, j \in I$, if $i < k < j$ then $k \in I$.

DEFINITION 2.3. A *Morse decomposition* of S, denoted by $\mathsf{M}(S) = \{M(i) | i \in (P, >)\}$, is a collection of mutually disjoint isolated invariant subsets of S, indexed by $(P, >)$, such that given $x \in S$, then either $x \in M(i)$ for some i, or there exists $i, j \in P$ such that $i > j$ and $x \in C(M(i), M(j))$. The invariant sets, $M(i)$, are called *Morse* sets.

Recall that in our problem we have a Lyapunov function, V. Hence, a natural Morse decomposition of S, the set of all bounded solutions, would consist of all the critical points of the system. Furthermore, an ordering on this Morse decomposition could be given by $i > j$ if $V(M(i)) > V(M(j))$.

Let I be an interval in P, then one can define a new isolated invariant set by

$$M(I) = \left(\bigcup_{i \in I} M(i) \right) \bigcup \left(\bigcup_{j i \in I} C(M(j), M(i)) \right)$$

Since $M(I)$ is an isolated invariant set $CH*(M(I))$ is defined.

Turning to the definition of a connection matrix, let

$$\Delta : \oplus_{i \in P} CH_*(M(i)) \to \oplus_{i \in P} CH_*(M(i))$$

be a linear map, then Δ can be written as a matrix of maps

$$\Delta = (\Delta(i, j))$$

where $\Delta(ij) : CH_*(M(i)) \to CH_*(M(j))$. For any interval I in P, let

$$\Delta(I) = (\Delta(i, j))_{i, j \in I} \, .$$

DEFINITION 2.4. Δ is called a *connection matrix* for $\mathsf{M}(S)$ if the following four properties are satisfied:

(i) If $i \not> j$ then $\Delta(i, j) = 0$. *(strict upper triangularity)*

(ii) $\Delta(i, j)(CH_k(M(i))) \subset CH_{k-1}(M(j))$. *(degree -1 map)*

(iii) $\Delta \circ \Delta = 0$.

(iv) $CH_*(M(I)) = \frac{\text{kernel } \Delta(I)}{\text{image } \Delta(I)}$ for every interval I. *(rank condition)*

Some of the basic theorems concerning connection matrices are as follows.

THEOREM 2.5. (Franzosa [F]) *Given a Morse decomposition there exists at least one connection matrix.*

THEOREM 2.6. (Franzosa [F]) *Let $\{i,j\}$ be an interval in P and assume that $\Delta(i,j) \neq 0$, then $C(M(i), M(j)) \neq \emptyset$.*

THEOREM 2.7. (McCord [Mc]) *Let $\mathsf{M}(S) = \{M(1), M(0)|1 > 0\}$ be a Morse decomposition consisting of hyperbolic critical points. Assume for $i = 0, 1$, that*

$$CH_n(M(i)) = \begin{cases} Z_2 & \text{if } n = k + i \\ 0 & \text{otherwise .} \end{cases}$$

Furthermore, let $C(M(1), M(0))$ consists of exactly q heteroclinic orbits which arise as the intersection of the stable and unstable manifolds of $M(0)$ and $M(1)$, respectively. Then $\Delta(1,0) = q \bmod 2$.

Up to this point, our discussion has been based on a given flow. However, the problem we wish to study consists of a parameterized family of flows. Thus we need to compare different connection matrices corresponding to the flows at different parameter values. With this in mind, consider the family of differential equations

$$x' = f(x, \lambda)$$

where $x \in \mathbf{R}^n$ and $\lambda \in \Lambda = \mathbf{R}$. The first assumption that needs to be made is that there exists an isolated invariant set S which continues over Λ (see [Sa]). For a fixed λ this isolated invariant set will be denoted by S^λ. Let $\mathsf{M}(S^0) = \{M^0(i)|i \in (P^0, >^0)\}$ and $\mathsf{M}(S^1) = \{M^1(i)|i \in (P^1, >^1)\}$ denote Morse decompositions of S^λ at $\lambda = 0$ and 1. Let Δ^0 and Δ^1 be connection matrices for these Morse decompositions.

DEFINITION 2.8. A *transition system* consists of an infinite sequence of systems of the form

$$\begin{aligned} x' &= f(x, \lambda) \\ \lambda' &= \varepsilon_n \lambda(\lambda - 1) \end{aligned}$$

where $1 >> \varepsilon_n > \varepsilon_{n+1}$ and $\lim_{n \to \infty} \varepsilon_n = 0$.

For each ε_n there exists an isolated invariant set with a Morse decomposition

$$\mathsf{M} = \{\bar{M}(k)|k \in P^0 \cup P^1 \text{ and } \bar{M}(k) = M^i(k) \text{ for } k \in P^i\}$$

and with the partial order relation

$$\begin{aligned} j > i \quad & \text{for } j \in P^1 \text{ and } i \in P^0 \\ i >^0 i' \quad & \text{for } i, i' \in P^0 \\ j >^1 j' \quad & \text{for } j, j' \in P^1. \end{aligned}$$

PROPOSITION 2.9. $CH_n(\bar{M}(i)) \approx CH_n(M^0(i))$

$$CH_n(\bar{M}(j)) \approx CH_{n-1}(M^1(j))$$

PROPOSITION 2.10. ([R], [M2]) *A connection matrix for* $\mathbf{M}$ *takes the form*

$$\begin{pmatrix} \Delta^0 & T^{10} \\ 0 & \Delta^1 \end{pmatrix}$$

where the entries Δ^0 *and* Δ^1 *are the connection matrices for* $\mathbf{M}(S^0)$ *and* $\mathbf{M}(S^1)$.

Remark 2.11. Notice, by Proposition 2.9 that the homology indices of $\bar{M}(j)$ are not in exact agreement with those of $M^1(j)$. However, the difference is only in the grading. Thus the Δ^1 referred to in Proposition 2.10 is obtained by shifting the grading of the connection matrix for $\mathbf{M}(S^1)$ up by one.

DEFINITION 2.12. T^{10} is called the *transition matrix* from $\lambda = 1$ to $\lambda = 0$.

Returning to the transition system. Notice that in the limit one obtains the parameterized system

$$\begin{aligned} x' &= f(x, \lambda) \\ \lambda' &= 0. \end{aligned}$$

The following theorem indicates that entries in the transition matrix correspond to connections for the system $x' = f(x, \lambda)$ for $\lambda \in (0,1)$.

THEOREM 2.13. (Reineck [R]) *Let* $T(i^1 j^0) : CH*(M^1(i)) \to CH*(M^0(j))$ *be an entry of the transition matrix* T^{10}. *If* $T(i^1 j^0) \neq 0$, *then there exists a finite sequence* $1 \geq \lambda_1 \geq \lambda_2 \geq \cdots \geq \lambda_k \geq 0$ *of parameter values and corresponding* $k_r \in P^{\lambda_r}$ *such that* $C(M(k_r), M(k_{r+1})) \neq \emptyset$ *under the flow defined by* $x' = f(x, \lambda_r)$.

Finally, the following theorem indicates that certain entries in the transition matrix can be determined by understanding how the individual Morse sets continue.

THEOREM 2.12. (Mischaikow [M1]) *Let the Morse set* $M(i)$ *continue over* $[0,1]$ *as an attractor or repeller. Then* $T(i^1, i^0)$ *is an isomorphism.*

3. The Critical Points. As was indicated in the introduction, the first step in applying the connection matrix techniques is to study the critical points of the system. Obviously, one obtains the critical points of $\mathbf{P}_U$ by solving the following system of equations:

$$\begin{aligned} 0 &= v \\ 0 &= -Uv - U^2(w - w_-) - p(w, \theta) + p_- \\ 0 &= \frac{\mu}{\alpha} U \left\{ -(\varepsilon(w,\theta) - \varepsilon_-) - (w - w_-)p_- + \frac{U^2}{2}(w - w_-)^2 + \frac{Av^2}{2} \right\}. \end{aligned} \tag{3.1}$$

Substituting in the definitions of $p(w,\theta)$ and $\varepsilon(w,\theta)$ one gets

(3.2a)

$$0 = -U^2(w-w_-) - \frac{R\theta}{w-b} + \frac{a}{w^2} + p_-$$

(3.2b)

$$0 = \frac{a}{w} - F(\theta) + \theta F'(\theta) + \varepsilon_- - (w-w_-)p_- + \frac{U^2}{2}(w-w_-)^2.$$

Define

$$f_1(w) = \frac{w-b}{R}\left\{-U^2(w-w_-) + \frac{a}{w^2} + p_-\right\}$$

and

$$f_2(w) = \frac{a}{w} + \varepsilon_- - (w-w_-)p_- + \frac{U^2}{2}(w-w_-)^2.$$

Clearly then (3.2) is equivalent to

(3.3a) $$\theta = f_1(w)$$

(3.3b) $$F(\theta) - \theta F'(\theta) = f_2(w).$$

Notice that $f_2'(w) = -\frac{R\theta}{w-b}f_1(w)$. Thus, differentiating (3.3b) with respect to w leads to

$$-\theta F''(\theta)\frac{d\theta}{dw} = -\frac{R\theta}{w-b}f_1(w)$$

or

(3.4) $$\frac{d\theta}{dw} = \frac{R}{w-b}\frac{1}{\theta F''(\theta)}f_1(w).$$

Hence, for $f_1(w) >)$, one has that $\frac{d\theta}{dw} < 0$, i.e. the graph of the curve defined by (3.2b) is monotone decreasing as long as $f_1(w)$ is positive. In particular, (3.2b) defines θ as a function of w for all $\theta > 0$. From now on this function will be denoted by g_2, i.e. $\theta = g_2(w)$.

Recall, from the introduction, that θ_- was chosen in such a manner as to insure that $p(w,\theta_-) = p_-$ has three solutions. Thus, setting $U = 0$ implies that the graph of $f_1(w)$ is as in Figure 3.1. Choosing $F = F^*$ and c_v large one obtains Figure 3.1(a). In particular, there are three values of w which correspond to critical points of $\mathbf{P}_U$ and the third one, w_+, is greater than w_β. If we consider arbitrary F (though insisting that $F'' < 0$) then it is possible to have the situation described in Figure 3.1(b). Notice, however, that w_+ is still well defined. On the other hand if one chooses $U > 0$, but small then one obtains the diagrams of Figure 3.2. Again, (a) corresponds to $F = F^*$ and (b) represents a possible graph for a more general F. And again, it is worth emphasizing that the critical point $(w_+, 0, \theta_+)$ is uniquely defined in both cases.

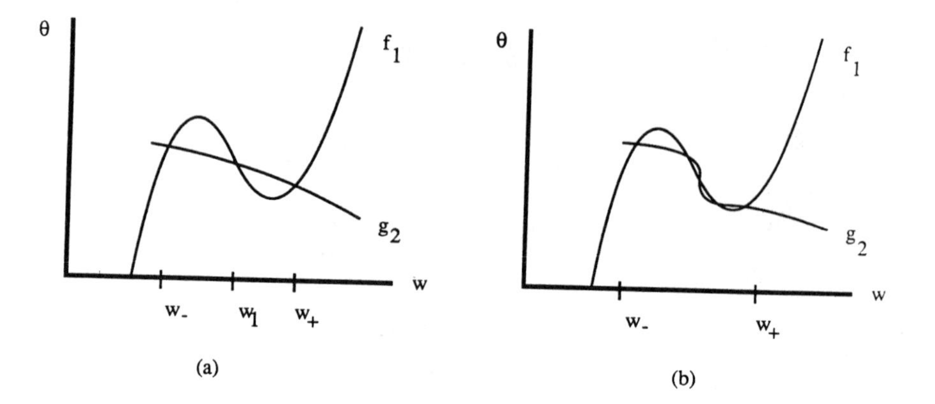

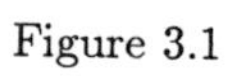

Figure 3.1

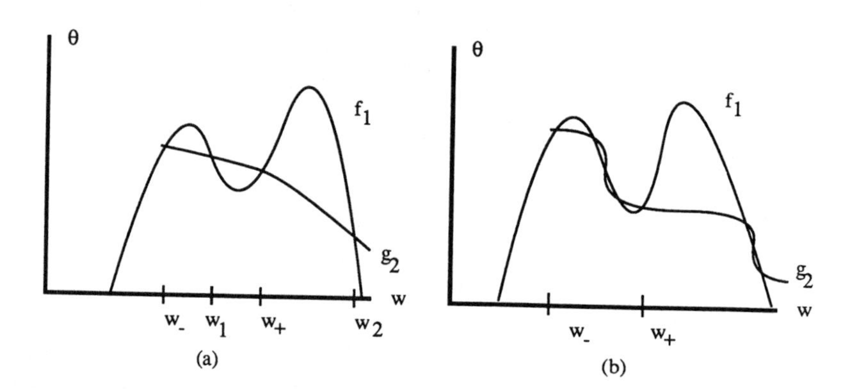

Figure 3.2

Before continuing we wish to introduce some notation. As was mentioned before we are primarily interested in the critical points $(w_-, 0, \theta_-)$ and $(w_+, 0, \theta_-)$. To simplify the notation we shall denote these critical points by $\mathbf{w}_\pm$. However, the other critical points shall also play a role in our analysis. If $F = F^*$ then we need only concern ourselves with the points $(w_i, 0, \theta_i)$ for $i = 1, 2$, however, for general F there may be more critical points. We can however define the Morse sets

$$\mathbf{W}_1 = \left(\bigcup_{w_\alpha < w_i < w_\beta} (w_i, 0, \theta_i) \right) \bigcup \left(\bigcup_{w_\alpha < w_i, w_j < w_\beta} C((w_1, 0, \theta_i), (w_j, 0, \theta_j)) \right)$$

and

$$\mathbf{W}_2 = \left(\bigcup_{w_+ < w_i} (w_1, 0, \theta_i) \right) \bigcup \left(\bigcup_{w_+ < w_i, w_j} C((w_i, 0, \theta_i), (w_j, 0, \theta_j)) \right)$$

As was indicated in the introduction U is treated as a free parameter. Thus the information of Figures 3.1 and 3.2 needs to be presented in terms of U and w. We begin with the following definitions. Let

$$U_0 = \min\{U \geq 0 | w_+ \text{ exists } \}$$

and

$$U_1 = \max\{U \geq 0 | w_+ \text{ exists } \}.$$

As is indicated in Figure 3.1, it is possible for $U_0 = 0$.

To understand how the curves in Figures 3.1 and 3.2 change as a function of U, consider f_1 and f_2 as functions of U and w, i.e. (3.3) becomes

$$\theta = f_1(w, U) \tag{3.5a}$$

$$F(\theta) - \theta F'(\theta) = f_2(w, U). \tag{3.5b}$$

Now differentiating by U gives

$$\frac{\partial \theta}{\partial U} = -2(w - b)(w - w_-)\frac{U}{R} > 0$$

and

$$-\theta F''(\theta)\frac{\partial \theta}{\partial U} = U(w - w_-)^2$$

$$\frac{\partial \theta}{\partial U} = -\frac{U(w - w_-)^2}{\theta F''(\theta)} > 0$$

for $w > w_-$. Thus, as U is increased the curve corresponding to f_1 moves down pointwise and g_2 moves up pointwise.

LEMMA 3.1. *There exists* $U_1 \in (0, \infty)$.

Proof. Consider $f_1(w, U) = 0$ as a function of U. This is equivalent to solving

$$-w^4 + \left[w_- + b + \frac{p_-}{U^2}\right] w^3 - b\left(w_- + \frac{p_-}{U^2}\right) w^2 + \frac{a}{U} w - \frac{ab}{U} = 0.$$

If we let $U \to \infty$, then we have that the zeroes are given by $0, b$, and w_-. Furthermore for $w > w_1, f(w, \infty) < 0$. Thus the right local maxima of f_1 tends to zero as $U \to \infty$. Therefore, there exists U^* such that the right local maximum of f_1 is tangent to g_2. Since f_1 is moving down pointwise and g_2 is moving up pointwise as a function of U for all $U > U^*, w_1$ does not exist. □

If we set $F = F^*$, then the above arguments can be summarized by the bifurcation diagrams of Figures 3.3 and Figures 3.4. (For general F one might have more wiggles).

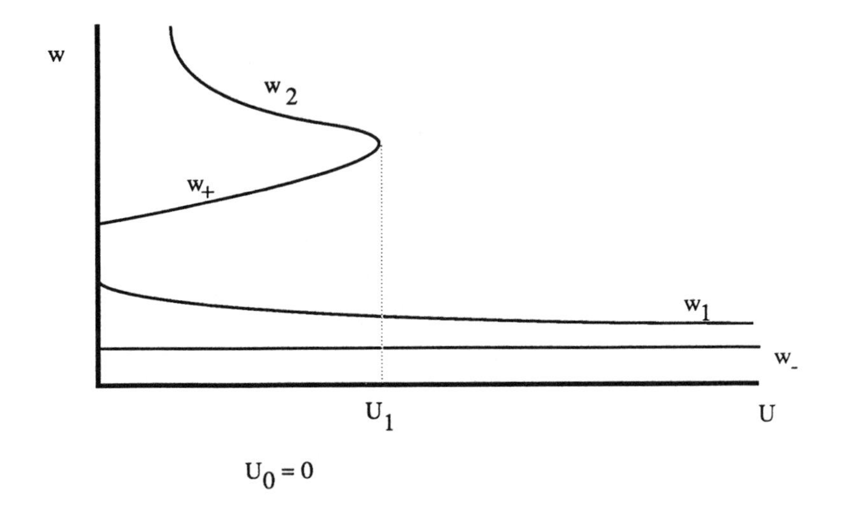

Figure 3.3

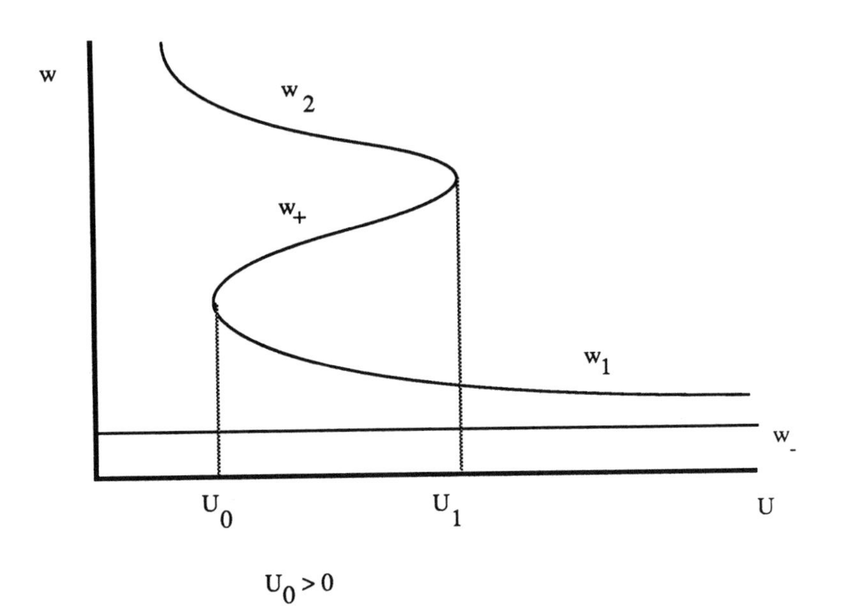

Figure 3.4

Our next step is to determine the Conley indices of the critical points. This is most easily done by assuming that $F = F^*$, linearizing about the critical points, and computing the characteristic equation. Doing this simple calculation yields:

$$\lambda^3+\left(\frac{\mu U c_v}{\alpha}+\frac{U}{A}\right)ld^2+\left(\frac{\mu U^2 c_v}{A\alpha}-\frac{R}{A(w-b)}f_1'(w)\right)\lambda-\frac{\mu U c_v}{\alpha}\frac{R}{A(w-b)}(f_1'(w)-f_2'(w))=0.$$

Solving for the roots explicitly one gets that if $f_1'(w)-f_2'(w)>0$, then there exists a unique positive solution and if $f_1'(w)-f_2'(w)<0$, then all the solutions have negative real parts. Therefore, we have the following lemma.

LEMMA 3.2. *$CH_*(\mathbf{w}\pm)\approx(0,Z_2,0,0,\dots)$ and $CH_*(\mathbf{W}_i)\approx(Z_2,0,0,\dots)$.*

Before considering this lemma in the context of more general F's, we shall make the following three assumptions.

A1 $V(\mathbf{w}_-)>V(\mathbf{w}_+)$

A2 If $\mathbf{w}^*$ is a critical point in $\mathbf{W}_i$ then $V(\mathbf{w}\pm)>V(\mathbf{w}^*)$.

A3 For $U\in[U^*,U^{**}],\theta_2>0$.

Since we are interested in obtaining connections of the form $\mathbf{w}_-\to\mathbf{w}_+$ and since V is a Lyapunov function for (**P**) it is obvious that **A1** is a necessary condition. For a proof that this assumption can be satisfied, at least in the case of $F = F^*$, we refer the reader to [G3]. Since it holds for F^* it must also hold for F near F^*.

The second assumption is considerably stronger than what is necessary, but clearly holds for $F = F^*$ and hence for F near F^*. It guarantees two things, first, that $\mathbf{W}_i$ are Morse sets and, two, that $CH_*(\mathbf{W}_i)$ remains the same as in Lemma 3.2. Therefore, Lemma 3.2 remains true for all F for which assumption **A2** holds.

Finally, **A3** is necessary on physical grounds since θ represents absolute temperature, and hence, must be positive.

4. The Set of Bounded Solutions. The second step in applying the Conley theory, is to study the set of bounded solutions. In our case this means understanding the set of bounded solutions for the parameter values $U^* < U < U^{**}$ and for which $w > b$ and $\theta > 0$. Thus in this section we define an appropriate isolating neighborhood, i.e. one which contains all the orbits of interest to us, and then compute the index of the isolated invariant.

Notice that one can think of the Lyapunov function V as a function of the parameter value. To emphasize this we shall write $V = V_U$. Define

$$K = \max_{U \in [U^*, U^{**}]} V_U(w_-, 0, \theta_-).$$

Let $N = \{(w, v, \theta) | w \geq \lambda, \theta > 0, V(w, v, \theta) \leq 2K\}$ and let S_U be the maximal invariant set in N under $\mathbf{P}_U$.

PROPOSITION 4.1. *There exists $\lambda \in (b, w_-)$ such that N is an isolating neighborhood for $\mathbf{P}_U$ for all $U \in [U^*, U^{**}]$.*

Proof. The first step is to show that N is compact. This follows from the following limits which hold for $w > b$, and $\theta > 0$;

$$\lim_{w \to \infty} V(w, v, \theta) = \infty$$
$$\lim_{|v| \to \infty} V(w, v, \theta) = \infty$$
$$\lim_{\theta \to 0} V(w, v, \theta) = \infty$$

and

$$\lim_{\theta \to \infty} V(w, v, \theta) = \infty.$$

Next one needs to show that there exists λ such that $S \cap \partial N = \varnothing$. Let $(w, v, \theta) \in \partial N$ with $w > \lambda$. Since $\frac{dV}{dt} < 0$ for these points, $(w, v, \theta) \notin S$. Thus we need only consider those points $(w, v, \theta) \in \partial N$ for which $w = \lambda$. If $v > 0$, then $(\lambda, v, \theta) \cdot [-t, 0] \not\subset N$ and if $v < 0$, then $(\lambda, v, \theta) \cdot [0, t] \not\subset N$. So assume $(\lambda, v, \theta) \in \partial N$ and $v = 0$. If $v' < 0$, then one has an external tangency. Let $(\lambda, 0, \bar{\theta}) \in \partial N$ for which $v' = 0$, then $\lim_{\lambda \to b} V(\lambda, 0, \bar{\theta}) = \infty$, i.e. there exists a λ such that $v' \neq 0$. One easily checks that in fact $v' < 0$. □

PROPOSITION 4.2. $CH_*(S) \approx (0,0,0,\dots.)$.

Proof. Notice that N is a contractible region. Furthermore $L = \{(\lambda, v, \theta) \in \partial N | v \leq 0\}$ is contractible and (N, L) is an index pair for S. □

In the following sections S_U will be the total isolated invariant set and its Morse decomposition will consist of the Morse sets $\mathbf{w}\pm$, and $\mathbf{W}_i, i = 0, 1$ with the ordering induced by V.

5. The Connection Matrices. We are now in the position to begin step three in the application of the Conley index theory, namely the computations of the connection matrices.

Recall that we are attempting to show that $\mathfrak{U} \neq \varnothing$. So choose $U^1, U^2, U^3 \notin \mathfrak{U}$ such that $0 \leq U^* < U^1 < U_0 < U^2 < U_1 < U^3 < U^{**}$ (obviously, if $U_0 = 0$, then U^1 cannot be defined). Let Δ^i denote the connection matrix for the isolated invariant set S_{U^i}.

LEMMA 5.1.

$$\Delta^1 = \begin{array}{c} \\ CH_0(\mathbf{w}_2) \\ CH_1(\mathbf{w}_-) \end{array} \begin{array}{c} \begin{array}{cc} CH_0(\mathbf{w}_2) & CH_1(\mathbf{w}_-) \end{array} \\ \begin{pmatrix} 0 & 1 \\ 0 & 0 \end{pmatrix} \end{array}$$

and

$$\Delta^3 = \begin{array}{c} \\ CH_0(\mathbf{w}_1) \\ CH_1(\mathbf{w}_-) \end{array} \begin{array}{c} \begin{array}{cc} CH_0(\mathbf{w}_1) & CH_1(\mathbf{w}_-) \end{array} \\ \begin{pmatrix} 0 & 1 \\ 0 & 0 \end{pmatrix} \end{array}$$

Proof. Consider Δ^1. In this case the Morse decomposition consists of $\{\mathbf{W}_2, \mathbf{w}_-\}$. Thus, $\Delta^1 : CH_*(\mathbf{w}_2) \oplus CH_*(\mathbf{w}_-) \to CH_*(\mathbf{w}_2) \oplus CH_*(\mathbf{w}_-)$. By Lemma 3.2, Δ^1 can be considered as a map on $CH_0(\mathbf{w}_2) \oplus CH_1(\mathbf{w}_-)$. Since a connection matrix is a degree -1 map Δ^1 must take the form

$$\begin{pmatrix} 0 & * \\ 0 & 0 \end{pmatrix}$$

where $*$ remains to be determined. By the rank condition, $*$ must be an isomorphism, which in $\mathbb{Z}_2$ coefficients implies that $* = 1$.

The proof for Δ^3 is similar. □

The same sorts of arguments, which rely only on the definition of the connection matrix give the following result.

LEMMA 5.2.

$$\Delta^2 = \begin{array}{c} \\ CH_0(\mathbf{w}_1) \\ CH_0(\mathbf{w}_2) \\ CH_1(\mathbf{w}_+) \\ CH_1(\mathbf{w}_-) \end{array} \begin{array}{c} \begin{array}{cccc} CH_0(\mathbf{w}_1) & CH_0(\mathbf{w}_2) & CH_1(\mathbf{w}_+) & CH_1(\mathbf{w}_-) \end{array} \\ \begin{pmatrix} 0 & 0 & a & b \\ 0 & 0 & c & d \\ 0 & 0 & 0 & 0 \\ 0 & 0 & 0 & 0 \end{pmatrix} \end{array}$$

and $ad + bc = 1$.

Since $U^i \notin \mathfrak{U}$, we shall use the transition matrices to show that for some value of U between U^1 and U^3 there exists an element of $\mathfrak{U}$. The transition matrices we shall use are those which relate Δ^1 to Δ^2 and Δ^2 to Δ^3 and we shall denote these by T^{12} and T^{23} respectively.

LEMMA 5.3.

$$\begin{pmatrix} \Delta^2 & T^{12} \\ 0 & \Delta^1 \end{pmatrix} =$$

$$\begin{array}{c} \\ CH_0(\mathbf{W}_1) \\ CH_0(\mathbf{W}_2) \\ CH_1(\mathbf{w}_+) \\ CH_1(\mathbf{w}_-) \\ CH_1(\mathbf{W}_2) \\ CH_2(\mathbf{w}_-) \end{array} \begin{array}{c} \begin{array}{cccccc} CH_0(\mathbf{W}_1) & CH_0(\mathbf{W}_2) & CH_1(\mathbf{w}_+) & CH_1(\mathbf{w}_-) & CH_1(\mathbf{W}_2) & CH_2(\mathbf{w}_-) \end{array} \\ \begin{pmatrix} 0 & 0 & a & b & 0 & 0 \\ 0 & 0 & c & d & 1 & 0 \\ 0 & 0 & 0 & 0 & 0 & \delta \\ 0 & 0 & 0 & 0 & 0 & 1 \\ 0 & 0 & 0 & 0 & 0 & 1 \\ 0 & 0 & 0 & 0 & 0 & 0 \end{pmatrix} \end{array}$$

where $a\delta + b = 0$ and $c\delta + d + 1 = 0$.

Proof. From the discussion of §2 it is clear that only the T^{12} entries need to be determined. Because T^{12} is a degree -1 map, one immediately obtains the fact that

$$T^{12} = \begin{pmatrix} \alpha & 0 \\ \beta & 0 \\ 0 & \delta \\ 0 & \gamma \end{pmatrix}.$$

To determine the values of the entries α, β, γ, and δ require a more careful analysis of the flow. Notice that $\mathbf{w}_-$ continues as a Morse set over the interval $[U^*, U^{**}]$ and furthermore, is a repeller for all parameter values U. Thus $\gamma = 1$. Similarly, $\mathbf{W}_2$ continues as an attractor over the interval $[U^1, U^2]$ and hence, $\beta = 1$ and $\alpha = 0$. The equalities involving the entries of the matrix follow from the condition that connection matrices square to zero. □

A similar argument results in:

LEMMA 5.4.

$$\begin{pmatrix} \Delta^2 & T^{23} \\ 0 & \Delta^3 \end{pmatrix} =$$

$$\begin{array}{c} \\ CH_0(\mathbf{W}_1) \\ CH_0(\mathbf{W}_2) \\ CH_1(\mathbf{w}_+) \\ CH_1(\mathbf{w}_-) \\ CH_1(\mathbf{W}_1) \\ CH_2(\mathbf{w}_-) \end{array} \begin{array}{c} \begin{array}{cccccc} CH_0(\mathbf{W}_1) & CH_0(\mathbf{W}_2) & CH_1(\mathbf{w}_+) & CH_1(\mathbf{w}_-) & CH_1(\mathbf{W}_1) & CH_2(\mathbf{w}_-) \end{array} \\ \begin{pmatrix} 0 & 0 & a & b & 1 & 0 \\ 0 & 0 & c & d & 0 & 0 \\ 0 & 0 & 0 & 0 & 0 & \gamma \\ 0 & 0 & 0 & 0 & 0 & 1 \\ 0 & 0 & 0 & 0 & 0 & 1 \\ 0 & 0 & 0 & 0 & 0 & 0 \end{pmatrix} \end{array}$$

where $a_\gamma + b + 1 = 0$ and $c_\gamma + d = 0$.

Notice that $\gamma = 0$ or 1 and $\delta = 0$ or 1. Thus we can solve the systems of equations

$$\begin{pmatrix} \Delta^2 & T^{12} \\ 0 & \Delta^1 \end{pmatrix}^2 = 0$$

and

$$\begin{pmatrix} \Delta^2 & T^{23} \\ 0 & \Delta^3 \end{pmatrix}^2 = 0$$

to obtain additional constraints on a, b, c, and d. Doing so one obtains the following lemma

LEMMA 5.5.

$$\Delta^2 = \begin{pmatrix} 0 & 0 & 1 & 1 \\ 0 & 0 & 1 & 0 \\ 0 & 0 & 0 & 0 \\ 0 & 0 & 0 & 0 \end{pmatrix}$$

or

$$\Delta^2 = \begin{pmatrix} 0 & 0 & 1 & 0 \\ 0 & 0 & 1 & 1 \\ 0 & 0 & 0 & 0 \\ 0 & 0 & 0 & 0 \end{pmatrix}$$

Remark 5.6. The result of Lemma 5.5 is independent of the fact that $U_0 > 0$. this follows from the fact that we can continue the system with $U_0 = 0$ to a system with $U_0 > 0$. Thus, given a connection matrix Δ^2 for $U_0 = 0$ it must be related to a Δ^2 for $U_0 > 0$ by a transition matrix, which only allows for the possibilities stated in Lemma 5.5.

6. Results. We are now ready to present the results. We remark that Theorems 6.1 and 6.4 can be found in [G3] while 6.2 is new. Hopefully the brevity and simplicity of the proofs will convince the reader that the previous sections were worth reading.

THEOREM 6.1. *Assume* **A1** *through* **A3**, *and that $0 \leq U^* < U_0 < U_1 < U^{**}$, then $\mathfrak{U} \neq \varnothing$, i.e. there exists $\bar{U} \in [U_0, U_1]$ for which there exists a heteroclinic orbit from $(w_-, 0, \theta_-)$ to $(w_+, 0, \theta_+)$ solving $\mathbf{P}_{\bar{U}}$.*

Proof. Recall from Lemmas 5.3 and 5.4 that $\delta a + b = 0$ and $\gamma a + b + 1 = 0$. If $\mathfrak{U} = \varnothing$, then $\delta = \gamma = 0$ and hence, $b = 0 = 1$. Therefore $\mathfrak{U} \neq \varnothing$. □

THEOREM 6.2. *Assume* **A1** *through* **A3**, *that $0 = U^* = U_0 < U_1 < U^{**}$, and that for some $U \in (0, U_1)$ there exists $\mathbf{w}_- \to \mathbf{W}_2$, then $\mathfrak{U} \neq \varnothing$.*

Proof. Lemma 5.5 determines the set of possible connection matrices for $\mathbf{P}_U$ with $U \in (0, U_1)$. In particular, either there exists a $\mathbf{w}_- \to \mathbf{W}_2$ or there exists a $\mathbf{w}_- \to \mathbf{W}_1$. Furthermore, since $\mathbf{W}_2$ is an attractor, the existence of a $\mathbf{w}_- \to \mathbf{W}_i$ orbit for $i = 1, 2$ implies that the corresponding connection matrix entry is 1. (If

$\mathbf{W}_i$ is a single attracting critical point then this follows from Theorem 2.7. Even if $\mathbf{W}_i$ is more complicated the proof of the result still applies.) Thus if we assume that $\mathbf{w}_- \to \mathbf{W}_2$ exists and we let Δ^2 denote the corresponding connection matrix, then by Lemma 5.4, $\gamma = 1$. therefore, there exists $\bar{U} \in (U, U_1)$ for which $\mathbf{w}_- \to \mathbf{w}_+$ exists. □

Remark 6.3. Since $\mathbf{w}_-$ has a one dimensional unstable manifold and since $\mathbf{W}_2$ is an attractor, checking the existence of a $\mathbf{w}_- \to \mathbf{W}_2$ orbit is numerically tractable.

We finish with a theorem which cannot be proved directly with our techniques. However, we feel it is important to generalize or extend the Conley index theory to the point that it can be applied to problems of this nature.

THEOREM 6.4. *Assume* **A1** *through* **A3**, *that* $0 = U^* = U_0 < U_1 < U^{**}$, *that* $F = F^*$, *that* c_v *is sufficiently large, and that* μ *is sufficiently small, then* $\mathfrak{U} \neq \emptyset$.

REFERENCES

[C] C. CONLEY, *Isolated Invariant Sets and the Morse Index*, Conf. Board Math. Sci. 38, AMS, Providence (1978).

[F] R. FRANZOSA, The connection matrix theory for Morse decompositions, Trans. AMS, Vol. 311, No. 2, Feb. 1989, pp. 781-803.

[HS] R. HAGEN AND M. SLEMROD, The viscosity-capillarity admissibility criterion for shocks and phase transitions, Arch. for Rat. Mech. and Anal, 83, pp. 333-361.

[G1] M. GRINFELD, *Topological Techniques in Dynamic Phase Transitions*, Ph.D. Thesis, Rensselar Polytechnic Institute (1986).

[G2] M. GRINFELD, Dynamic phase transitions: existence of "cavitation" waves, Proc. of the Royal Soc. of Edin., 107A (1987), pp. 153-163.

[G3] M. GRINFELD, Nonisothermal dynamic phase transitions, Quart. of Applied Math., Vol 47, No. 1 (March 1989), pp. 71-84.

[Mc] C. MCCORD, *The connection map for attractor-repeller pairs*, To appear in Trans. A.M.S.

[M1] K. MISCHAIKOW, *Existence of generalized homoclinic orbits for one-parameter families of flows*, Proc. A.M.S., May 1988.

[M2] K. MISCHAIKOW, *Transition systems*, Proceedings of the Royal Society of Edinburgh, 112 A (1989), pp. 155-175.

[R] J. REINECK, *Connecting orbits in one-parameter families of flows*, J. Erg. thy. and Dyn. Sys., Vol 8* (1988), pp. 359-374.

[S] D. SALAMON, *Connected simple systems and the Conley index of isolated invariant sets*, Trans. AMS, 291(1) (1985).

[Sl1] M. SLEMROD, *Admissibility criteria for propagating phase boundaries in a van der Waals fluid*, Arch. Rat. Mech. Anal., Vol. 81, No 4, pp. 301-315.

[Sl2] M. SLEMROD, *Dynamic phase transitions in a van der Waals fluid*, Journal of Differential Equations, Vol. 52, No. 1, pp. 1-23.

[Sm] J. SMOLLER, *Shock Waves and Reaction Diffusion Equations*, Springer-Verlag, New York, 1983.

This paper is dedicated to Daniel D. Joseph
on the occasion of his 60th birthday

A WELL-POSED BOUNDARY VALUE PROBLEM FOR SUPERCRITICAL FLOW OF VISCOELASTIC FLUIDS OF MAXWELL TYPE

MICHAEL RENARDY*

Abstract. For a class of viscoelastic fluids with differential constitutive laws of Maxwell type, we investigate the existence and uniqueness of steady flows. We consider small perturbations of uniform flow transverse to a strip. A well-posed boundary value problem is formulated for the case when the velocity of the fluid exceeds the speed of propagation of shear waves.

Key words. viscoelastic fluids, boundary conditions, change of type

AMS(MOS) subject classifications. 35M05,76A10

1. Introduction. While the study of existence and uniqueness results for steady flows of Newtonian fluids is well advanced, relatively little is known about viscoelastic fluids with memory. For such fluids, the nature of boundary conditions leading to well-posed problems is in general different from the Newtonian case. There are two reasons for this:

1. The memory of the fluid implies that what happens in the domain under consideration is dependent on the deformation history of the fluid before it entered the domain. Information about this deformation history must therefore be given in the form of boundary conditions at inflow boundaries. The precise nature of such inflow conditions is dependent on the constitutive relation; for example, fluids of Maxwell type [4] are different from fluids of Jeffreys type [5].
2. For fluids of Maxwell type, there is a change of type in the governing equations when the velocity of the fluid exceeds the propagation speed of shear waves (cf. [1], [2], [7], [8]). This necessitates a change in the nature of boundary conditions. If boundary conditions which would be correct in the subcritical case are imposed in a supercritical situation, the problem becomes ill-posed in a similar fashion as the Dirichlet problem for the wave equation (see [6]).

In the following, v denotes the velocity, p the pressure, $\mathbf{T}$ the extra stress tensor, ρ the density and f a given body force. The equation of motion is

$$\rho(v \cdot \nabla)v = \text{div } \mathbf{T} - \nabla p + f, \tag{1}$$

and the incompressibility condition is

$$\text{div } v = 0. \tag{2}$$

*Department of Mathematics and ICAM, Virginia Tech, Blacksburg, VA 24061-0123. This research was completed while I was visiting the Institute for Mathematics and its Applications at the University of Minnesota. Financial support from the IMA and from the National Science Foundation under Grant No. DMS-8796241 is gratefully acknowledged.

We assume a Maxwell-type constitutive relation of the following form:

$$((v\cdot\nabla)+\lambda)T_{ij}-\frac{\partial v_i}{\partial x_k}T_{kj}-T_{ik}\frac{\partial v_j}{\partial x_k}+P_{ik}(\mathbf{T})(\frac{\partial v_k}{\partial x_j}+\frac{\partial v_j}{\partial x_k})+P_{jk}(\mathbf{T})(\frac{\partial v_k}{\partial x_i}+\frac{\partial v_i}{\partial x_k})$$

$$+g_{ij}(\mathbf{T})=\mu(\frac{\partial v_i}{\partial x_j}+\frac{\partial v_j}{\partial x_i}). \tag{3}$$

Here λ and μ are positive constants, and the matrix-valued functions $\mathbf{P}$ and $\mathbf{g}$ are assumed to be smooth; moreover, $\mathbf{P}$, $\mathbf{g}$ and the first derivatives of $\mathbf{g}$ vanish at $\mathbf{T}=\mathbf{0}$. Equation (3) includes a number of popular rheological models (cf. e.g. [3]).

The domain on which we want to solve (1)-(3) is the strip bounded by the planes $x_1=0$ and $x_1=1$. In the x_2- and x_3-directions, we assume periodicity with periods L and M. The solutions we seek are small perturbations of the uniform flow $v=(V,0,0)$, $p=0$, $\mathbf{T}=0$. The given body force and the imposed boundary conditions are assumed to satisfy smallness conditions consistent with this. In [4], we considered this problem under the assumption that $\rho V^2<\mu$. A well-posed boundary value problem was obtained by prescribing the velocities at both boundaries plus additional stress conditions at the inflow boundary $x_1=0$. In two dimensions it is possible to prescribe the diagonal components T_{11} and T_{22}. In three dimensions a correct choice of inflow stress conditions was obtained as follows. We expand each stress component in a Fourier series, e.g.

$$T_{11}(0,x_2,x_3)=\sum_{k,l}t_{11}^{kl}\exp(2\pi i(kx_2/L+lx_3/M)). \tag{4}$$

Then one can, for example, prescribe the following inflow conditions:

$$t_{11}^{kl},\ t_{22}^{kl},\ t_{13}^{kl},\ t_{33}^{kl}\text{ if }|k|>>|l|,$$

$$t_{11}^{kl},\ t_{22}^{kl},\ t_{12}^{kl},\ t_{33}^{kl}\text{ if }|l|>>|k|,$$

$$t_{11}^{kl},\ t_{13}^{kl},\ t_{23}^{kl},\ t_{33}^{kl}\text{ if }|k|\text{ and }|l|\text{ are comparable,}$$

$$t_{11}^{kl},\ t_{22}^{kl},\ t_{23}^{kl},\ t_{33}^{kl}\text{ if }k=l=0. \tag{5}$$

If, on the other hand, $\rho V^2>\mu$, then this choice of boundary conditions does not lead to a well-posed problem [6]. We shall show that, in this case, one can prescribe the following conditions: the inflow stresses as above, the normal velocity at both boundaries, plus the vorticity and its normal derivative (in two dimensions), or, respectively, the second and third components of the vorticity and their normal derivatives (in three dimensions) at the inflow boundary. The analysis for two space dimensions will be carried out in Section 2; the modifications needed for three dimensions will be discussed in Section 3.

2. The two-dimensional case. We apply the operation $(v\cdot\nabla)+\lambda+(\nabla v)^T$ to the equation of motion (1) and we use equation (3) to reexpress $((v\cdot\nabla)+\lambda)\mathbf{T}$. After some algebra, this yields an equation of the following form (written in components)

$$\rho((v\cdot\nabla)+\lambda)(v\cdot\nabla)v_i = \mu\Delta v_i + (T_{kj}-P_{kj}(\mathbf{T}))\frac{\partial^2 v_i}{\partial x_j\partial x_k}$$

$$-P_{ik}(\mathbf{T})\frac{\partial^2 v_k}{\partial x_j^2} - P_{jk}(\mathbf{T})\frac{\partial^2 v_k}{\partial x_i\partial x_j} - \frac{\partial q}{\partial x_i}$$

$$+((v\cdot\nabla)+\lambda)f_i + \frac{\partial v_j}{\partial x_i}f_j + h_i(v,\nabla v,\mathbf{T},\nabla\mathbf{T}). \tag{6}$$

Here we have set $q=((v\cdot\nabla)+\lambda)p$. The term h is a complicated expression which we do not write out explicitly; it contains only quadratic and higher order terms.

Next we introduce a streamfunction-vorticity formulation. We set

$$v_1 = -\frac{\partial\psi}{\partial x_2},\ v_2 = \frac{\partial\psi}{\partial x_1},\ \zeta = \frac{\partial v_2}{\partial x_1} - \frac{\partial v_1}{\partial x_2}, \tag{7}$$

so that the incompressibility condition (2) is automatically satisfied and

$$\Delta\psi = \zeta. \tag{8}$$

We take the curl of equation (6), which results in

$$\rho((v\cdot\nabla)+\lambda)(v\cdot\nabla)\zeta = \mu\Delta\zeta + (T_{kj}-P_{kj}(\mathbf{T}))\frac{\partial^2\zeta}{\partial x_j\partial x_k} - P_{22}(\mathbf{T})\frac{\partial^2\zeta}{\partial x_1^2} - P_{11}(\mathbf{T})\frac{\partial^2\zeta}{\partial x_2^2}$$

$$+(P_{12}(\mathbf{T})+P_{21}(\mathbf{T}))\frac{\partial^2\zeta}{\partial x_1\partial x_2} + r(v,\nabla v,\nabla^2 v,\mathbf{T},\nabla\mathbf{T},\nabla^2\mathbf{T},f,\nabla f,\nabla^2 f). \tag{9}$$

Here r is again a complicated expression which we do not write out explicitly.

In the following, we shall solve (1)-(3) subject to the following boundary conditions:

$$\psi = -Vx_2+\psi_0,\ \text{on } x_1=0,\ \psi=-Vx_2+\psi_1,\ \text{on } x_1=1,$$

$$\zeta=\zeta_0,\ \text{on } x_1=0,\ \frac{\partial\zeta}{\partial x_1}=\eta_0,\ \text{on } x_1=0,$$

$$T_{11}=t_1,\ \text{on } x_1=0,\ T_{22}=t_2 \text{ on } x_1=0. \tag{10}$$

We note that prescribing ψ on both boundaries is equivalent to prescribing the normal velocity on these boundaries as well as the total flow rate in the x_2-direction.

We denote by H^s the space of all functions on the strip $0\le x_1\le 1$ which are periodic with period L in the x_2-direction and have s derivatives which are square integrable over one period. Sobolev spaces of periodic functions depending only on x_2 are denoted by $H^{\langle s\rangle}$. The corresponding norms are denoted by $\|\cdot\|_s$ and $\|\cdot\|_{\langle s\rangle}$. Moreover, $\|\cdot\|_{k,l}$ denotes the norm in $W^{k,\infty}([0,1];H^{\langle l\rangle})$.

The goal of this section is the following existence and uniqueness result:

THEOREM. *Assume that* $\|f\|_4$, $\|\psi_0\|_{\langle 9/2\rangle}$, $\|\psi_1\|_{\langle 9/2\rangle}$, $\|\zeta_0\|_{\langle 3\rangle}$, $\|\eta_0\|_{\langle 2\rangle}$, $\|t_1\|_{\langle 3\rangle}$ *and* $\|t_2\|_{\langle 3\rangle}$ *are sufficiently small. Then there is a solution of (1)-(3) which satisfies the boundary conditions (10) and the regularity* $\psi + Vx_2 \in H^5$, $\mathbf{T} \in H^3$. *Moreover, this solution is the only one for which* $\|\psi + Vx_2\|_5$, *and* $\|\mathbf{T}\|_3$ *are small.*

The construction of the solution is based on an iterative scheme. As a starting value for the iteration we use the uniform flow

$$\psi^0 = -Vx_2,\ \zeta^0 = 0,\ \mathbf{T}^0 = 0. \tag{11}$$

Given ψ^n, ζ^n and $\mathbf{T}^n$, we define v^n by

$$v_1^n = -\frac{\partial \psi^n}{\partial x_2},\ v_2^n = \frac{\partial \psi^n}{\partial x_1}. \tag{12}$$

Next, we determine $\mathbf{T}^{n+1}$ from the equation

$$((v^n\cdot\nabla)+\lambda)T_{ij}^{n+1} - \frac{\partial v_i^n}{\partial x_k}T_{kj}^n - T_{ik}^n\frac{\partial v_j^n}{\partial x_k} + P_{ik}(\mathbf{T}^n)(\frac{\partial v_k^n}{\partial x_j} + \frac{\partial v_j^n}{\partial x_k}) + P_{jk}(\mathbf{T}^n)(\frac{\partial v_k^n}{\partial x_i} + \frac{\partial v_i^n}{\partial x_k})$$

$$+g_{ij}(\mathbf{T}^n) = \mu(\frac{\partial v_i^n}{\partial x_j} + \frac{\partial v_j^n}{\partial x_i}), \tag{13}$$

subject to the following initial conditions at $x_1 = 0$:

$$T_{11}^{n+1} = t_1,\ T_{22}^{n+1} = t_2, \tag{14}$$

$$\rho \text{ curl } ((v^n\cdot\nabla)v^n) = \text{curl } (\text{div } \mathbf{T}^{n+1} + f), \tag{15}$$

Then we determine ζ^{n+1} from the initial-value problem

$$\rho((v^n\cdot\nabla)+\lambda)(v^n\cdot\nabla)\zeta^{n+1} = \mu\Delta\zeta^{n+1} + (T_{kj}^{n+1} - P_{kj}(\mathbf{T}^{n+1}))\frac{\partial^2\zeta^{n+1}}{\partial x_j \partial x_k}$$

$$-P_{22}(\mathbf{T}^{n+1})\frac{\partial^2\zeta^{n+1}}{\partial x_1^2} - P_{11}(\mathbf{T}^{n+1})\frac{\partial^2\zeta^{n+1}}{\partial x_2^2} + (P_{12}(\mathbf{T}^{n+1}) + P_{21}(\mathbf{T}^{n+1}))\frac{\partial^2\zeta^{n+1}}{\partial x_1\partial x_2}$$

$$+r(v^n, \nabla v^n, \nabla^2 v^n, \mathbf{T}^{n+1}, \nabla\mathbf{T}^{n+1}, \nabla^2\mathbf{T}^{n+1}, f, \nabla f, \nabla^2 f). \tag{16}$$

$$\zeta^{n+1} = \zeta_0, \text{ on } x_1 = 0,\ \frac{\partial\zeta^{n+1}}{\partial x_1} = \eta_0, \text{ on } x_1 = 0. \tag{17}$$

Finally, we obtain ψ^{n+1} from the Dirichlet problem

$$\Delta\psi^{n+1} = \zeta^{n+1},$$

$$\psi^{n+1} = -Vx_2 + \psi_0, \text{ on } x_1 = 0, \ \psi^{n+1} = -Vx_2 + \psi_1, \text{ on } x_1 = 1. \tag{18}$$

We define

$$X(M) = \{(\psi, \zeta, \mathbf{T}) \mid \|\psi + Vx_2\|_5 + \|\zeta\|_3 + \|\mathbf{T}\|_3 \le M\}. \tag{19}$$

The space $X(M)$ is complete under the metric

$$d((\psi, \zeta, \mathbf{T}), (\hat{\psi}, \hat{\zeta}, \hat{\mathbf{T}})) = \|\psi - \hat{\psi}\|_4 + \|\zeta - \hat{\zeta}\|_2 + \|\mathbf{T} - \hat{\mathbf{T}}\|_2. \tag{20}$$

We choose M small relative to 1, but sufficiently large relative to the norms of the prescribed data. In order to prove the theorem, it is sufficient to show that the mapping defined by the iteration (12)-(18) is a contraction in $X(M)$. We begin by showing that the iteration maps $X(M)$ into itself. Let us assume that $(\psi^n, \zeta^n, \mathbf{T}^n)$ lies in $X(M)$.

We first discuss the solution of (13)-(15). A rearrangement of (15) yields

$$\frac{\partial^2}{\partial x_1 \partial x_2}(T_{22}^{n+1} - T_{11}^{n+1}) + (\frac{\partial^2}{\partial x_1^2} - \frac{\partial^2}{\partial x_2^2})T_{12}^{n+1} = \rho \text{ curl } ((v^n \cdot \nabla)v^n) - \text{ curl } f. \tag{21}$$

We can use (13) to express x_1-derivatives of the stresses, i.e.

$$\frac{\partial}{\partial x_1} T_{ij}^{n+1} = \frac{1}{v_1^n}\left[-v_2^n \frac{\partial}{\partial x_2} T_{ij}^{n+1} - \lambda T_{ij}^{n+1} \ ...\right]. \tag{22}$$

After substituting (22) in (21), we obtain an ODE from which we can determine T_{12}^{n+1} at the inflow boundary $x_1 = 0$. We denote this boundary value by t^{n+1}. The following estimate is immediate

$$\|t^{n+1}\|_{\langle 3 \rangle} \le C(\|t_1\|_{\langle 3 \rangle} + \|t_2\|_{\langle 3 \rangle} + \|w^n\|_4 + \|w^n\|_4 \|\mathbf{T}^n\|_3 + \|\mathbf{T}^n\|_3^2). \tag{23}$$

Here we have set $w^n = v^n - (V, 0)$. After determining t^{n+1}, we have a full set of initial conditions to solve (13). Using standard energy estimates for hyperbolic equations (see [4] for some more details) we obtain a unique solution which satisfies an estimate of the form

$$\|\mathbf{T}^{n+1}\|_{3,0} + \|\mathbf{T}^{n+1}\|_{2,1} + \|\mathbf{T}^{n+1}\|_{1,2} + \|\mathbf{T}^{n+1}\|_{0,3}$$

$$\le C(\|t_1\|_{\langle 3 \rangle} + \|t_2\|_{\langle 3 \rangle} + \|t^{n+1}\|_{\langle 3 \rangle} + \|w^n\|_4 + \|\mathbf{T}^n\|_3 \|w^n\|_4 + \|\mathbf{T}^n\|_3^2). \tag{24}$$

From equation (13), we can see that an expression like the one on the right hand side of (24) also provides an upper bound for $\|(v^n \cdot \nabla)\mathbf{T}^{n+1}\|_3$.

For the solution of the initial-value problem (16), (17), one readily obtains the estimate

$$\|\zeta^{n+1}\|_{0,2} + \|\zeta^{n+1}\|_{1,1} + \|\zeta^{n+1}\|_{2,0} \le C(\|\zeta_0\|_{\langle 2 \rangle} + \|\eta_0\|_{\langle 1 \rangle} + \|r^n\|_{1,0} + \|r^n\|_{0,1}). \tag{25}$$

This is insufficient because we need to estimate third order derivatives of ζ^{n+1}. Results available in the literature would require the existence of a higher order derivative of r^n either with respect to x_1 or with respect to x_2. We cannot use such an assumption because of the dependence of r on second derivatives of $\mathbf{T}^{n+1}$. However, because of the bound for $\|(v^n \cdot \nabla)\mathbf{T}^{n+1}\|_3$, we can get bounds on $\|(v^n \cdot \nabla)r^n\|_1$. Instead of differentiating (16) with respect to either x_1 or x_2, which is what is usually done, we can apply the operation $(v^n \cdot \nabla)$ to it. By doing this and deriving energy estimates in the usual fashion, we obtain an estimate of the form

$$\|\zeta\|_3 \leq C(\|\zeta_0\|_{\langle 3\rangle} + \|\eta_0\|_{\langle 2\rangle} + \|r^n\|_{0,1} + \|r^n\|_{1,0} + \|(v^n \cdot \nabla)r^n\|_1.) \tag{26}$$

By taking into account the form of r, we can estimate the last three terms in (26) by a constant times

$$\|f\|_4 + (\|w^n\|_4 + \|\mathbf{T}^{n+1}\|_{3,0} + \|\mathbf{T}^{n+1}\|_{2,1} + \|\mathbf{T}^{n+1}\|_{1,2} + \|\mathbf{T}^{n+1}\|_{0,3} + \|(v^n \cdot \nabla)\mathbf{T}^{n+1}\|_3)^2. \tag{27}$$

Finally, from (18) we immediately obtain

$$\|\psi^{n+1} + Vx_2\|_5 \leq C(\|\psi_0\|_{\langle 9/2\rangle} + \|\psi_1\|_{\langle 9/2\rangle} + \|\zeta^{n+1}\|_3). \tag{28}$$

The claim that the iteration maps $X(M)$ into itself now follows easily by combining the estimates (23)-(28).

The derivation of estimates to show that the mapping defined by the iteration is a contraction is fairly straightforward, and we shall only demonstrate one step. From (16), (17) we obtain

$$\begin{aligned}
&\rho((v^n \cdot \nabla) + \lambda)(v^n \cdot \nabla)(\zeta^{n+1} - \zeta^n) = \mu\Delta(\zeta^{n+1} - \zeta^n) \\
&+(T_{kj}^{n+1} - P_{kj}(\mathbf{T}^{n+1}))\frac{\partial^2(\zeta^{n+1} - \zeta^n)}{\partial x_j \partial x_k} - P_{22}(\mathbf{T}^{n+1})\frac{\partial^2(\zeta^{n+1} - \zeta^n)}{\partial x_1^2} \\
&-P_{11}(\mathbf{T}^{n+1})\frac{\partial^2(\zeta^{n+1} - \zeta^n)}{\partial x_2^2} + (P_{12}(\mathbf{T}^{n+1}) + P_{21}(\mathbf{T}^{n+1}))\frac{\partial^2(\zeta^{n+1} - \zeta^n)}{\partial x_1 \partial x_2} \\
&-\rho\Big[((v^n \cdot \nabla) + \lambda)(v^n \cdot \nabla) - ((v^{n-1} \cdot \nabla) + \lambda)(v^{n-1} \cdot \nabla)\Big]\zeta^n \\
&+[(T_{kj}^{n+1} - P_{kj}(\mathbf{T}^{n+1})) - (T_{kj}^n - P_{kj}(\mathbf{T}^n))]\frac{\partial^2 \zeta^n}{\partial x_j \partial x_k} \\
&-[P_{22}(\mathbf{T}^{n+1}) - P_{22}(\mathbf{T}^n)]\frac{\partial^2 \zeta^n}{\partial x_1^2} - [P_{11}(\mathbf{T}^{n+1}) - P_{11}(\mathbf{T}^n)]\frac{\partial^2 \zeta^n}{\partial x_2^2} \\
&+[(P_{12}(\mathbf{T}^{n+1}) + P_{21}(\mathbf{T}^{n+1})) - (P_{12}(\mathbf{T}^n) + P_{21}(\mathbf{T}^n))]\frac{\partial^2 \zeta^n}{\partial x_1 \partial x_2} + r^n - r^{n-1},
\end{aligned} \tag{29}$$

$$\zeta^{n+1} - \zeta^n = 0, \text{ on } x_1 = 0, \quad \frac{\partial(\zeta^{n+1} - \zeta^n)}{\partial x_1} = 0, \text{ on } x_1 = 0. \tag{30}$$

Energy estimates now yield

$$\|\zeta^{n+1} - \zeta^n\|_{0,2} + \|\zeta^{n+1} - \zeta^n\|_{1,1} + \|\zeta^{n+1} - \zeta^n\|_{2,0}$$

(31)

$$\leq C(\|\zeta^n\|_3(\|v^n - v^{n-1}\|_3 + \|\mathbf{T}^{n+1} - \mathbf{T}^n\|_2) + \|r^n - r^{n-1}\|_{0,0} + \|(v^n \cdot \nabla)(r^n - r^{n-1})\|_0).$$

We note that a bound on $\|\zeta^n\|_3$ has already been obtained. In order to deal with the last term in (31), we note that

$$(32)\quad (v^n \cdot \nabla)(r^n - r^{n-1}) = (v^n \cdot \nabla)r^n - (v^{n-1} \cdot \nabla)r^{n-1} + ((v^n - v^{n-1}) \cdot \nabla)r^{n-1}.$$

By taking into account the form of r and equation (13), it is easy to estimate the terms on the right hand side of (32).

3. Modifications in three dimensions. The basic iteration scheme used to construct solutions and the function spaces chosen for the analysis will be as in the two-dimensional case, and we shall therefore confine the following discussion to those points where modifications are needed. One of these changes is that $\zeta = \operatorname{curl} v$ is now a vector, and the equation analogous to (9), written in components, is

$$\rho((v \cdot \nabla) + \lambda)(v \cdot \nabla)\zeta_i = \mu \Delta \zeta_i + (T_{kj} - P_{kj}(\mathbf{T}))\frac{\partial^2 \zeta_i}{\partial x_j \partial x_k} - \epsilon_{mli}\epsilon_{krs}P_{mk}(\mathbf{T})\frac{\partial^2 \zeta_s}{\partial x_l \partial x_r}$$

$$(33)\qquad + r_i(v, \nabla v, \nabla^2 v, \mathbf{T}, \nabla \mathbf{T}, \nabla^2 \mathbf{T}, f, \nabla f, \nabla^2 f).$$

This is a system of PDE's for the components of ζ, and in order to make it symmetric hyperbolic, we add the assumption that the matrix $\mathbf{P}$ is symmetric. In the iteration, we use the following equation which is analogous to (16):

$$\rho((v^n \cdot \nabla) + \lambda)(v^n \cdot \nabla)\zeta_i^{n+1} = \mu \Delta \zeta_i^{n+1} + (T_{kj}^{n+1} - P_{kj}(\mathbf{T}^{n+1}))\frac{\partial^2 \zeta_i^{n+1}}{\partial x_j \partial x_k}$$

$$-\epsilon_{mli}\epsilon_{krs}P_{mk}(\mathbf{T}^{n+1})\frac{\partial^2 \zeta_s^{n+1}}{\partial x_l \partial x_r}$$

$$(34)\qquad + r_i(v^n, \nabla v^n, \nabla^2 v^n, \mathbf{T}^{n+1}, \nabla \mathbf{T}^{n+1}, \nabla^2 \mathbf{T}^{n+1}, f, \nabla f, \nabla^2 f).$$

The velocity can be determined in terms of the vorticity if we prescribe the normal velocity on both boundaries and the mean flux in the y- and z-directions. Unfortunately, we shall not be able to guarantee that all the iterates satisfy $\operatorname{div} \zeta^{n+1} = 0$, and hence we cannot simply use the equation $\operatorname{curl} v^{n+1} = \zeta^{n+1}$. Let Π denote the orthogonal projection (in L^2) onto the subspace

$$(35)\qquad V = \{\zeta \in L^2 \mid \operatorname{div} \zeta = 0, \int_0^L \int_0^M \zeta_1(\cdot, x_2, x_3)\, dx_3\, dx_2 = 0\}.$$

The set of equations determining v^{n+1} is

$$\text{curl } v^{n+1} = \Pi\zeta^{n+1}, \text{ div } v^{n+1} = 0,$$

$$v_1^{n+1} = V + a(x_2, x_3), \text{ at } x_1 = 0, \; v_1^{n+1} = V + b(x_2, x_3) \text{ at } x_1 = 1,$$

$$\int_0^1 \int_0^L \int_0^M v_2^{n+1} \, dx_3 \, dx_2 \, dx_1 = \alpha, \; \int_0^1 \int_0^L \int_0^M v_3^{n+1} \, dx_3 \, dx_2 \, dx_1 = \beta. \tag{36}$$

Here the numbers α and β as well as the functions a and b are prescribed and $|\alpha| + |\beta| + \|a\|_{\langle 7/2 \rangle} + \|b\|_{\langle 7/2 \rangle}$ is assumed to be small. Moreover, we have to assume the compatibility condition

$$\int_0^L \int_0^M a(x_2, x_3) \, dx_3 \, dx_2 = \int_0^L \int_0^M b(x_2, x_3) \, dx_3 \, dx_2. \tag{37}$$

It can easily be shown along the same lines as in Section 2 that by combining (34), (36) with (13) and (small) inflow data for $\mathbf{T}$, ζ and $\partial\zeta/\partial x_1$, we obtain a convergent iteration. However, there are two problems:

1. It is not guaranteed that the limit of the iteration satisfies $\zeta = \text{curl } v$, or, equivalently, $\zeta = \Pi\zeta$.
2. It is not guaranteed that the original equation of motion (1) holds. This is because in proceeding from (1) to (6) we have applied the operation $(v \cdot \nabla) + \lambda + (\nabla v)^T$. In order to reverse this step and go from (6) to (1), we have to assume that (1) holds on the inflow boundary $x_1 = 0$ (cf. [4]). In two dimensions we imposed this condition as equation (15).

In order to remove these two difficulties, we must restrict the inflow data; i.e., only part of these data can be prescribed, and the rest have to be determined at each step of the iteration.

We take the divergence of equation (33). If ζ were equal to curl v, we would get 0 (to see this, recall that (33) was derived by taking the curl of (6)). Hence we find (we set curl $v = \omega$)

$$\begin{aligned} &\frac{\partial}{\partial x_i} \Big[\rho((v \cdot \nabla) + \lambda)(v \cdot \nabla)(\zeta_i - \omega_i) - \mu\Delta(\zeta_i - \omega_i) \\ &-(T_{kj} - P_{kj}(\mathbf{T})) \frac{\partial^2(\zeta_i - \omega_i)}{\partial x_j \partial x_k} + \epsilon_{mli}\epsilon_{krs} P_{mk}(\mathbf{T}) \frac{\partial^2(\zeta_s - \omega_s)}{\partial x_l \partial x_r} \Big] = 0. \end{aligned} \tag{38}$$

After some algebra, this yields

$$\begin{aligned} &\Big[\rho((v \cdot \nabla) + \lambda)(v \cdot \nabla) - \mu\Delta - (T_{kj} - P_{kj}(\mathbf{T})) \frac{\partial^2}{\partial x_j \partial x_k} \Big] \text{ div } \zeta \\ &= D^2(v, \nabla v, \nabla^2 v, \mathbf{T}, \nabla \mathbf{T})(\zeta - \omega). \end{aligned} \tag{39}$$

Here D^2 is a second order differential operator with coefficients depending on the arguments indicated. Let d_1 denote the value of div ζ at $x_1 = 0$, and let d_2 denote the value of $\frac{\partial}{\partial x_1}$ div ζ at $x_1 = 0$. From (39), we obtain the estimate

$$\|\text{div } \zeta\|_{0,1} + \|\text{div } \zeta\|_{1,0} \leq C(\|d_1\|_{\langle 1\rangle} + \|d_2\|_{\langle 0\rangle} + (\|w\|_4 + \|\mathbf{T}\|_3)\|\zeta - \omega\|_2). \tag{40}$$

As before, w denotes $v - (V, 0, 0)$. We note that $\|w\|_4 + \|\mathbf{T}\|_3$ is small. Moreover, (36) yields the estimate

$$\|\zeta - \omega\|_2 \leq C\|\text{div } \zeta\|_1 + \left|\int_0^L \int_0^M \zeta_1(0, x_2, x_3)\, dx_3\, dx_2\right|. \tag{41}$$

Inflow conditions are now handled as follows. We prescribe arbitrary data for ζ_2, ζ_3 and their normal derivatives. The initial datum for $\frac{\partial \zeta_1}{\partial x_1}$ is then determined by requiring that div $\zeta = 0$ at $x_1 = 0$. Finally, the initial datum for ζ_1 cannot be determined a priori, but must be computed at each step of the iteration. We require that, at $x_1 = 0$,

$$\frac{\partial^2 \zeta_1^{n+1}}{\partial x_1^2} + \frac{\partial^2 \zeta_2^{n+1}}{\partial x_2 \partial x_1} + \frac{\partial^2 \zeta_3^{n+1}}{\partial x_3 \partial x_1} = s_{n+1}, \tag{42}$$

where s_{n+1} is a constant to be determined. We then solve (42) for $\partial^2 \zeta_1^{n+1}/\partial x_1^2$ and substitute into the first equation of (34). This yields an elliptic problem from which we can uniquely determine the inflow datum for ζ_1^{n+1} up to an arbitrary constant, as well as the constant s_{n+1}. Finally, the arbitrary constant in ζ_1^{n+1} is fixed by the requirement that

$$\int_0^L \int_0^M \zeta_1^{n+1}(0, x_2, x_3)\, dx_3\, dx_2 = 0. \tag{43}$$

For the limit of the iteration, this obviously insures that $d_1 = 0$, $d_2 = s$, and $\int_0^L \int_0^M \zeta_1(0, x_2, x_3)\, dx_3\, dx_2 = 0$. To determine the constant s, we take the first equation of (33), set $x_1 = 0$, and integrate over x_2 and x_3. If ζ is replaced by $\omega =$ curl v, we obtain an expression which vanishes identically (recall that (33) was derived by taking the curl of (6)). Hence we find

$$\begin{aligned}\int_0^L \int_0^M \Big[&\rho((v \cdot \nabla) + \lambda)(v \cdot \nabla)(\zeta_1 - \omega_1) - \mu\Delta(\zeta_1 - \omega_1)\\ &-(T_{kj} - P_{kj}(\mathbf{T}))\frac{\partial^2(\zeta_1 - \omega_1)}{\partial x_j \partial x_k}\\ &+\epsilon_{ml1}\epsilon_{krs}P_{mk}(\mathbf{T})\frac{\partial^2(\zeta_s - \omega_s)}{\partial x_l \partial x_r}\Big](0, x_2, x_3)\, dx_3\, dx_2 = 0.\end{aligned} \tag{44}$$

Next we integrate by parts in all terms which involve second order derivatives of $\zeta - \omega$ such that one of the differentiations is with respect to x_2 or x_3. This leads to terms which can be estimated by a constant times $(\|\mathbf{T}\|_3 + \|w\|_4)\|\zeta - \omega\|_2$. The term which remains is the integral of

$$\rho v_1^2 \frac{\partial^2}{\partial x_1^2}(\zeta_1 - \omega_1) + \lambda \rho v_1 \frac{\partial}{\partial x_1}(\zeta_1 - \omega_1) - \mu \frac{\partial^2}{\partial x_1^2}(\zeta_1 - \omega_1)$$

$$-(T_{11} - P_{11}(\mathbf{T}))\frac{\partial^2}{\partial x_1^2}(\zeta_1 - \omega_1). \tag{45}$$

We now note that

$$\frac{\partial}{\partial x_1}(\zeta_1 - \omega_1) = -\frac{\partial}{\partial x_2}(\zeta_2 - \omega_2) - \frac{\partial}{\partial x_3}(\zeta_3 - \omega_3),$$

$$\frac{\partial^2}{\partial x_1^2}(\zeta_1 - \omega_1) = -\frac{\partial^2}{\partial x_2 \partial x_1}(\zeta_2 - \omega_2) - \frac{\partial^2}{\partial x_3 \partial x_3}(\zeta_3 - \omega_3) + s. \tag{46}$$

By using this, we obtain again terms which can, after an integration by parts, be estimated by a constant times $(\|\mathbf{T}\|_3 + \|w\|_4)\|\zeta - \omega\|_2$, plus s times the integral of $\rho v_1^2 - \mu - T_{11} + P_{11}(\mathbf{T})$. As a result, s can be estimated by a constant times $(\|\mathbf{T}\|_3 + \|w\|_4)\|\zeta - \omega\|_2$. In conjunction with (40) and (41) this yields that div ζ is indeed zero.

To make sure that (1) is satisfied, we proceed as in [4]. At each step of the iteration, q^{n+1} is determined by the relation

$$\nabla q^{n+1} = \Sigma\Big[((v^n \cdot \nabla) + \lambda + (\nabla v^n)^T)(\text{div } \mathbf{T}^n + f - \rho(v^n \cdot \nabla)v^n)\Big]. \tag{47}$$

Here Σ is the orthogonal projection of L^2 onto the subspace of vectorfields with vanishing curl. From (47), q^{n+1} is determined up to an arbitrary constant; we may fix this constant by requiring that

$$\int_0^1 \int_{-L/2}^{L/2} \int_{-M/2}^{M/2} q^{n+1}(x_1, x_2, x_3)\, dx_3\, dx_2\, dx_1 = 0. \tag{48}$$

We note that q^{n+1} is not necessarily periodic in the x_2 and x_3-directions, but may contain a part which is linear in x_2 and x_3. At $x_1 = 0$, we impose the condition

$$\rho(v^n \cdot \nabla)v^n = \text{div } \mathbf{T}^{n+1} - \nabla p^{n+1} + f. \tag{49}$$

This condition, in conjunction with (13) and the equation

$$(v^n \cdot \nabla)p^{n+1} + \lambda p^{n+1} = q^{n+1} \tag{50}$$

can be used to express some of the inflow data for $\mathbf{T}$ in terms of others. For details we refer to [4]. Specifically we can prescribe the stress components specified in (5) and solve for the rest.

Let us summarize the iteration scheme. We prescribe the following boundary data a priori: the normal velocities on both boundaries and the total flux in the x_2- and x_3-directions according to (35); the second and third components of the vorticity and their normal derivatives at the inflow boundary,

$$\zeta_2(0,x_2,x_3) = \zeta_2^0(x_2,x_3),\ \zeta_3(0,x_2,x_3) = \zeta_3^0(x_2,x_3),$$

$$\frac{\partial \zeta_2}{\partial x_1}(0,x_2,x_3) = \zeta_2^1(x_2,x_3),\ \frac{\partial \zeta_3}{\partial x_1}(0,x_2,x_3) = \zeta_3^1(x_2,x_3),$$

$$\frac{\partial \zeta_1}{\partial x_1}(0,x_2,x_3) = -\frac{\partial \zeta_2^0}{\partial x_2}(x_2,x_3) - \frac{\partial \zeta_3^0}{\partial x_3}(x_2,x_3); \tag{51}$$

and the inflow stresses according to (5). We denote this prescribed part of the stress by $\mathbf{T}_p$. We start the iteration by setting $\mathbf{T}=0$, $\zeta=0$, $v=(V,0,0)$. At each step of the iteration, we first determine q^{n+1} from (47), (48). Then we calculate the inflow boundary value of $\mathbf{T}^{n+1}$ from (5), (49), (50) and (13). We can now determine $\mathbf{T}^{n+1}$ from (13). Next we use (42), (43) and (34) to determine the inflow value of ζ_1^{n+1}. Then we determine ζ^{n+1} from (34) and v^{n+1} from (36).

The existence theorem thus obtained is the following:

THEOREM:. *Assume that* $\|f\|_4$, $\|a\|_{\langle 7/2\rangle}$, $\|b\|_{\langle 7/2\rangle}$, $\|\zeta_2^0\|_{\langle 3\rangle}$, $\|\zeta_3^0\|_{\langle 3\rangle}$, $\|\zeta_2^1\|_{\langle 2\rangle}$, $\|\zeta_3^1\|_{\langle 2\rangle}$, $\|\mathbf{T}_p\|_{\langle 3\rangle}$, $|\alpha|$ *and* $|\beta|$ *are sufficiently small. Assume, moreover, that the matrix function* $\mathbf{P}$ *has symmetric values. Then there is a solution of (1-3) which satisfies the boundary conditions given by (5), (35) and (51) and the regularity* $v \in H^4$, $\mathbf{T} \in H^3$. *Moreover, this solution is the only one for which* $\|v-(V,0,0)\|_4$ *and* $\|\mathbf{T}\|_3$ *are small.*

REFERENCES

[1] D.D. Joseph, M. Renardy and J.C. Saut, *Hyperbolicity and change of type in the flow of viscoelastic fluids*, Arch. Rat. Mech. Anal., 87 (1985), pp. 213–251.

[2] M. Luskin, *On the classification of some model equations for viscoelasticity*, J. Non-Newt. Fluid Mech., 16 (1984), pp. 3–11.

[3] J.G. Oldroyd, *Non-Newtonian effects in steady motion of some idealized elastico-viscous liquids*, Proc. Roy. Soc. London, A 245 (1958), pp. 278–297.

[4] M. Renardy, *Inflow boundary conditions for steady flow of viscoelastic fluids with differential constitutive laws*, Rocky Mt. J. Math., 18 (1988), pp. 445–453.

[5] *Recent advances in the mathematical theory of steady flows of viscoelastic fluids*, J. Non-Newt. Fluid Mech., 29 (1988), pp. 11–24.

[6] *Boundary conditions for steady flows of non-Newtonian fluids*, *Proc. Xth Int. Congr. Rheology (ed. P.H.T. Uhlherr), Vol. 2*, Sydney, 1988, pp. 202–204.

[7] I.M. Rutkevich, *The propagation of small perturbations in a viscoelastic fluid*, J. Appl. Math. Mech., 34 (1970), pp. 35–50.

[8] J.S. Ultman and M.M. Denn, *Anomalous heat transfer and a wave phenomenon in dilute polymer solutions*, Trans. Soc. Rheol., 14 (1970), pp. 307–317.

LOSS OF HYPERBOLICITY IN YIELD VERTEX PLASTICITY MODELS UNDER NONPROPORTIONAL LOADING

DAVID G. SCHAEFFER* AND MICHAEL SHEARER†

§0. Introduction

Several authors [5,8,10,12] have shown that the dynamic partial differential equations arising from continuum models for granular flow may be linearly ill-posed. In a typical theory with shear-strain hardening, the equations are linearly well-posed for small deformations but become linearly ill-posed at some critical deformation which occurs before the maximum shear stress is achieved. Before this critical deformation the dynamic equations are hyperbolic, but after this point the equations are of no definite type; rather they resemble $u_{tt} = u_{xx} - u_{yy}$, the wave equation with a rotated time axis, in that the possible uncontrolled growth of a plane wave depends on its direction of propagation.

The analyses to date have been restricted to the case of proportional loading. This term means that at every step the principal axes of the applied stress increment are parallel to those of the current stress; in other words, the stress axes are fixed in the material and never rotate. However, for yield vertex models such as in [3], the governing equations are fully nonlinear so that their type depends on the loading path.

In this paper we determine, to lowest order in the rotation rate, how the type of the equations is changed by nonproportional loading. Even for proportional loading, our analysis extends previous work by including the full Jaumann derivative. We also apply this analysis to explain the preferred orientation of fracture in shearing between parallel plates [1].

In §1 we present the equations to be analyzed, including a discussion of constitutive assumptions. Our results are formulated in §2 and proved in §3. (The application to shearing between parallel plates appears in §2.2.)

§1. Constitutive assumptions

1.1. Formulation of the equations. We begin with several caveats: (i) we restrict our attention to two space dimensions; (ii) we consider the equations without reference to boundary conditions that might be imposed; and (iii) we neglect elastic deformations. (It would be only a technical complication to include elastic effects.)

The unknowns describing granular flow consist of the density ρ, the velocity v_i, and the (Cauchy) stress tensor T_{ij} (with compressive stresses regarded as positive). These six unknowns are subject to conservation of mass and momentum,

$$\text{(1.1)} \qquad (a) \quad d_t\rho + \rho\partial_j v_j = 0 \qquad (b) \quad \rho d_t v_i + \partial_j T_{ij} = 0,$$

*Research supported under NSF Grants DMS 86-04141 and 88-04592. The latter includes funds from AFOSR. Department of Mathematics, Duke University, Durham, NC 27706

†Research supported under NSF Grant DMS 87-01348 and ARO Grant DAAL 03-88-K-0080.
Department of Mathematics, North Carolina State University, Raleigh, NC 27695

where $d_t = \partial_t + v_j\partial_j$ is the material derivative and we employ the summation convention. The unknowns are also subject to appropriate constitutive laws that we will formulate after the introduction of some notation.

The deformation rate tensor is defined by

$$V_{ij} = -\frac{1}{2}(\partial_i v_j + \partial_j v_i) \tag{1.2}$$

(note the minus sign), and the Jaumann stress rate is defined by

$$\nabla_t T_{ij} = d_t T_{ij} - \omega_{ki}T_{jk} - \omega_{kj}T_{ik} \tag{1.3}$$

where

$$\omega_{\ell m} = \frac{1}{2}(\partial_\ell v_m - \partial_m v_\ell) \tag{1.4}$$

is the antisymmetric part of the velocity gradient. As explained in Chapter VIII, §1 of [9], the final two terms on the right in (1.3) allow for changes in stress due to rotation of the material, yielding an objective measure of the rate of change of the stress.

Given any 2×2 matrix A, we can split it into its deviatoric and spherical parts,

$$A = \text{dev}A + (\frac{1}{2}trA)I. \tag{1.5}$$

We identify the space of symmetric 2×2 matrices with $\mathbb{R}^3$ using the coordinates

$$\left(\frac{A_{11} - A_{22}}{2}, A_{12}, \frac{A_{11} + A_{22}}{2}\right), \tag{1.6}$$

so that (1.5) results from the decomposition

$$\mathbb{R}^3 \cong \mathbb{R}^2 \oplus \mathbb{R}^1. \tag{1.7}$$

Let $\pi : O(2) \to \mathcal{L}(\mathbb{R}^3)$ denote the natural action of the orthogonal group on symmetric 2×2 matrices,

$$\pi(R) \cdot A = RAR^T. \tag{1.8}$$

Note that the decomposition (1.7) is invariant under this action. Applying (1.5) to the stress tensor, we obtain the mean stress and (scalar) shear stress

$$(1.9) \qquad \begin{aligned} (a) \qquad & \sigma = \frac{1}{2}trT = \frac{1}{2}(\sigma_1 + \sigma_2) \\ (b) \qquad & \tau = |\text{dev}T| = \frac{1}{2}(\sigma_1 - \sigma_2), \end{aligned}$$

where σ_1, σ_2 are the eigenvalues of T_{ij}, with $\sigma_1 > \sigma_2 > 0$, and $|\cdot|$ is the Euclidean norm,

$$|A| = (A, A)^{1/2} = \left\{ \frac{1}{2} \sum_{i,j} A_{ij}^2 \right\}^{1/2}. \tag{1.10}$$

Under conditions of continued loading, the ratio

$$\mu = \tau / \sigma \tag{1.11}$$

is called the coefficient of mobilized friction. In a typical constitutive test, this quantity varies with the total shear strain as sketched in Figure 1.1, but in the present context μ cannot be expressed as a function of the shear strain since the constitutive laws that we consider are path dependent. It follows from (1.9) that $\mu < 1$ if, as we always assume, both principal stresses are compressive.

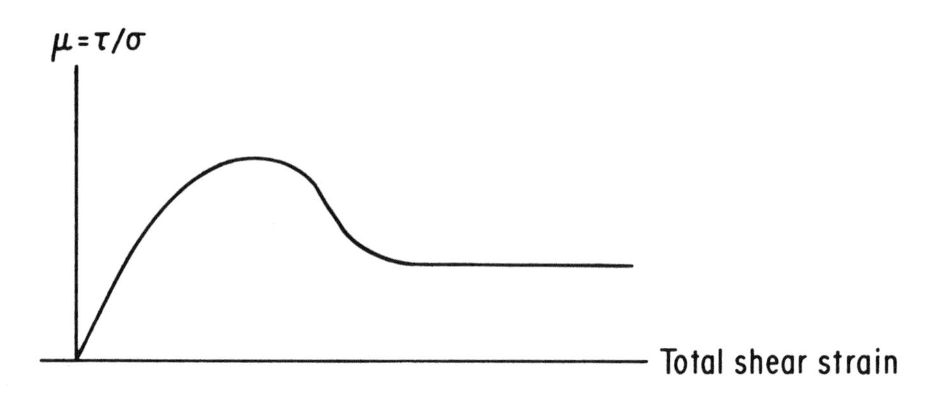

Figure 1.1: An illustration of shear strain hardening

The constitutive laws relate the stress and deformation rates. Assuming continued loading, we may write these laws as

$$\begin{aligned} &(a) \qquad \text{dev}V = \Psi(T, \nabla_t(\sigma^{-1}\text{dev}T)) \\ &(b) \qquad \frac{1}{2}\text{tr}V = -\beta|\text{dev}V| \end{aligned} \tag{1.12}$$

where β is a real parameter ($|\beta| < 1$) and $\Psi : \mathbf{R}^3 \times \mathbf{R}^2 \to \mathbf{R}^2$ is a smooth function (Note that $\nabla_t(\sigma^{-1}\text{dev}\, T)$ is trace free so the second argument of Ψ belongs to $\mathbf{R}^2$) with the following two properties:

(1) *Rate independence*: Ψ is homogeneous of degree 1 in its second argument, and

(2) *Isotropy*: For any rotation matrix $R \in O(2)$,

$$\Psi(\pi(R)\cdot T,\ \pi(R)\cdot A) = \pi(R)\cdot\Psi(T,A). \tag{1.13}$$

(In our notation we suppress the dependence of β and Ψ on history parameters such as the total shear strain. Inclusion of such parameters would not change the princiapl part, and hence the hyperbolicity or nonhyperbolicity, of the equations under study.) A simple example of a nonlinear function satisfying these hypotheses can be obtained by a slight modification, to make Ψ smooth, of a constitutive relation proposed in [16]; viz.,

$$\Psi(T,A) = \frac{\sigma}{G_p}\left\{PA + \alpha\frac{|(I-P)A|^2}{|A|}\frac{devT}{|devT|}\right\} + \frac{\sigma}{G_r}(I-P)A \tag{1.14}$$

where $P:\mathbb{R}^2 \to \mathbb{R}^2$ is the projection operator along the direction $devT$,

$$PA = \frac{1}{|devT|^2}(devT, A)devT$$

(cf. Figure 1.2), and G_p, G_r, and α are constants. The subscripts p and r are mnemonic for "proportional" and "rotating", respectively, and the constants G_p and G_r are the strength moduli of the material with respect to such loading. This terminology should be clarified by the discussion in §1.2 below. The appearance of $\nabla_t(\sigma^{-1}devT)$ in the second argument of Ψ will also be addressed there. Here we focus on working out the consequences of the above two assumptions.

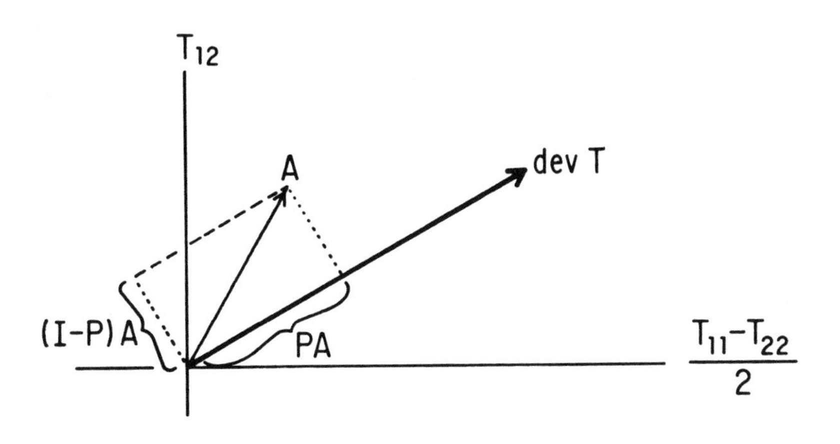

Figure 1.2: Projection onto the stress deviator

Because of the isotropy condition (1.13), we may rotate coordinates so that, at one specific point in space and time, the stress tensor T is diagonal and $T_{11} > T_{22} > 0$. In such a coordinate system we compute from (1.3) that

$$\nabla_t T = \begin{pmatrix} d_t(\frac{T_{11}-T_{22}}{2}) \\ d_t T_{12} - \tau(\partial_1 v_2 - \partial_2 v_1) \\ d_t(\frac{T_{11}+T_{22}}{2}) \end{pmatrix} \tag{1.15}$$

where τ is given by (1.9b). Thus

$$\nabla_t(\sigma^{-1}\text{dev}T) = \sigma^{-1}\begin{pmatrix} X_1 - \mu X_3 \\ X_2 \end{pmatrix} \tag{1.16}$$

where X_1, X_2, X_3 are the three components on the right in (1.15) and μ is given by (1.11). We conclude from homogeneity that

$$\Psi(T_{(d)}, \nabla_t(\sigma^{-1}\text{dev}T)) = \sigma^{-1}\Psi(T_{(d)}, (X_1 - \mu X_3, X_2)); \tag{1.17}$$

the supscript (d) on the first argument of Ψ is a reminder that (1.17) holds at a point where the stress tensor is diagonal. Incidentally, in proportional loading we have $X_2 = 0$, expressing the condition of no rotation in the stress axes; and, at least before the maximum shearing stress in Figure 1.1, we have

$$X_1 - \mu X_3 > 0. \tag{1.18}$$

We now apply the isotropy condition (1.13) using the reflection matrix

$$R = \begin{pmatrix} 1 & 0 \\ 0 & -1 \end{pmatrix}.$$

Since T is diagonal in the chosen coordinate system, $\pi(R) \cdot T = T$. On the other hand

$$\pi(R) \cdot \begin{pmatrix} X_1 \\ X_2 \\ X_3 \end{pmatrix} = \begin{pmatrix} X_1 \\ -X_2 \\ X_3 \end{pmatrix}.$$

Expressing (1.12a) in components

$$\frac{V_{11} - V_{22}}{2} = \sigma^{-1}\Psi_1(T, (X_1 - \mu X_3, X_2))$$
$$V_{12} = \sigma^{-1}\Psi_2(T, (X_1 - \mu X_3, X_2))$$

and applying (1.13), we conclude that Ψ_1 is even in X_2 and Ψ_2 is odd. Note that functions of the form

$$\begin{aligned} &(a) \quad \Psi_1(T, (X_1 - \mu X_3, X_2)) = (X_1 - \mu X_3)\psi_1(T, s^2) \\ &(b) \quad \Psi_2(T, (X_1 - \mu X_3, X_2)) = X_2\psi_2(T, s^2), \end{aligned} \tag{1.19}$$

where

$$s = \frac{X_2}{X_1 - \mu X_3} \tag{1.20}$$

is the slope of the stress path, satisfy the requirements of homogeneity and isotropy; since s is odd under the reflection R, only the square of this quantity appears in (1.19). Indeed, provided (1.18) holds, any function Ψ satisfying the above two conditions may be expressed in the form (1.19). Our analysis below will be on a small neighborhood of the proportional loading path in which (1.18) is satisfied. (Remark: The dependence of Ψ on history, which our notation suppresses, could invalidate the derivation of (1.19). Specifically, to obtain (1.19) we must assume that either (i) the history of the sample under study is invariant under the reflection R or (ii) Ψ depends on history through quantities such as the total shear stress γ which are invariant under reflections.)

Later we shall need the Taylor expansion of (1.19) near $X_2 = 0$. Retaining only $\mathcal{O}(s^2)$ terms in ψ_1 and $\mathcal{O}(1)$ terms in ψ_2, we derive the constitutive relations

$$(1.21) \qquad \begin{aligned} &(a) \quad \frac{V_{11} - V_{22}}{2} = \frac{1}{G_p}(X_1 - \mu X_3)\left\{1 + \alpha\left(\frac{X_2}{X_1 - \mu X_3}\right)^2\right\} \\ &(b) \quad V_{12} = \frac{1}{G_r} X_2, \end{aligned}$$

where the constants G_p, G_r, and α may depend on T. (Remark: We shall see in § 3 that including an $\mathcal{O}(s^2)$ term in ψ_2 would not change condition (2.9) below.) Equations (1.21) merely express (1.14) in the chosen coordinate system.

1.2. Discussion of the constitutive assumptions.

(a) The flow rule (1.12a). The classical flow theory of plasticity for a strain hardening granular material is formulated in terms of a yield condition. If γ is the total shear stress (defined by the ordinary differential equation $d_t\gamma = |devV|$), the yield condition typically has the form

$$(1.22) \qquad \frac{\tau}{\sigma} = \mu(\gamma)$$

for some function $\mu(\gamma)$ such as illustrated in Figure 1.1. (Strictly speaking (1.22) should be an inequality; equality holds if the material is deforming plastically, which we always assume.) As sketched in Figure 1.3, the yield surface (1.22) is a cone in stress space; the figure shows both the full three dimensional space and a representative two dimensional cross section in a deviatoric plane, $\{\sigma = const\}$. This conical shape reflects the fact that friction between the grains is the dominant mechanism controlling the deformation of a granular material; i.e., the shearing stress required to overcome friction is proportional to the mean stress. These ideas have been formalized [4] in the notion of "psammic" material ($\psi\alpha\mu\mu o s$ is Greek for sand); such materials are defined to have no material properties with the dimensions of stress, so that stresses in constitutive relations can appear only as ratios.

1.3: A smooth yield surface exhibiting isotropy

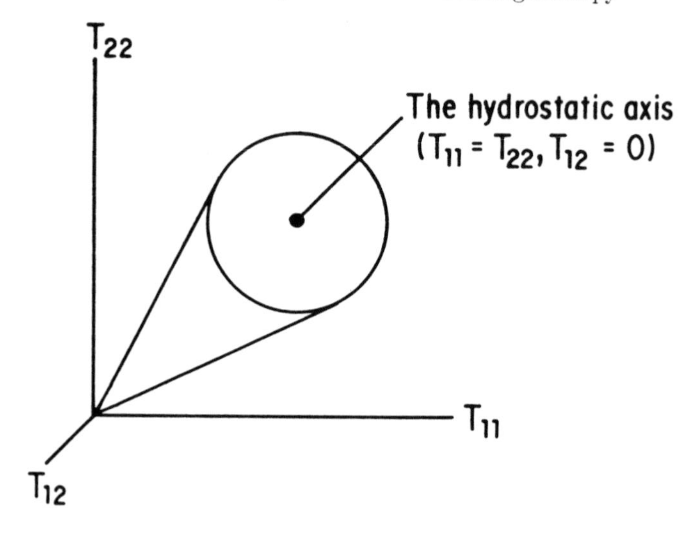

Figure 1.3 (a) The full three dimensional stress space

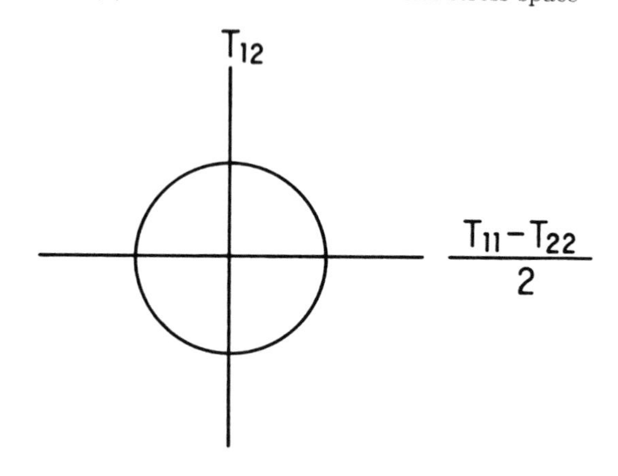

Figure 1.3 (b) Intersection with a deviatoric plane $\{\sigma = const.\}$

The constitutive function $\mu(\gamma)$ can be measured in a biaxial test. Figure 1.4 shows schematically the apparatus for such a test. Starting from a state of hydrostatic stress $\sigma_1 = \sigma_2 = \sigma_*$, the lateral pressure σ_2 is held constant while the top plate is moved slowly down. The pressure σ_1 required to achieve this (as well as the lateral displacement) is continuously monitored, giving a graph such as in Figure 1.1. To extract this constitutive information one must assume that stresses and deformations are uniform across the sample.

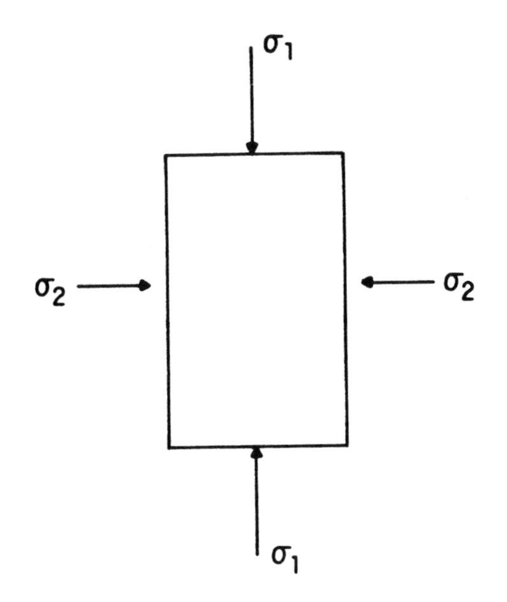

Figure 1.4: Schematic of a biaxial test

As shown for example in [12], the yield condition (1.22) can be combined with a flow rule to derive a constitutive relation of the form (1.12a). The constitutive function Ψ so derived is linear in its second argment; in symbols,

$$\Psi(T, A) = \frac{\sigma}{G_p} PA, \tag{1.23}$$

the special case of (1.14) obtained by setting $\alpha = 0$, $G_r = \infty$.

The prediction that $G_r = \infty$ discredits this theory. To see why, consider a uniform sample subjected to a stress history such as the circular arc BC in Figure 1.5, the mean stress σ being held constant. In such a history the principal stress axes rotate relative to material axes, but the principal stresses σ_i remain constant. Equation (1.23) predicts that no deformation will occur along such a stress path since $\nabla_t(\sigma^{-1}\, dev\, T)$ is everywhere orthogonal to T. Unlike the biaxial test, tests which rotate the stress axes are difficult to perform. Nevertheless, there is general agreement [2,6,13,14] on one point: *plastic deformation accompanies rotation of stress axes* when the material is at yield.

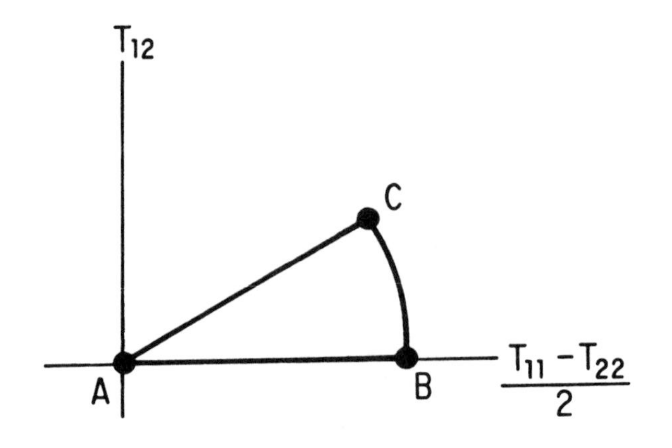

Figure 1.5: Some informative stress histories

The deformation theory of plasticity [7] predicts a finite value for G_r that can be determined using just data from biaxial tests. For stress paths close to proportional loading one obtains the constitutive relation (1.14) with

$$G_p = \sigma \frac{d\mu}{d\gamma}, \qquad G_r = \sigma \frac{\mu}{\gamma}, \qquad \alpha = 0. \tag{1.24}$$

However, this theory also suffers from a serious defect regarding rotation of the stress axes. Consider subjecting an isotropic material to two stress histories moving from A to C in Figure 1.5, one moving directly along AC, the other indirectly along ABC. The total work in the deformation theory is independent of path, so equal work is performed during both histories. On the other hand, equal work is performed along AC and AB since one stress history may be transformed to the other by a change in the coordinate system. Thus, although plastic deformation occurs along BC, *no work is performed during this deformation.* This is a most unsatisfactory prediction. To obtain positive work with the constitutive law (1.14), one must require that the coefficient α satisfy

$$\alpha > 0. \tag{1.25}$$

This derivation of (1.25) is flawed because we used (1.14) to describe the material response along the stress path BC. Since the tangent to BC makes an angle of 90° with the current stress, BC differs greatly from proportional loading. However, (1.14) is an approximation valid for nearly proportional loading with limited justification elsewhere. Despite these reservations, we shall assume (1.25). Our hypothesis is that, at the first possible occurrence in the Taylor expansion, there is a term which increases the rate of plastic work $T_{ij}V_{ij}$ relative to a linear theory.

Indeed, we shall augment (1.25) by assuming that the dimensionless constant α is at least of the order of unity.

Let us note in passing that (1.25) implies that Ψ cannot be derived as the gradient of a plastic potential. If Ψ is a gradient, then the first nonlinear term in (1.21a) is $\mathcal{O}(X_2^4)$. Such cases will be studied in a future publication. Here we assume (1.25), noting that at least in some cases (e.g. [13]) the experimentally determined Ψ cannot be derived from a potential.

In our formulation of equations in §1.1 we made no mention of a yield condition. The only role of a yield surface in the present theory is to separate loading from unloading. Since plastic yielding accompanies rotation of the stress axes, stress histories such as BC in Figure 1.5 must lie outside the yield surface. Thus the yield surface must have a vertex at the current stress such as is shown in Figure 1.6. Note that, consistent with the assumption of psammic material, all cross sections with a deviatoric plane are geometrically similar to Figure 1.6(b). Thus in the full three dimensional stress space, the "vertex" in "yield vertex models" is really a line of corners on the yield surface.

Figure 1.6: A yield surface with a corner exhibiting induced anisotropy

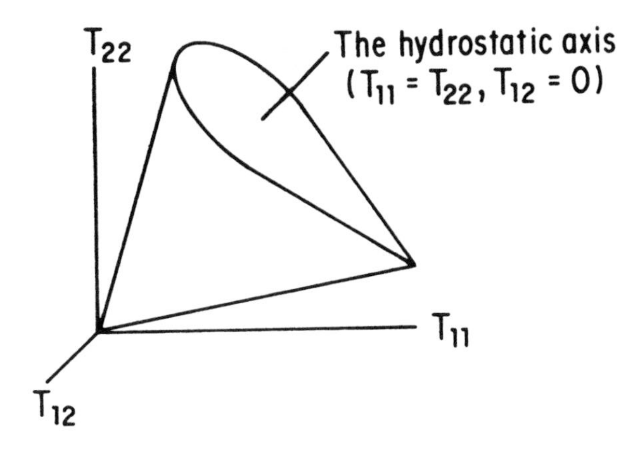

Figure 1.6 (a) The full three dimensional stress space

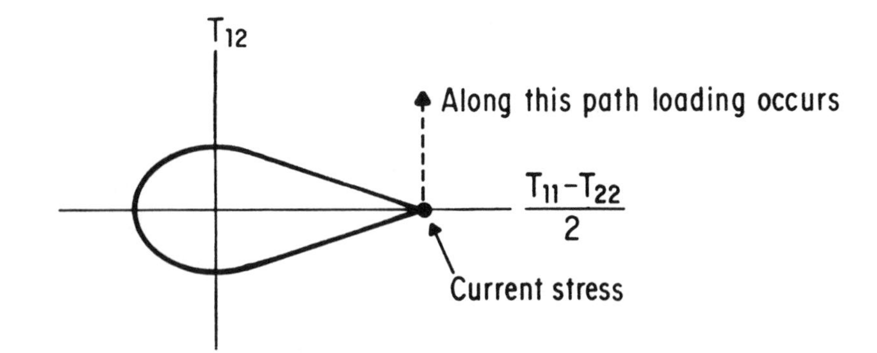

Figure 1.6 (b) Intersection with a deviatoric plane $\{\sigma = const.\}$, including an illustrative stress path

The yield surface in Figure 1.6 exhibits anisotropy. Specifically, consider unloading the sample to a hydrostatic stress (the origin in Figure 1.6(b)) and then reloading until yielding occurs, the latter being performed with proportional loading. As illustrated in Figure 1.7, much less reloading is needed to achieve yielding if the principal stress axes for reloading are rotated relative to the original axes. Experiments [17] confirm this prediction of anisotropy induced by the stress history of the sample.

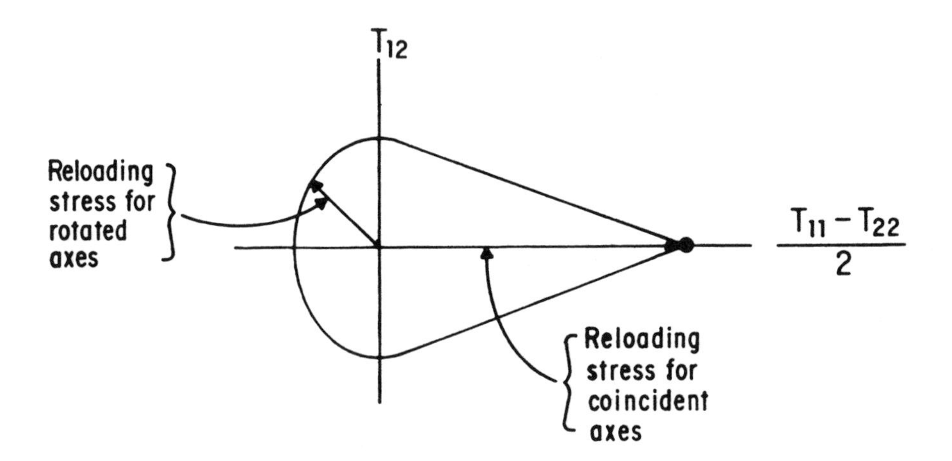

Figure 1.7: Consequences of anisotropy: reloading stresses depend on direction

The hardening rule assumes that, as long as loading continues, the vertex moves so as to coincide with the current stress. We refer to [3] for further information on such models.

We conclude by remarking that $\nabla_t(\sigma^{-1}\text{dev}T)$ appears in the second argument in (1.12a) as a reflection of the assumption of psammic material—only changes in *ratios* of stresses are significant. In fact the dependence of Ψ on its first argument must also be through ratios of stresses, but since this fact is irrelevant for our analysis, we do not incorporate it explicitly into our notation.

(b) The dilatancy condition (1.12b). It is well documented [15] that a densely packed granular material must dilate in order to deform. As an extreme example of this, a close-packed array of disks on a hexagonal lattice is rigid unless dilation is permitted. When $\beta > 0$, equation (1.12b) expresses the condition that any shearing is accompanied by a corresponding dilation. When $\beta < 0$, this relation describes loosely packed material consolidating under deformation.

In general the parameter β depends on the strain history of the material, perhaps through the total shear strain. Since only the current value of β is relevant for our analysis, we do not indicate this dependence. We always assume that $|\beta| < 1$.

In the governing equations, (1.1) and (1.12a) contain time derivatives but (1.12b) contains no such derivative. In this respect (1.12b) resembles the incompressibility condition of the Navier-Stokes equation; indeed, incompressibility emerges as a special case of (1.12b) when $\beta = 0$. Because there are no time derivatives in (1.12b), the speed of longitudinal sound waves is infinite, and consequently such waves do not appear explicitly in an eigenvalue analysis. The loss of hyperbolicity studied below occurs when the speed of *transverse* sound waves becomes imaginary.

§2. Hyperbolicity and its consequences

2.1. Conditions for hyperbolicity. For convenient reference we recall from §1 the equations governing granular flow:

$$\begin{aligned}
&(a)\quad d_t\rho + \rho\partial_j v_j = 0\\
&(b)\quad \rho d_t v_i + \partial_j T_{ij} = 0\\
&(c)\quad \text{dev}\, V = \Psi(T, \nabla_t(\sigma^{-1}\text{dev}\, T))\\
&(d)\quad \frac{1}{2}tr\ V = -\beta|\text{dev}\, V|.
\end{aligned} \tag{2.1}$$

In this paper we derive conditions for (2.1) to be hyperbolic. Let us use the shorthand $U = (\rho, v, T)$ for the state vector. To test for hyperbolicity, we linearize (2.1) by expanding the solution $U = U^{(0)} + \bar{U}$ about a base state $U^{(0)}$ and retaining only terms that are linear in $\bar{U}$; drop all lower order terms (i.e., not containing a derivative of $\bar{U}$); freeze coefficients at a point x_0, t_0 in the terms which remain; and look for solutions with exponential dependence

$$e^{i(\xi,x)-i\omega t} \tag{2.2}$$

where $\xi \in \mathbb{R}^2$, $(\cdot,\cdot)$ is the inner product on $\mathbb{R}^2$, and $\omega \in \mathbb{C}$ is to be determined. Roughly speaking, we call (2.1) *hyperbolic* with respect to $U^{(0)}$ at a given point x_0, t_0 if the frequencies $\omega(\xi)$ so obtained are all real. (In case the eigenvalue problem for

$\omega(\xi)$ has multiple roots, this definition of hyperbolicity needs to be augmented by the technical condition that the maximum number of eigenvectors exist. Although we find multiple eigenvalues in analyzing (2.1), there always are enough eigenvectors. When the system (2.1) loses hyperbolicity, it does so by the straightforward mechanism of frequencies becoming complex , not the subtle one of eigenvectors degenerating.) Hyperbolicity is a necessary and sufficient condition for (2.1) to be linearly well-posed; i.e., for the growth rates $-i\omega(\xi)$ of solutions (2.2) to satisfy

$$\sup_{\xi} Re[-i\omega(\xi)] < \infty. \tag{2.3}$$

Since (2.1) is fully nonlinear, the hyperbolicity of this system depends on derivatives of the base state $U^{(0)}$ as well as the values of $U^{(0)}$ at the point in question. In §3 we shall show that hyperbolicity depends on these derivatives only through the combination

$$s^{(0)} = \frac{X_2^{(0)}}{X_1^{(0)} - \mu X_3^{(0)}} = \frac{d_t T_{12}^{(0)} - \tau(\partial_1 v_2^{(0)} - \partial_2 v_1^{(0)})}{d_t(\frac{T_{11}^{(0)} - T_{22}^{(0)}}{2}) - \mu d_t(\frac{T_{11}^{(0)} + T_{22}^{(0)}}{2})} \tag{2.4}$$

where now $d_t = \partial_t + v_j^{(0)} \partial_j$. Actually in (2.4), τ and μ also depend on the base state $U^{(0)}$ but we reserve the superscript zero for quantities depending on derivatives of the base state.

Our first result concerns the hyperbolicity of (2.1) in case $s^{(0)} = 0$; i.e., hyperbolicity with respect to proportional loading. Related results, but neglecting the Jaumann derivative, have been obtained by several authors [5,8, 10,12]. The parameters G_p and G_r in the theorem are defined by (1.21).

THEOREM 2.1. *Equations (2.1) are hyperbolic with respect to a state satisfying* $s^{(0)} = 0$ *if and only if*

$$\begin{aligned} &(a) \quad G_r > 2\tau \quad \text{and} \\ &(b) \quad G_p > (\mu - \beta)\tau + \frac{G_r}{2}\left\{1 - \mu\beta - \sqrt{(1-\mu^2)(1-\beta^2)(1-(\frac{2\tau}{G_r})^2)}\right\} \end{aligned} \tag{2.5}$$

Although we do not accept the deformation theory of plasticity, nonetheless we refer to (1.24) for representative values for G_p and G_r. In particular, the modulus G_p for proportional loading tends to zero as the maximum shear stress is approached. Since the RHS of (2.5b) is positive, this inequality fails before the maximum is reached, resulting in nonhyperbolic equations. By contrast, according to (1.22,24), $G_r = \gamma^{-1}\tau$. Since constitutive tests rarely proceed beyond 10% deformation, the factor γ^{-1} is substantially larger than 2. Thus (2.5a) will not be violated, so we focus on (2.5b).

The term $(\mu - \beta)\tau$ and the factor $1 - (2\tau/G_r)^2$ in (2.5b) result from including the full Jaumann derivative. Because of the latter, even in the case of normality (i.e., $\mu = \beta$), the threshold for ill-posedness occurs before G_p goes to zero at the maximum stress in Figure 1.1.

The complexity of (2.5b) makes some approximations desirable. Since $2\tau/G_r << 1$ we may use the binomial theorem to expand $[1 - (2\tau/G_r)^2]^{1/2}$. Thus the RHS of (2.5b) is approximately equal to

$$(\mu - \beta)\tau + \frac{\tau^2}{G_r} + \frac{G_r}{2}\left\{1 - \mu\beta - \sqrt{(1-\mu^2)(1-\beta^2)}\right\}. \tag{2.6}$$

In fact τ^2/G_r is typically much smaller than $(\mu - \beta)\tau$, so we will drop the middle term in (2.6) altogether. It is not difficult to show that the third term in (2.6) is nonnegative and vanishes when $\mu = \beta$. Thus this quantity is $\mathcal{O}((\mu-\beta)^2)$; specifically

$$1 - \mu\beta - \sqrt{(1-\mu^2)(1-\beta^2)} = \frac{2(\mu-\beta)^2}{4-(\mu+\beta)^2} + \mathcal{O}((\mu-\beta)^4).$$

Combining the above, we obtain the following approximation for (2.5b):

$$G_p > (\mu - \beta)\tau + \frac{(\mu-\beta)^2}{4-(\mu+\beta)^2}G_r. \tag{2.7}$$

If we take representative values for the parameters, say

$$\mu = .7, \qquad \beta = .3, \qquad G_r = \tau/\gamma \quad \text{where} \quad \gamma = .06, \tag{2.8}$$

then the first term in (2.7) equals $.4\tau$ and the second equals $(8/9)\tau$, or approximately $.9\tau$.

The following theorem, the main analytical result of the paper, gives the first order correction to (2.5b) if $s^{(0)} \neq 0$.

THEOREM 2.2. *For small $s^{(0)}$, equations (2.1) are hyperbolic if and only if*

$$G_p > G^* + CG_r|s^{(0)}| + \mathcal{O}(s^{(0)^2}) \tag{2.9}$$

where G^* is the RHS of (2.5b) and C is a dimensionless positive constant.

Remark. C depends on $\alpha, \beta, \mu, G_p/G_r$, and τ/G_r. Equation (3.32) below gives an explicit formula for C.

Observe from comparison of (2.5b) and (2.9) that the effect of taking small $s^{(0)} \neq 0$ is destablizing; i.e., hyperbolicity is lost sooner if $s^{(0)}$ is nonzero. This behavior is illustrated in Figure 2.1. We chose $1/G_p$ as the abscissa in the figure so that the quasistatic evolution of the parameters in a constitutive test is described by motion to the right in the figure.

2.2. An application to shearing between parallel plates. The hyperbolic/nonhyperbolic boundary in Figure 2.1 is the union of two curves, a situation which arises as follows. At the critical point for proportional loading, hyperbolicity is lost because the frequencies of plane waves in certain directions go to zero and become complex. Specifically, this happens for plane waves in two directions located symmetrically with respect to the axes, as illustrated in Figure 2.2. Plane waves in these two directions respond differently to rotations in the stress of the base state. Hence there are two curves in Figure 2.1. The dotted curves are continuations of the boundary for one direction into the region where the frequency of the other direction becomes complex first.

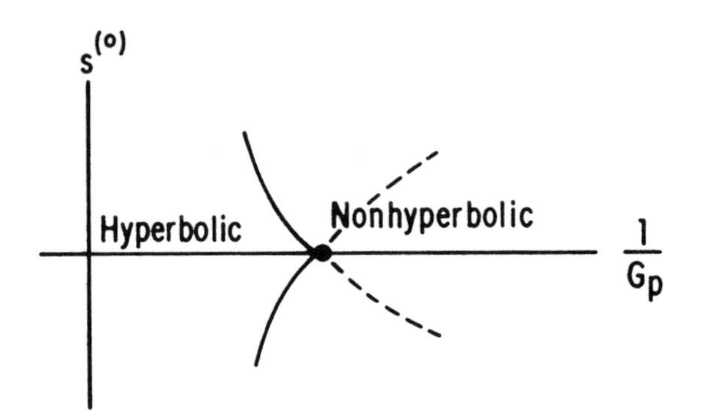

Figure 2.1: Regions of hyperbolicity in parameter space

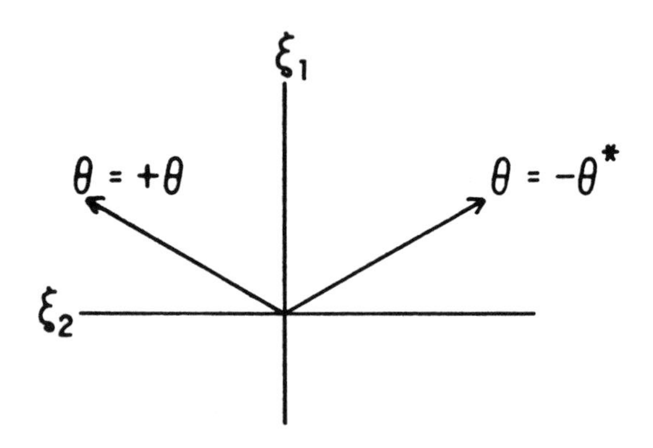

Figure 2.2: Directions of wave vectors of plane waves whose frequencies become complex

(Note: The ξ_1-axis, corresponding to the major principal stress axis, is shown as vertical to facilitate comparison with Figure 2.3.)

This structure has consequences for experiments. It is felt [11] that shear bands form in material when the equations lose hyperbolicity, the shear band being approximately normal to the direction ξ of the plane wave whose frequency becomes complex. In the biaxial test, which has proportional loading, the frequencies for both directions in Figure 2.2 become complex at the same time. Moreover, as illustrated in Figure 2.3, shear bands in such experiments are observed to have either orientation. By contrast, in shearing between parallel plates (cf. Figure 2.4(a)), the rotation $s^{(0)}$ of the base state is nonzero, so the frequency for one direction becomes complex before that of the other. This leads to a preferred orientation for the shear band. We shall argue in §3.4 that this mechanism selects shear bands of the orientation shown in Figure 2.4(c), the orientation invariably seen in experiments [1].

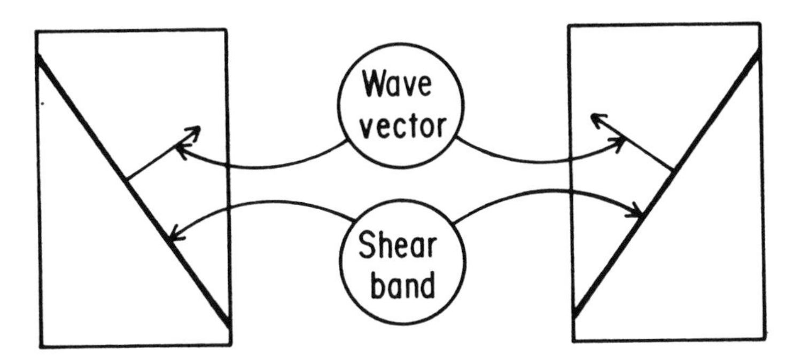

Figure 2.3: Shear bands in the biaxial test

§3. Proofs

3.1. Formulation of the eigenvalue problem. Our first task is to compute the principal part of the linearization of (2.1). For (2.1a) we obtain the equation

$$d_t\bar{\rho} + \rho^{(0)}\partial_j\bar{v}_j = 0 \tag{3.1}$$

where $d_t = \partial_t + v_j^{(0)}\partial_j$ and for (2.1b),

$$\begin{aligned} &(a)\quad \rho^{(0)}d_t\bar{v}_1 + \partial_1\left(\frac{\bar{T}_{11}-\bar{T}_{22}}{2}\right) + \partial_2\bar{T}_{12} + \partial_1\left(\frac{\bar{T}_{11}+\bar{T}_{22}}{2}\right) = 0 \\ &(b)\quad \rho^{(0)}d_t\bar{v}_2 - \partial_2\left(\frac{\bar{T}_{11}-\bar{T}_{22}}{2}\right) + \partial_1\bar{T}_{12} + \partial_2\left(\frac{\bar{T}_{11}+\bar{T}_{22}}{2}\right) = 0. \end{aligned} \tag{3.2}$$

Stresses in the latter equation are written in terms of the coordinates (1.6) for symmetric tensors. Recalling (1.17), we compute that the principal part of the linearization of (2.1c) is

$$\begin{aligned} \operatorname{dev}\bar{V} = \frac{1}{\sigma^{(0)}}\Bigg\{&\frac{\partial\Psi}{\partial X_1}\left[d_t\left(\frac{\bar{T}_{11}-\bar{T}_{22}}{2}\right) - \mu d_t\left(\frac{\bar{T}_{11}+\bar{T}_{22}}{2}\right)\right] \\ &+ \frac{\partial\Psi}{\partial X_2}[d_t\bar{T}_{12} - \tau^{(0)}(\partial_1\bar{v}_2 - \partial_2\bar{v}_1)]\Bigg\}. \end{aligned} \tag{3.3}$$

To vectorize the notation, define a 2×2 matrix

$$B = \frac{1}{\sigma^{(0)}} \begin{pmatrix} \frac{\partial \Psi_1}{\partial X_1} & \frac{\partial \Psi_1}{\partial X_2} \\ \frac{\partial \Psi_2}{\partial X_1} & \frac{\partial \Psi_2}{\partial X_2}. \end{pmatrix} \tag{3.4}$$

Equation (3.3) may be rewritten in this notation as

(3.5)

$$Bd_t \begin{pmatrix} \frac{\bar{T}_{11}-\bar{T}_{22}}{2} \\ \bar{T}_{12} \end{pmatrix} - \mu d_t \left(\frac{\bar{T}_{11}+\bar{T}_{22}}{2} \right) B_1 - \tau^{(0)}(\partial_1 \bar{v}_2 - \partial_2 \bar{v}_1) B_2 + \begin{pmatrix} \frac{\partial_1 \bar{v}_1 - \partial_2 \bar{v}_2}{2} \\ \frac{\partial_1 \bar{v}_2 + \partial_2 \bar{v}_1}{2} \end{pmatrix} = 0$$

where B_i, $i = 1, 2$ denotes the ith column of B. Finally, equation (2.1d) linearizes to

(3.6)

$$\frac{\beta}{|devV^{(0)}|} \left\{ \left(\frac{V_{11}^{(0)} - V_{22}^{(0)}}{2} \right) (\partial_1 \bar{v}_1 - \partial_2 \bar{v}_2) + V_{12}^{(0)} (\partial_1 \bar{v}_2 + \partial_2 \bar{v}_1) \right\} + \partial_1 \bar{v}_1 + \partial_2 \bar{v}_2 = 0.$$

If $V^{(0)}$ were diagonal, then we would have $V_{12}^{(0)} = 0$ and

$$\frac{V_{11}^{(0)} - V_{22}^{(0)}}{2} = |devV^{(0)}|,$$

so that (3.6) would simplify to

$$(1+\beta)\partial_1 \bar{v}_1 + (1-\beta)\partial_2 \bar{v}_2 = 0.$$

Of course when $s^{(0)} \neq 0$, the deformation rate $V^{(0)}$ is not diagonal, even though $T^{(0)}$ is diagonal. However if $s^{(0)}$ is small, then $V^{(0)}$is approximately diagonal, resulting in partial simplification of (3.6). Specifically, recalling (2.4), we deduce from (1.21) that

$$\frac{V_{12}^{(0)}}{\frac{V_{11}^{(0)} - V_{22}^{(0)}}{2}} = \frac{G_p}{G_r} s^{(0)} + \mathcal{O}((s^{(0)})^2).$$

Thus

$$|devV^{(0)}| = \frac{V_{11}^{(0)} - V_{22}^{(0)}}{2} + \mathcal{O}((s^{(0)})^2).$$

Using this information in (3.6) yields

$$(1+\beta)\partial_1 \bar{v}_1 + \beta s^{(0)} \frac{G_p}{G_r} (\partial_1 \bar{v}_2 + \partial_2 \bar{v}_1) + (1-\beta)\partial_2 \bar{v}_2 = 0, \tag{3.7}$$

modulo an error that is $\mathcal{O}((s^{(0)})^2)$.

Next we look for exponential solutions

$$e^{i(\xi,x) - i\omega t} \bar{U}$$

of the linearized equations, where $\bar{U}$ is a 6-component vector of constants. This exponential satisfies (3.1,2,5,7) if and only if

$$\lambda = \omega - (v^{(0)}, \xi) \tag{3.8}$$

solves the generalized eigenvalue problem (written in blocks corresponding to the decomposition $\bar{U} = (\bar{\rho}, \bar{v}, dev\bar{T}, \bar{\sigma})$

$$
(3.9) \qquad \begin{pmatrix} 0 & \rho^{(0)}\xi^T & 0 & 0 \\ 0 & 0 & \Xi^T & \xi \\ 0 & \frac{1}{2}\Xi + \tau B_2 \xi^{\natural} & 0 & 0 \\ 0 & \ell(\xi)^T & 0 & 0 \end{pmatrix} \bar{U} = \lambda \begin{pmatrix} 1 & 0 & 0 & 0 \\ 0 & \rho^{(0)} I & 0 & 0 \\ 0 & 0 & B & -\mu B_1 \\ 0 & 0 & 0 & 0 \end{pmatrix} \bar{U},
$$

where

$$
(3.10) \qquad (a) \quad \Xi = \begin{pmatrix} \xi_1 & -\xi_2 \\ \xi_2 & \xi_1 \end{pmatrix},
$$

$$
(b) \quad \ell(\xi) = \begin{pmatrix} (1+\beta)\xi_1 + \beta s^{(0)} \frac{G_p}{G_r} \xi_2 \\ \beta s^{(0)} \frac{G_p}{G_r} \xi_1 + (1-\beta)\xi_2 \end{pmatrix},
$$

ξ is a 2-component column vector, and $\xi^{\natural}$ is the row vector $(\xi_2, -\xi_1)$. (The symbol $\natural$ may be regarded as a contraction of the symbol $\perp$ for orthogonal and T for transpose.)

Equations (2.1) are hyperbolic if and only if the frequency ω is real for all ξ. Since the term $(v^{(0)}, \xi)$ in (3.8) is real, ω is always real if and only if the eigenvalue λ in (3.9) is always real.

3.2. Analysis of the eigenvalue problem (3.9). We claim that zero is a double eigenvalue of (3.9). Now $\bar{U} = (1,0,0,0)^T$ is one eigenvector of (3.9) associated with $\lambda = 0$. Since the 2×3 matrix (Ξ^T, ξ) has rank at most 2, there is another null eigenvector of (3.9) of the form $(0, 0, \bar{U}_3, \bar{U}_4)$. Thus zero is a double eigenvalue of (3.9), and moreover there are two linearly independent eigenvectors. (As we will see below, under certain circumstances the multiplicity may be higher than two.)

Apart from one zero eigenvalue, the eigenvalues of (3.9) coincide with those of the 5×5 problem

$$
(3.11) \qquad \begin{pmatrix} 0 & \Xi^T & \xi \\ \frac{1}{2}\Xi + \tau B_2 \xi^{\natural} & 0 & 0 \\ \ell(\xi)^T & 0 & 0 \end{pmatrix} W = \lambda \begin{pmatrix} \rho^{(0)} I & 0 & 0 \\ 0 & B & -\mu B_1 \\ 0 & 0 & 0 \end{pmatrix} W.
$$

Since our interest is in nonzero eigenvalues of (3.9), we confine our attention to (3.11). First we premultiply and postmultiply (3.11) by the invertible matrices

$$
\begin{pmatrix} (1/\rho^{(0)}) I & 0 & 0 \\ 0 & B^{-1} & 0 \\ 0 & 0 & I \end{pmatrix}, \qquad \begin{pmatrix} I & 0 & 0 \\ 0 & I & \mu e_1 \\ 0 & 0 & I \end{pmatrix},
$$

respectively, where e_i, $\quad i = 1, 2$ is a unit vector along the ith axis, to reduce (3.11) to a 5×5 generalized eigenvalue problem in standard form

$$
(3.12) \qquad \begin{pmatrix} 0 & \Xi^T/\rho^{(0)} & (\xi + \mu \Xi^T e_1)/\rho^{(0)} \\ \frac{1}{2} B^{-1}\Xi + \tau e_2 \xi^{\natural} & 0 & 0 \\ \ell(\xi)^T & 0 & 0 \end{pmatrix} \begin{pmatrix} W_1 \\ W_2 \\ W_3 \end{pmatrix} = \lambda \begin{pmatrix} W_1 \\ W_2 \\ 0 \end{pmatrix}
$$

For purposes of the following lemma, let us abstract the structure of (3.12) as

$$\begin{pmatrix} 0 & M & m \\ L & 0 & 0 \\ \ell^T & 0 & 0 \end{pmatrix} \begin{pmatrix} W_1 \\ W_2 \\ W_3 \end{pmatrix} = \lambda \begin{pmatrix} W_1 \\ W_2 \\ 0 \end{pmatrix} \tag{3.13}$$

where L and M are 2×2 matrices and ℓ and m are 2-component column vectors. In the lemma we use the notation (m, ℓ) for the inner product on $\mathbf{R}^2$ and

$$m^{\perp} = \begin{pmatrix} m_2 \\ -m_1 \end{pmatrix}$$

for the orthogonal vector.

LEMMA 3.1. *Provided*

$$(m, \ell) \neq 0, \tag{3.14}$$

(3.13) has precisely three eigenvalues, one of which is zero. The other two eigenvalues are real if and only if

$$(m, \ell)^{-1}(m^{\perp}, ML\ell^{\perp}) \geq 0. \tag{3.15}$$

Remark. Although (3.13) involves 5×5 matrices, there are only three eigenvalues. As regards the original system (3.12), this deficit is because the speed of longitudinal sound waves is infinite and does not appear as an eigenvalue of (3.12). The two nonzero eigenvalues of (3.12), equal in magnitude and opposite in sign, are the speeds of transverse sound waves.

Proof. Solving the first component of (3.13) for W_1 and substituting the result into the latter two components reduces (3.13) to a 3×3 generalized eigenvalue problem

$$\begin{pmatrix} LM & Lm \\ \ell^T M & \ell^T m \end{pmatrix} \begin{pmatrix} W_2 \\ W_3 \end{pmatrix} = \lambda^2 \begin{pmatrix} W_2 \\ 0 \end{pmatrix} \tag{3.16}$$

Since by assumption $\ell^T m \neq 0$, the second component of (3.16) may be solved for W_3 and substituted into the first, yielding the 2×2 ordinary eigenvalue problem

$$(LM - (m, \ell)^{-1} Lm\ell^T M)W_2 = \lambda^2 W_2. \tag{3.17}$$

As noted above, one eigenvalue of (3.13), hence also of (3.17), is zero. The other eigenvalue of (3.17), let's call it c, gives rise to a pair of eigenvalues of (3.13) of the form $\lambda = \pm\sqrt{c}$. Thus (3.13) admits three eigenvalues. These eigenvalues are real if and only if $c \geq 0$, which happens if and only if the trace of (3.17) is nonnegative; in symbols,

$$tr(LM - (m, \ell)^{-1} Lm\ell^T M) \geq 0. \tag{3.18}$$

Using the identity $tr(AB) = tr(BA)$ to rearrange both terms in (3.18), we rewrite this inequality as

$$tr(MLP) \geq 0 \tag{3.19}$$

where

$$P = I - (m, \ell)^{-1} m \ell^T.$$

Observe that P is the projection from $\mathbb{R}^2$ onto $\ell^\perp$ relative to the basis $\{n_1, n_2\}$ where

$$n_1 = m, \qquad n_2 = \ell^\perp; \tag{3.20}$$

i.e.,

$$Pn_1 = 0, \qquad Pn_2 = n_2.$$

(By (3.14), n_1 and n_2 are linearly independent so (3.20) is a basis for $\mathbb{R}^2$.) The dual basis to (3.20) is

$$n_1^* = (m, \ell)^{-1} \ell, \qquad n_2^* = (m, \ell)^{-1} m^\perp.$$

Now

$$tr\ MLP = \sum_{i=1}^{2} (n_i^*, MLPn_i) = (n_2^*, MLn_2).$$

The proof is complete.

To apply the lemma, we substitute from (3.12) and (3.10) to compute that

$$\begin{aligned} \rho^{(0)}(m, \ell) &= (\xi + \mu \Xi^T e_1, \ell(\xi)) \\ &= (1 + \mu)(1 + \beta)\xi_1^2 + 2\beta s^{(0)} \frac{G_p}{G_r} \xi_1 \xi_2 + (1 - \mu)(1 - \beta)\xi_2^2. \end{aligned}$$

Recall that $0 \leq \mu < 1$ and $|\beta| < 1$. Thus, at least for small $s^{(0)}$, this quadratic form is positive definite. Therefore condition (3.14) of the lemma is verified. Moreover, since $(m, \ell) > 0$, condition (3.15) simplifies to

$$(m^\perp, ML\ell^\perp) \geq 0. \tag{3.21}$$

We claim that, modulo terms that are $\mathcal{O}((s^{(0)})^2)$, the LHS of (3.21) equals $1/2(\rho^{(0)})^2$ times the quantity

$$\begin{aligned} &(1 + \mu)(1 + \beta)(G_r - 2\tau)\xi_1^4 + (1 - \mu)(1 - \beta)(G_r + 2\tau)\xi_2^4 \\ &\qquad + 2(2G_p - (1 - \mu\beta)G_r - 2(\mu - \beta)\tau)\xi_1^2\xi_2^2 + 2s^{(0)}\xi_1\xi_2 p(\xi) \end{aligned} \tag{3.22}$$

where

(3.23)

$$p(\xi) = 2\alpha G_r[(1 + \beta)\xi_1^2 - (1 - \beta)\xi_2^2] + \beta \frac{G_p^2}{G_r}(\xi_1^2 + \xi_2^2) - 2\beta \frac{G_p}{G_r} \tau[(1 + \mu)\xi_1^2 - (1 - \mu)\xi_2^2].$$

To summarize the above analysis, this claim yields the following conclusion: *The system (2.1) is hyperbolic if and only if the homogeneous quartic form (3.22) is positive definite.*

The calculation verifying (3.22) may be shortened somewhat by the following trick. Taking transposes to shift M to the other factor in (3.21) and substituting data from (3.12), we rewrite the LHS of (3.21) as $1/2(\rho^{(0)})^2$ times the quantity

$$(\Xi(\xi + \mu\Xi^T e_1)^\perp, (B^{-1}\Xi + 2\tau e_2 \xi^\natural)\ell(\xi)^\perp). \tag{3.24}$$

For any $w \in \mathbb{R}^2$, we have $w^\perp = Rw$ where R is the rotation matrix

$$R = \begin{pmatrix} 0 & 1 \\ -1 & 0 \end{pmatrix}.$$

We simplify the vector on the left in (3.24) as follows

$$\Xi(\xi+\mu\Xi^T e_1)^\perp = \Xi R(\xi+\mu\Xi^T e_1) = \Xi(R\xi+\mu\Xi^T Re_1) = \Xi(\xi^\perp-\mu\Xi^T e_2) = \Xi\xi^\perp-\mu|\xi|^2 e_2.$$

Here we have used the fact, which may be seen from (3.10(a)), that Ξ is a multiple of a rotation matrix; hence (i) Ξ^T commutes with R and (ii) $\Xi\Xi^T = |\xi|^2 I$.

To continue we need a formula for B^{-1} where B was defined by (3.4). Recall from (1.17) that $\Psi(T, X)$ equals σ times the RHS of (1.21). Differentiating (1.21) we find that, modulo terms which are $\mathcal{O}((s^{(0)})^2)$,

$$B = \begin{pmatrix} \frac{1}{G_p} & \frac{2\alpha s^{(0)}}{G_p} \\ 0 & \frac{1}{G_r} \end{pmatrix}.$$

This, together with formulas (3.10), completes the data needed to verify (3.22).

Incidentally, to prove a claim made in §1 following (1.21), let us consider the effect of including the $\mathcal{O}((s^{(0)})^2)$ term in (1.21b). Such a term would be proportional to X_2^3/X_1^2, so it would perturb the 2,1-entry of B by $\mathcal{O}((s^{(0)})^3)$, the 2,2-entry by $\mathcal{O}((s^{(0)})^2)$. Thus this term is irrelevant for the present $\mathcal{O}(s^{(0)})$ calculation.

3.3. Analysis of quartic forms.

(a) Case 1: $s^{(0)} = 0$. We now prove Theorem 2.1. We must test whether (3.22) is positive definite. Note that for the present case the final term in (3.22) vanishes. A quartic form

$$a\xi_1^4 + 2b\xi_1^2\xi_2^2 + c\xi_2^4$$

is positive definite if and only if

$$\begin{aligned} &(a) \quad a > 0 \\ &(b) \quad c > 0 \\ &(c) \quad b > -\sqrt{ac}. \end{aligned} \tag{3.25}$$

When this test is applied to the above form, (3.25a) yields (2.5a), (3.25b) is vacuous, and (3.25c) yields (2.5b). The proof is complete.

Case 2: $s^{(0)} \neq 0$. Next we prove Theorem 2.2 The framework for this proof is the following observation. In the borderline case when $s^{(0)} = 0$ and $G_p = G^*$ where G^* equals the RHS of (2.5b), the form (3.22) is nonnegative, but it vanishes along the two lines in the ξ- plane through the origin and making angles $\pm\theta^*$ with the ξ_1-axis, where

$$\tan\theta^* = \left\{ \frac{(1+\mu)(1+\beta)(1-\frac{2\tau}{G_r})}{(1-\mu)(1-\beta)(1+\frac{2\tau}{G_r})} \right\}^{1/4}. \tag{3.26}$$

The perturbing term in (3.22), $2s^{(0)}\xi_1\xi_2 p(\xi)$, has opposite signs at these two directions. When $s^{(0)} \neq 0$, hyperbolicity will be lost slightly earlier near the direction for which this term is negative, slightly later near the direction for which it is positive.

To make this quantitative, let us write $Q(\xi; G_p, s^{(0)})$ for the expression (3.22); we indicate explicitly the parameters that will be varied in the following analysis. We define a function

$$F(\theta; \delta, s^{(0)}) = Q((\cos\theta, \sin\theta); G^* + \delta, s^{(0)}).$$

If $s^{(0)} \neq 0$, then (2.1) loses hyperbolicity at $G_p = G^* + \delta$ where δ is defined implicitly by

$$\min_\theta F(\theta; \delta, s^{(0)}) = 0. \tag{3.27}$$

The minimum will be obtained at a critical point; i.e., at a solution of the system

$$\begin{array}{ll} (a) & F(\theta; \delta, s^{(0)}) = 0 \\ (b) & F_\theta(\theta; \delta, s^{(0)}) = 0. \end{array} \tag{3.28}$$

This system is to be solved for θ and δ as functions of $s^{(0)}$. When $s^{(0)} = 0$, (3.28) has the two solutions

$$\theta = \pm\theta^*, \qquad \delta = 0.$$

When $s^{(0)} \neq 0$, we have to lowest order

$$\delta \approx \frac{d\delta}{ds}(0)s^{(0)}. \tag{3.29}$$

To evaluate the derivative in (3.29), we invert the relation

$$\begin{pmatrix} F_\theta & F_\delta \\ F_{\theta\theta} & F_{\theta\delta} \end{pmatrix} \begin{pmatrix} \frac{d\theta}{ds} \\ \frac{d\delta}{ds} \end{pmatrix} = - \begin{pmatrix} F_s \\ F_{\theta s} \end{pmatrix}$$

which is obtained by differentiating (3.28) with respect to $s^{(0)}$. Taking advantage of the fact that $F_\theta(\theta^*, 0, 0) = 0$, we obtain

$$\frac{d\delta}{ds}(0) = -\frac{F_s}{F_\delta}, \tag{3.30}$$

provided $F_\delta F_{\theta\theta}$ is nonzero, which is easily checked. Referring to (3.22) to evaluate the derivatives in (3.30) and substituting into (3.29) we find

$$\delta \approx \mp \frac{p(\cos\theta^*, \sin\theta^*)}{2\cos\theta^* \sin\theta^*} s^{(0)}, \tag{3.31}$$

the minus sign holding near $\theta = +\theta^*$ and the plus sign near $\theta = -\theta^*$. Since both signs occur, the correction to G_p in (2.5b) needed to guarantee hyperbolicity must be positive. On substituting (3.23) into (3.31) and recalling (3.26), we verify (2.9) with C equal to $[(1-\mu)^2(1-\beta^2)(1-(\frac{2\tau}{G_r})^2)]^{-1/4}$ times the absolute value of

(3.32)

$$\alpha\sqrt{1-\beta^2}\{\sqrt{(1+\mu)(1-\beta)(1-\frac{2\tau}{G_r})} - \sqrt{(1-\mu)(1+\beta)(1+\frac{2\tau}{G_r})}\}$$

$$+\beta\sqrt{1-\mu^2}\frac{G^*}{G_r} \quad \frac{\tau}{G_r}\{\sqrt{(1+\mu)(1-\beta)(1+\frac{2\tau}{G_r})} - \sqrt{(1-\mu)(1+\beta)(1-\frac{2\tau}{G_r})}\}$$

$$-\frac{1}{2}\beta(\frac{G^*}{G_r})^2\left\{\sqrt{(1+\mu)(1+\beta)(1-\frac{2\tau}{G_r})} + \sqrt{(1-\mu)(1-\beta)(1+\frac{2\tau}{G_r})}\right\}.$$

3.4. Application to shearing between parallel plates. Consider homogeneous shearing of granular material between two infinite parallel plates (cf. Figure 2.4(a)). We claim that during shearing the major principal stress in the material will be tilted relative to the normal as indicated in Figure 2.4(b). To see this, recall that yield vertex constitutive relations are elaborations of a cruder theory in which the eigenvectors of T and V are required to be coaxial. If the shearing is isochoric (i.e., at constant volume), then the eigenvectors of V have slope $\pm 45°$, contraction occurring along the direction with slope $-45°$. Thus in the cruder theory the major principal stress axis would have slope $-45°$. The qualitative conclusion, that the major principal stress slopes downward, survives the transition to a yield vertex model with possible dilatant deformations.

It is believed that shear bands form in such an experiment when (2.1) loses hyperbolicity, the shear band being approximately orthogonal to the direction ξ along which hyperbolicity is lost. As indicated in Figure 2.4(b), there are two competing orientations for shear bands, making angles of approximately $\pm(90° - \theta^*)$ with the x_1-axis of a coordinate system aligned with respect to the current principal stress axes. If $s^{(0)} \neq 0$, one orientation will be favored, the orientation along whose normal hyperbolicity is lost earlier (i.e., at a larger value of G_p).

Figure 2.4: Shearing between parallel plates

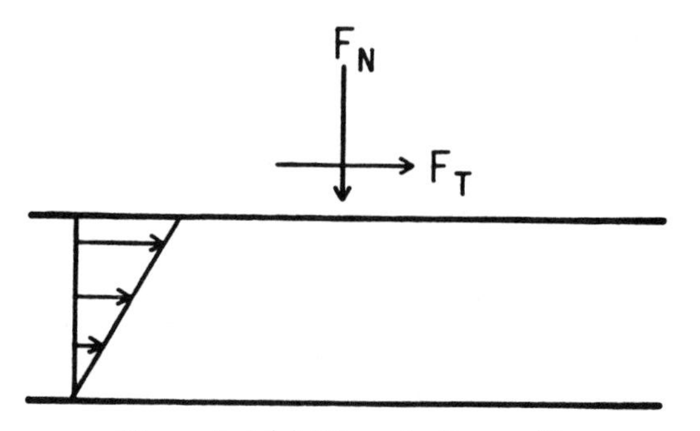

Figure 2.4 (a) The velocity profile

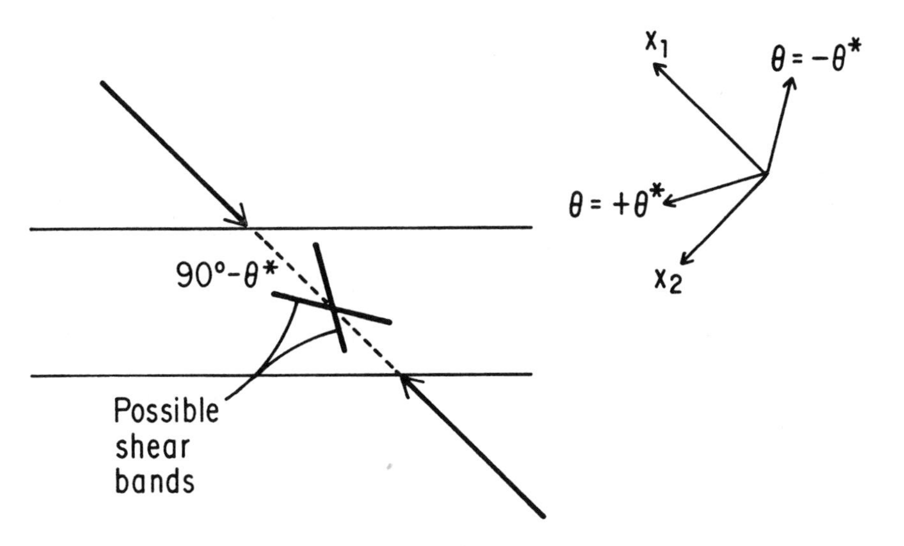

Figure 2.4 (b) Two possible orientations of shear bands

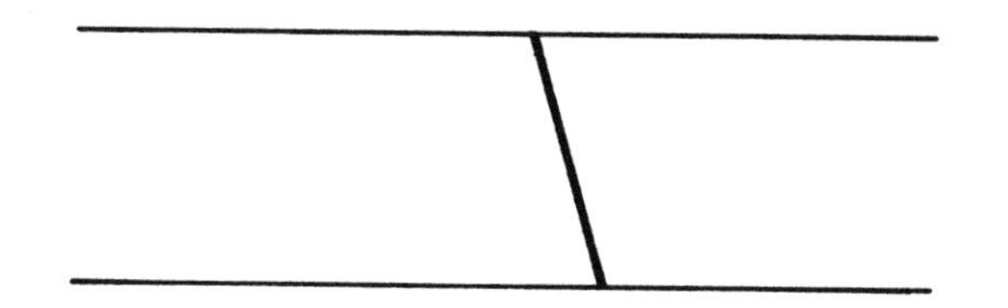

Figure 2.4 (c) The favored orientation of shear bands

As discussed at the beginning of §3.3(b), at least for small $s^{(0)}$, hyperbolicity is lost earlier along the ray $\theta = \pm\theta^*$ for which the perturbing term $2s^{(0)}\xi_1\xi_2 p(\xi)$ in (3.22) is negative. Below we will argue that

$$s^{(0)} > 0 \tag{3.33}$$

and

$$\xi_1 \xi_2 p(\xi) \quad \text{is negative along} \quad \theta = +\theta^*. \tag{3.34}$$

Combining these ideas, we conclude that shear bands with the orientation shown in Figure 2.4(c) are favored. This prediction is subject to the caveat that in an experiment $s^{(0)}$ need not be small, but we believe that the mechanism which selects orientation at small $s^{(0)}$ will operate in the same way at large $s^{(0)}$. Experiments [1] consistently bear out this prediction.

Regarding (3.33), we deduce from (1.16), (1.18), and (1.20) that the sign of $s^{(0)}$ equals that of $\nabla_t T_{12}^{(0)}$, where

$$\nabla_t T_{12}^{(0)} = d_t T_{12}^{(0)} - \tau^{(0)}(\partial_1 v_2^{(0)} - \partial_2 v_1^{(0)}). \tag{3.35}$$

The first term in (3.35) is nonzero when the principal stress axes rotate (relative to a frame fixed in space), while the second is nonzero when axes fixed in the material rotate (relative to a frame fixed in space). The second term in (3.35) is definitely nonzero because material axes rotate when shearing occurs. In fact the second term is positive; this may be seen mathematically by calculating *curl* $v^{(0)}$ or physically by observing that material axes rotate clockwise. In contrast, the first term in (3.35) should be small. Indeed, as discussed above, for isochoric deformation with a constitutive relation requiring coaxiality, this term would vanish identically. In this way we verify (3.33).

As a convenient point of reference in verifying (3.34), observe that the sign of $\xi_1 \xi_2 p(\xi)$ along the ray $\theta = -\theta^*$ equals the sign of the expression (3.32). Let us argue that one should expect (3.32) to be positive, which will prove (3.34). Our argument has the following outline:

(1) The first term in (3.32) is dominant and

(2) This first term is positive because

(i) α is positive by (1.25) and

(ii) the difference in square roots is positive because usually μ is significantly larger than β and much larger than τ/G_r.

Point (2) and its subclauses are self-explanatory. The idea behind point (1) is that the second and third terms in (3.32) are much smaller than the first because of the factors of τ/G_r and G^*/G_r; moreover these two terms have opposite signs, so partial cancellation further reduces their contribution to (3.32). By way of illustration, if we take parameter values as given by (2.8) and $G_p = 1.3\tau$, then (3.32) equals approximately

$$.345\alpha - .0009.$$

Because the second term here is negative, in general we need to augment (1.25) by the condition that α is not excessively small in order to verify (3.34). However, for the special case of isochoric deformation, we have $\beta = 0$, so the latter two terms in (3.32) are strictly zero; thus we obtain (3.34) without qualification.

Acknowledgements. We are grateful to I. Vardoulakis for numerous discussions regarding this problem, especially for suggesting the application of Theorem 2.2 to shearing between parallel plates. Indeed his paper [16] observes that a constitutive relation of the type (1.14) will lead to a preferred orientation for cracks, although he does not attempt to calulate which orientation will be preferred. Our thanks also go to Andrew Drescher, Tomasz Hueckel, and Alan Needleman for calling various references to our attention. We are indebted to the Institute of Mathematics and its Applications for their hospitality – the bulk of this research was performed while both authors were visiting the Institute.

REFERENCES

[1] V. ANNIN, A. REVUZHENKO, AND E. SHEMYAKIN, *Deformed solid mechanics at the Siberian branch, Academy of Sciences of the USSR*, J. Appl. Mech. and Tech. Physics, 28 (1987), pp. 531–549.

[2] J. ARTHUR, K. CHUA, AND T. DUNSTAN, *Induced anisotropy in a sand*, Geotechnique, 27 (1977), pp. 13–30.

[3] J. CHRISTOFFERSEN AND J. HUTCHINSON, *A class of phenomenological corner theories of plasticity*, J. Mech. Phys. Solids, 27 (1979), pp. 465–487.

[4] T. DIETRICH, *A comprehensive mechanical model of sand at low stress levels, in*, Constitutive equations of soils, Preprints of Speciality Section 9, IXth Int'l Conf. in Soil and Found. Eng'g., Tokyo (1977), pp. 33–44.

[5] R. HILL, *Acceleration waves in solids*, J. Mech. Phys. Solids, 10 (1962), pp. 1–16.

[6] K. ISHIHARA AND I. TOWHATA, *Sand response to cyclic rotation of principal stress directions as induced by wave loads*, Soils Fdns., 23 (1983), pp. 11–26.

[7] L. KACHANOV, *Foundations of the Theory of Plasticity*, American Elsevier (1971).

[8] J. MANDEL, *Condition de stabilité et postulat de Drucker*, in Pheology and Soil Mechanics (J. Kravtchenko and P. Sirieys, eds.), IUTAM Symposium at Grenoble, Springer (1966), pp. 58–68.

[9] W. PRAGER, *Introduction to the Mechanics of Continua*, Ginn (1961).

[10] J. RICE, *The localization of plastic deformation*, in Proc. 14th IUTAM Congress (W. Koiter, ed.), Delft (1976), pp. 207-220.

[11] J. RUDNICKI AND J. RICE, *Conditions for the localization of deformation in pressure-sensitive dilatant materials*, J. Mech. Phys. Solids, 23 (1975), pp. 371–394.

[12] D. SCHAEFFER, *Instability and ill-posedness in the deformation of granular materials*, Int'l. J. Num. Anal. Methods Geomechanics, (to appear).

[13] S. STURE, M. ALAWI AND H.-Y. KO, *True triaxial and directional shear cell experiments on dry sand,*, Final report to U. S. Army Corps of Engineers of Contract DACW39-85-C-0080, November (1988).

[14] M. SYMES, A. GENS AND D. HIGHT, *Undrained anisotropy and principal stress rotation in saturated sand*, Geotechnique, 34 (1984), pp. 11–27.

[15] I. VARDOULAKIS, *Rigid granular plasticity model and bifurcation in the triaxial test*, Acta Mech, 43 (1983), pp. 57–79.

[16] I. VARDOULAKIS AND B. GRAF, *Calibration of constitutive models for granular materials from biaxial experiments*, Geotechnique, 35 (1985), pp. 299–317.

[17] R. WONG AND J. ARTHUR, *Induced and inherent anisotropy in sand*, Geotechnique, 35 (1985), pp. 471–481.

UNDERCOMPRESSIVE SHOCKS IN SYSTEMS OF CONSERVATION LAWS

MICHAEL SHEARER*† AND STEPHEN SCHECTER†

1. Introduction. In this paper, we describe recent progress in our understanding of Riemann problems that involve undercompressive shock waves for 2×2 systems of nonstrictly hyperbolic conservation laws. A 2×2 system of conservation laws

$$U_t + F(U)_x = 0, \tag{1.1}$$

$U = U(x,t) \in \mathbf{R}^2, F : \mathbf{R}^2 \longrightarrow \mathbf{R}^2$, is *nonstrictly hyperbolic* if the eigenvalues $\lambda_1(U) \leq \lambda_2(U)$ of $dF(U)$ are real, but not distinct for every U. As defined in [4], system (1.1) has an *umbilic point* at $U = U^*$ if $dF(U^*)$ is a multiple of the identity. Hyperbolic equations with an isolated umbilic point can be classified locally according to properties of the quadratic map $d^2F(U^*)$. Since linear changes of coordinates do not affect the shocks or rarefaction waves for quadratic nonlinearities F, the general family of quadratic nonlinearities F with a unique umbilic point can be reduced to a two parameter family, which we write as

$$Q(u,v) = d(au^3/3 + bu^2v + uv^2), \quad a \neq 1 + b^2, \tag{1.2}$$

where d denotes gradient with respect to $U = (u,v)$.

Systems of nonstrictly hyperbolic conservation laws support two types of nonclassical shock waves that we term *overcompressive* and *undercompressive*. Recall that a *Lax* shock, or *compressive* shock, has three characteristics $dx/dt = \lambda_k(U(x,t))$ entering the shock, while one characteristic leaves the shock. An overcompressive shock has all four characteristics converging on the shock. This type of shock is the superposition of a slow Lax shock (for which the slower characteristics converge on the shock from both sides) and a fast Lax shock. To describe undercompressive shocks, we first describe noncompressive shocks. These are shocks for which only one characteristic enters the shock on each side. That is, both families of characteristics pass through the shock. Noncompressive shocks have two unpleasant properties: they are linearly unstable, and Riemann problems have nonunique solutions if the class of admissible shocks is widened to include all noncompressive shocks. However, some Riemann problems that fail to have solutions with only compressive discontinuities, have been solved uniquely by allowing all shocks possessing viscous profiles. The requirement that shocks possess viscous profiles selects some but not all noncompressive shocks. Those noncompressive shocks with viscous profiles are termed undercompressive.

---

*Research supported by National Science Foundation grant DMS 8701348 and by Army Research Office grant DAAL03-88-K-0080.

†Department of Mathematics, North Carolina State University, Raleigh, North Carolina 27695.

Viscous profiles are travelling wave solutions

$$U = U((x - st)/\epsilon) \tag{1.3}$$

of the parabolic system

$$U_t + F(U)_x = \epsilon(DU_x)_x \tag{1.4}$$

that approximate the shock wave as $\epsilon \longrightarrow 0+$.

In this paper, we generally take the diffusion matrix D to be the identity. Specifically, a centered shock wave

$$U(x,t) = \begin{cases} U_-, & if \quad x < st \\ U_+, & if \quad x > st \end{cases} \tag{1.5}$$

is *admissible* if the system

$$U_t + F(U)_x = \epsilon U_{xx} \tag{1.6}$$

has a travelling wave solution (1.3) satisfying the boundary conditions

$$U(\pm\infty) = U_\pm, \quad U'(\pm\infty) = 0, \tag{1.7}$$

where $'$ denotes $d/d\xi$ and $\xi = (x - st)/\epsilon$. Substituting of (1.3) into (1.6), integrating from $\xi = -\infty$, and using the boundary conditions (1.7) at $\xi = -\infty$, leads to the following autonomous system of ordinary differential equations with parameters s and U_-:

$$U' = F(U) - F(U_-) - s(U - U_-). \tag{1.8}$$

The set $H(U_-)$ of points U that are equilibria for (1.8) for some value of s is referred to as the **Hugoniot locus**. A viscous profile corresponds to a trajectory of (1.8) from the equilibrium U_- to a second equilibrium U_+. Thus, if such a trajectory exists, then the shock (1.5) is admissible. It is easy to see that admissible Lax shocks correspond to node to saddle connections (for slow shocks), or saddle to node connections (for fast shocks). These connections are structurally stable to perturbations of F, U_-, s, or D. Undercompressive shocks correspond to saddle to saddle connections, and are thus nongeneric. In fact, the family of smooth vector fields in the plane possessing a saddle to saddle connection has codimension one.

When (1.8) is a quadratic gradient, it is known by a striking result of Chicone [1] that all saddle to saddle connections lie on straight invariant lines. For $F = Q$ given by (1.2), this leads (cf.[8]) not only to a description of undercompressive shocks, but to a complete solution of the Riemann problem for Case I ($a < 3b^2/4$) of the classification described in [4]. However, if U_- lies on an invariant straight line, then that line is invariant for all values of s, so that a saddle to saddle connection $U_- \longrightarrow U_+$ along the line is not broken by changing s. This situation is not generic, and does not persist when higher order terms are added to Q.

REMARK. For quadratic functions F, Isaacson, Marchesin and Plohr [3] have found a useful characterisation of invariant straight lines that can be used to study admissibility in the above sense when the matrix D in (1.4) is not a multiple of the identity.

In this paper, we study perturbations of the quadratic nonlinearity (1.2) by cubic terms. We show how the unique admissible solution of certain Riemann problems is established using a constructive method. In sections 2 and 3 we describe the problem precisely and outline results concerning saddle to saddle connections. These results are then used in section 4 to construct solutions of Riemann problems.

2. The Quadratic Case I. The Riemann problem for system (1.1), with quadratic F given by (1.2), was solved in a series of papers [5,7,8]. For the special case $a = -1, b = 0$, for which the Riemann problem was first solved, there is a high degree of symmetry that simplifies the analysis considerably. In particular, all Lax shocks have viscous profiles, and all saddle to saddle connections lie on lines of symmetry. We use this symmetric case as the starting point for our study, as it brings together various features of nonstrictly hyperbolic systems, which will then be separated by perturbation. In particular, lines of symmetry for the special case coincide with inflection loci, on which genuine nonlinearity fails; and when U_- lies on a line of symmetry the Rankine Hugoniot compatibility condition

$$F(U) - F(U_-) - s(U - U_-) = 0 \tag{2.1}$$

experiences a secondary bifurcation.

Undercompressive shocks for nonstrictly hyperbolic equations were discovered in connection with solving the Riemann problem for the prototype system ($a = -1, b = 0$)

$$\begin{aligned} u_t - (u^2 - v^2)_x &= 0 \\ v_t + (2uv)_x &= 0. \end{aligned} \tag{2.2}$$

System (2.1) has three lines of symmetry, $v = 0, v = \pm\sqrt{3}u$. These lines are invariant both for the hyperbolic system (2.2), and for the associated parabolic system

$$\begin{aligned} u_t - (u^2 - v^2)_x &= \epsilon u_{xx} \\ v_t + (2uv)_x &= \epsilon v_{xx}. \end{aligned} \tag{2.3}$$

It follows from a result of Chicone [1] that if (1.5) is an undercompressive shock for (2.2), then U_- and U_+ must lie on one of the lines of symmetry. To describe these undercompressive shocks in more detail, consider U_- and U_+ on the u-axis. Let $U_- = (u_-, 0)$. For $u_- > 0$, (1.5) is not an undercompressive shock for any choice of U_+. If $u_- < 0$ on the other hand, then the shock wave

$$(u, v)(x, t) = \begin{cases} (u_-, 0), & if \quad x < st \\ (u_+, 0), & if \quad x > st, \end{cases} \tag{2.4}$$

with speed $s = -(u_+ + u_-)$, is undercompressive if

$$(2.5) \qquad -u_-/3 < u_+ < -3u_-.$$

For $u_+ > -3u_-$, (2.4) is a slow shock, and for $u_- < u_+ < -u_-/3$, (2.4) is a fast shock.

Now let $U_- = (u_-, v_-)$, with $u_- < 0$ and $v_- \neq 0$ small. Then (1.5) cannot be an undercompressive shock for any choice of U_+. However, the portions of the Hugoniot locus $H(U_-)$ corresponding to admissible slow and fast Lax shocks vary smoothly with U_-.

The above description of the relative positions in (s, U_-, U_+) space of admissible Lax shocks and undercompressive shocks is not stable to perturbation of the nonlinearities in (2.2). To see this, recall that each undercompressive shock corresponds to a saddle-to-saddle connection for the corresponding vector field (1.8). However, vector fields with saddle-to-saddle connections are codimension one. But for (2.2), we have identified a one parameter family of vector fields, depending on s, each of which has a saddle-to-saddle connection . A generic family of vector fields depending on a single parameter s would have saddle-to-saddle connections at isolated values of s.

There are various perturbations of (2.2) that lead to a generic picture of the relation between undercompressive shocks and Lax shocks. In this paper, we consider equation (1.1) with the flux function F given by

$$(2.6) \qquad F(u,v) = d(-u^3/3 + bu^2v + uv^2) + \alpha \begin{pmatrix} 0 \\ u^3 \end{pmatrix}.$$

The special form in the cubic is chosen in this paper for simplicity, and because this term dominates contributions from other possible cubic and higher order terms in the description of the results. In a forthcoming paper [6], we show how other cubic terms affect the results (i.e., minimally), and we discuss the sense in which the results for $\alpha \neq 0$ do not depend on higher order terms.

Fix $u_- < 0$. For $U_- = (u_-, v_-)$, consider the vector field

$$(2.7) \qquad U' = F(U) - F(U_-) - s(U - U_-),$$

with F given by (2.6). For $b = \alpha = 0$, (2.7) has saddle-to-saddle connections along the $u-$ axis when $v_- = 0$, and s lies in the interval

$$(2.8) \qquad 2u_- \leq s \leq -2u_-/3.$$

The endpoints of this interval represent transitions to Lax shocks and correspond to trajectories between a saddle point and a degenerate saddle point.

The problem we address comes in two parts:

(i) Determine the phase portraits of the vector field (2.6),(2.7) for small values of the parameters v_-, b, α, with $\alpha \neq 0$, as s varies in a neighborhood of the interval (2.8). (The phase portraits for $\alpha = 0$ are well understood, using Chicone's result, cf.,[8].)

(ii) Interpret the unfolding results from (i) in terms of admissible shocks and solutions of Riemann problems.

3. The Vector Field. Let $G = (G_1, G_2)(U, s, \mu)$ denote the right hand side of (2.7), with F given by (2.6), $\mu = (v_-, b, \alpha), U = (u, v), U_- = (u_-, v_-), u_- < 0$.

We restrict s to lie in a neighborhood I of the range (2.8):

$$I = (2u_- - \delta, -2u_-/3 + \delta), \qquad \delta > 0 \quad \text{small.} \tag{3.1}$$

For $\mu = 0$, we have the vector field

$$U' = G(U, s, 0), \qquad \text{for which:}$$

The line $v = 0$ is invariant;

There are equilibria at $(u_-, 0)$ and $(-u_- - s, 0)$, and $u_- < 0 < -u_- - s$. The equilibria have eigenvectors $(1, 0), (0, 1)$;

For $s \in I$:

$(u_-, 0)$ has a positive eigenvalue associated with $(1, 0)$; the eigenvalue for eigenvector $(0, 1)$ is positive for $s < 2u_-$ and negative for $s > 2u_-$;

$(-u_- - s, 0)$ has a negative eigenvalue associated with $(1, 0)$; the eigenvalue for eigenvector $(0, 1)$ is positive for $s < -2u_-/3$ and negative for $s > -2u_-/3$;

A trajectory from $(u_-, 0)$ to $(-u_- - s, 0)$ along the line $v = 0$ is given by $\gamma_s(t) = (u_s(t), 0)$, with

$$u_s(t) = \left[e^{(2u_- + s)t} + 1\right]^{-1} \left[e^{(2u_- + s)t} u_- - (u_- + s)\right].$$

We first describe, for s in the interior of the interval (2.8), a simple use of Melnikov's method, as μ varies away from the origin, to describe saddle-to-saddle connections near the trajectory $\gamma_s(t)$ associated with the undercompressive shocks (2.3)-(2.5). For s near the endpoints of the interval (2.8), the use of Melnikov's method is more complicated, so we content ourselves here with sketching the ideas and outlining the results. Full details will be given in a forthcoming paper[6].

(i) Let $\beta > 0$ be small. For $s \in (2u_- + \beta, -2u_-/3 - \beta), \mu = 0$, the heteroclinic trajectory γ_s joins hyperbolic saddle points (i.e., with nonzero real eigenvalues of opposite sign). For (s, μ) in a neighborhood N of $(2u_- + \beta, -2u_-/3 - \beta) \times \{0\}$ in $\mathbf{R} \times \mathbf{R}^3$, there are unique hyperbolic equilibria near $(u_-, 0)$ and $(-u_- - s, 0)$. There is also a smooth function (the **Melnikov function**) $d : N \longrightarrow \mathbf{R}$ such that $d(s, \mu) = 0$ if and only if the equilibria are joined by a heteroclinic trajectory near γ_s. From above, we have $d(s, 0) = 0$. Let

$$\begin{aligned} J_s(t) &= exp\left[-\int_0^t divG(\gamma_s(t), s, 0)dt\right] \\ &= e^{2st}. \end{aligned}$$

(div means divergence with respect to (u, v)).

Let m be any component of μ. Then $\frac{\partial d}{\partial m}(s,0)$ is given by Melnikov's integral:

$$\begin{aligned}\frac{\partial d}{\partial m}(s,0) &= \int_{-\infty}^{\infty} J_s(t)G(\gamma_s(t),s,0)\wedge\frac{\partial G}{\partial m}(\gamma_s(t),s,0)dt\\ &= \int_{-\infty}^{\infty} e^{2st}\dot{u}_s(t)\frac{\partial G_2}{\partial m}((u_s(t),0),0,0)dt,\end{aligned}$$

where $\wedge$ means determinant.

These integrals can be evaluated explicitly, as explained in [6]. Let

$$\zeta(s) = \begin{cases} \frac{\pi s(s-2u_-)}{\sin\left(\frac{2u_-+3s}{2u_-+s}\right)\pi}, & if \quad s\in(2u_-,-2u_-/3), \quad s\neq 0;\\ 2u_-^2, & if \quad s=0.\end{cases}$$

Then

$$\frac{\partial d}{\partial v_-}(s,0) = 2\zeta(s) \tag{3.2a}$$

$$\frac{\partial d}{\partial b}(s,0) = \frac{2}{3}u_-\zeta(s) \tag{3.2b}$$

$$\frac{\partial d}{\partial \alpha}(s,0) = \frac{1}{3}u_-(s+2u_-)\zeta(s). \tag{3.2c}$$

Since

$$\zeta(s)\neq 0 \quad \text{for} \quad s\in(2u_-,-2u_-/3),$$

by the implicit function theorem, the set $d=0$ is a codimension one manifold.

When $\alpha=0$, there is a degeneracy in the problem, in that $d(s,v_-,b,0)=0$ identically in s along a curve

$$v_- = \phi(b) = -\frac{1}{3}u_-b+O(b^2) \tag{3.3}$$

Indeed, when $d=0$ and (v_-,b) satisfies (3.3), there is a line near the u-axis that is invariant for (2.7) for all s. Both hyperbolic equilibria lie on the line, as does the heteroclinic trajectory joining them.

We claim that the only simultaneous solutions of

$$d(s,v_-,b,\alpha)=0 \tag{3.4a}$$

$$\frac{\partial d}{\partial s}(s,v_-,b,\alpha)=0 \tag{3.4b}$$

in N are given by (3.3), with $\alpha=0, s$ arbitrary. The proof of this statement comes from (3.2):

$$\begin{aligned} d(s,v_-,b,\alpha) &= 2\zeta(s)v_- + \frac{2}{3}u_-\zeta(s)b+\frac{1}{3}u_-(s+3u_-)\zeta(s)\alpha\\ &\quad + O(|(v_-,b,\alpha)|^2).\end{aligned} \tag{3.5a}$$

$$\begin{aligned}\frac{\partial d}{\partial s}(s,v_-,b,\alpha) &= 2\zeta'(s)v_- + \frac{2}{3}u_-\zeta'(s)b+\frac{1}{3}u_-((s+3u_-)\zeta'(s)+\zeta(s))\alpha\\ &\quad + O(|(v_-,b,\alpha)|^2)\end{aligned} \tag{3.5b}$$

Since $\zeta(s)\neq 0$ for $s\in(2u_-,-2u_-/3)$, the gradients of (3.5a,b) on the s-axis are linearly independent. By the implicit function theorem, the zeroes of (3.5) form a two-dimensional surface. But then the surface must coincide with the surface of known solutions $\alpha=0, s$ arbitrary, $v_-=\phi(b)$.

In particular, we have proved the following result.

PROPOSITION 3.1. *Let β be small, and let N be the neighborhood of zero in $\mathbb{R}^3$ discussed above. If $(v_-, b, \alpha) \in N$ and $\alpha \neq 0$, then there is at most one $s \in [2u_- + \beta, -2u_-/3 - \beta]$ such that $d(s, v_-, b, \alpha) = 0$.*

To treat the end points of the interval (2.8), we have to deal with the complications of the bifurcation of equilibria. We use the bifurcation theory approach of Golubitsky and Schaeffer [2], in which the bifurcation parameter s is treated in a distinguished manner. Thus, a singularity corresponds to a nongeneric bifurcation diagram for equilibria. Similarly, an unfolding of a singularity, with unfolding parameters, has a bifurcation diagram for each choice of the unfolding parameters.

The parameters $\mu = (v_-, b, \alpha)$ play the role of unfolding parameters here. Since the unfolding is well understood for $\alpha = 0$, our goal is to describe the situation for $\alpha \neq 0$. We describe the endpoints $s = 2u_-, \quad s = -\frac{2}{3}u_-$ separately.

(ii) When $\mu = 0$, there is a subcritical pitchfork primary bifurcation at $s = 2u_-$. Since the trivial solution $U = U_-$ is preserved for all values of the parameters, the universal unfolding of the singularity (that preserves the trivial solution) has one unfolding parameter. In particular, there is a transcritical primary bifurcation and a limit point bifurcation for each value of μ, except for μ on a two dimensional surface corresponding to subcritical primary bifurcation. As shown in [5], this occurs precisely when the slow characteristic speed loses genuine nonlinearity at U_-:

$$v_- = -bu_- + O(\alpha). \tag{3.6}$$

Because the left state U_- undergoes a bifurcation, for certain parameter values, there are several equilibria that might be left hand states for saddle-to-saddle connections . We therefore define two Melnikov functions, $d(s, \mu)$ for U_-, and a separate function d_1 whose zeroes represent saddle-to-saddle connections $\tilde{U}_- \longrightarrow U_+$ with $\tilde{U}_- \neq U_-$. This is done roughly as follows. The universal unfolding of the primary bifurcation of equilibria at $\mu = 0, s = 2u_-$ may be written

$$x(x^2 + \lambda + \beta x) = 0, \tag{3.7}$$

where $x = v - v_- + h.o.t., \lambda = s - 2u_- + h.o.t.$, and β, the unfolding parameter is a function of μ that is zero at $\mu = 0$. Here, *h.o.t.* signifies higher order terms in U, s, μ. Now solutions of (3.7) are $x = 0$, corresponding to the trivial solution $U = U_-$, and nontrivial solutions that may be parameterized smoothly by x and β, with $\lambda = -(x^2 + \beta)$. The latter correspond to equilibria $\tilde{U}_- \neq U_-$. Thus, the function $d(s, \mu)$ is a smooth extension of that defined in subsection (i), and the function d_1 depends smoothly upon v, v_-, b, α.

As in Proposition 2.1, we have that $d(s, \mu) = 0$ for at most one value of s for fixed μ with $\alpha \neq 0$. The corresponding result for $d_1 = 0$ states that for each choice of the parameters v_-, b, α there is at most one v with $d_1(v, v_-, b, \alpha) = 0$. Since s is a smooth function of v, there is at most one s for each choice of the parameters.

Calculating derivatives of d and d_1 with respect to their variables at the origin leads to a description of the sets $d = 0, d_1 = 0$ for s near $2u_-$, μ near zero.

(iii) When $\mu = 0$, there is a supercritical secondary bifurcation of equilibria at $s = -\frac{2}{3}u_-$. The universal unfolding of this singularity requires two parameters, and the resulting bifurcation diagrams have transcritical secondary bifurcation on a codimension one manifold M. The unfolding takes the form

$$x^3 - \lambda x + \gamma x^2 + \beta = 0, \tag{3.8}$$

where $x = v - v_- + h.o.t.$, $\lambda = s + \frac{2}{3}u_- + h.o.t.$, and γ and β are unfolding parameters.

From (3.8), we express $\beta = \lambda x - x^3 - \gamma x^2$ as a smooth function of the other variables. Thus, the equilibria are parameterised smoothly by x, λ, γ. Then the Melnikov function d is a smooth function of $x, \lambda, \gamma, \alpha$, and the set $d = 0$ is a smooth surface in the space of these variables. Therefore, β is a smooth function on this surface. Next, we calculate derivatives of d at the origin, project out x and finally change back to the original parameters s, v_-, b, α. This process leads to a description of the set $d = 0$ for s near $-\frac{2}{3}u_-$, μ near zero.

The next step in the analysis is to put the pieces together from subsections (i)-(iii). Rather than describe this, together with the structure of the sets $d = 0, d_1 = 0$ in detail, we describe only the bifurcations of the vector field that are significant for shock admissibility. For example, consider bifurcation of equilibria. If none of the equilibria involved in the bifurcation is connected to U_-, then the bifurcation is not significant for shocks.

Now bifurcations of the vector field $G(\cdot, s, \mu)$ occur at certain values of (s, μ). However, s is treated as a distinguished parameter, so that we are really concerned with singular bifurcation diagrams for the family of vector fields $G(\cdot, \cdot, \mu)$ The following surfaces, in addition to $\alpha = 0$, describe the values of μ for which $G(\cdot, \cdot, \mu)$ has a singular bifurcation diagram that affects admisible shocks.

F_1 : *Inflection locus for slow family.* Primary bifurcation is subcritical at $s = \lambda_1(U_-)$ when μ lies on F_1.

$$F_1 : \quad v_- = -u_- b - \frac{1}{2}u_-^2 \alpha + h.o.t.$$

B : *Secondary bifurcation locus.*

$$v_- = -\frac{1}{3}u_- b - \frac{7}{18}u_-^2 \alpha + h.o.t.$$

F: $d(s, \mu) = 0$ at value of s for which there is a limit point for fast family *not* connected to U_-.

$U_- \longrightarrow U_+$ is a saddle-to-saddle connection with speed s, and $s = \lambda_2(U_+^*)$ for some equilibrium $U_+^* \neq U_\pm$.

E: $d(s, \mu) = 0$ at value of s for which there is a limit point for fast family connected to U_-.

$U_- \longrightarrow U_+$ is a saddle-to-saddle connection with speed $s = \lambda_2(U_+)$.

Equations for F and E coincide with the equation for B up to linear terms in μ.

E_1: $d(s, \mu) = 0$ at the primary bifurcation point $s = \lambda_1(U_-)$.

$$v_- = -\frac{1}{3}u_- b - \frac{5}{6}u_-^2 \alpha + h.o.t.$$

F_2: Remote saddle-to-saddle connection at $s = \lambda_1(U_-)$.

$$v_- = -\frac{4}{3}u_- b - \frac{1}{3}u_-^2 \alpha + h.o.t.$$

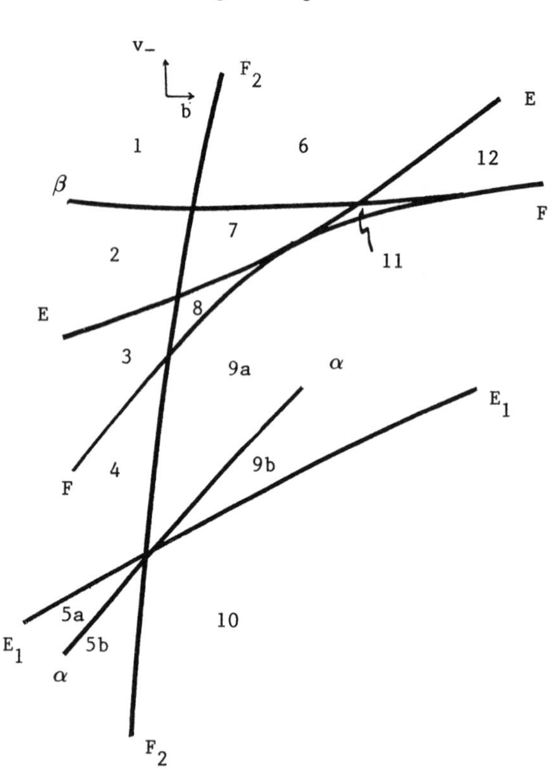

Figure 1. The (b, v_-) plane for $\alpha > 0$.

In Figure 1, we show the curves schematically (i.e., not to scale) in the (b, v_-) plane for fixed $\alpha > 0$. Note that F has tangential intersection with E and B but the curvatures are different. In Figure 2, we give the corresponding bifurcation diagrams for (b, v_-) in each region of Figure 1. Note that part of the bifurcation curve B plays no role in shock admissibility, and is therefore omitted from Figures 1 and 2. The curve F_1 similarly has a lesser role in understanding admissibility of fast shocks, so we show only that portion affecting the analysis of slow shocks in Figure 1 and omit F_1 from Figure 2. In the bifurcation diagrams of Figure 2, some transitions from one bifurcation diagram to its neighbor involve a bifurcation of equilibria that is not represented because it corresponds to inadmissible shocks. In Figure 2, an $\times$ represents a value of (s, U) for which there is a connection $U_- \longrightarrow U$ with speed s, with $d = 0$. A vertical arrow indicates points (s, U) and (s, U') for which $d_1 = 0$ and the corresponding connection with speed s is $U \longrightarrow U'$. Figures 1 and 2 contain the crucial information for solving Riemann problems, as explained in the next section.

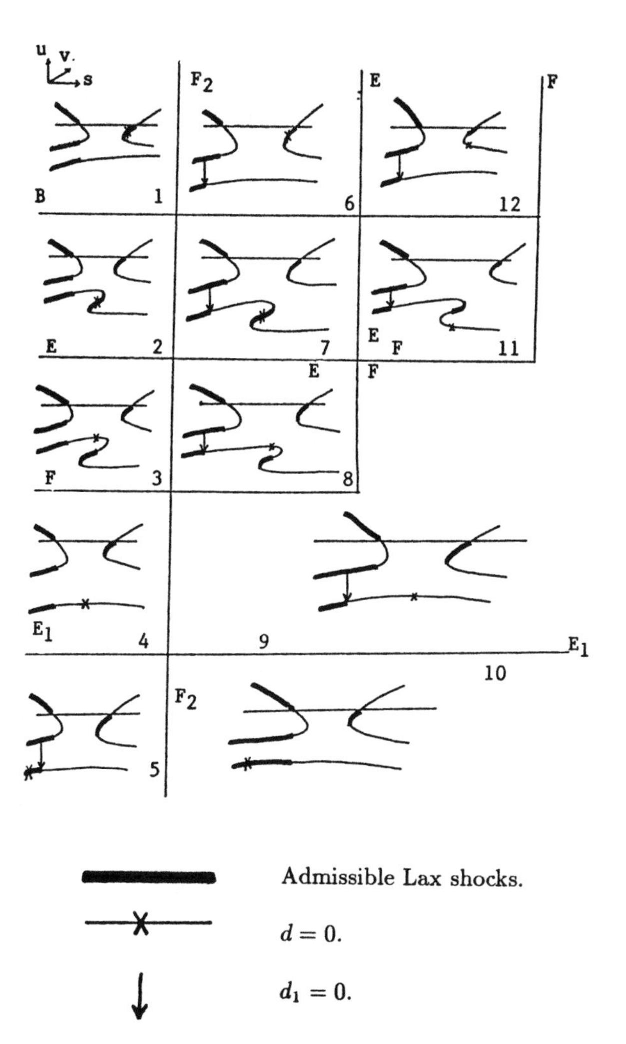

Figure 2. Bifurcation Diagrams

4. The Riemann Problem. Let $F(U)$ be given by (2.6) with $U = (u, v)$. The Riemann problem is

$$U_t + F(U)_x = 0 \tag{4.1}$$

$$U(x,0) = \begin{cases} U_L, & if \quad x < 0 \\ U_R, & if \quad x > 0. \end{cases} \tag{4.2}$$

For $\alpha = 0$, (4.1),(4.2) was solved uniquely in [8] for all b, U_L, U_R. Here, we show how the results of section 3 apply to the perturbed problem with $\alpha \neq 0$. To treat the global problem, with U_L, U_R in some neighborhood of $U = 0$, the umbilic point, would require unfolding a vector field with a high degree of degeneracy. In section 3, we considered a somewhat simpler degeneracy that captures the main features of the problem of admissibility of shocks for nonstrictly hyperbolic systems. Specifically, when $\alpha = 0$ and $b = 0$, the u-axis represents the coincidence of three phenomena:

(i) $v = 0$ is an inflection locus;

(ii) $v = 0$ is a secondary bifurcation curve;

(iii) $v = 0$ is an invariant straight line. This is associated with undercompressive shocks.

Varying b from zero splits items (i) and (ii), but (ii) coincides with (iii) even for $b \neq 0$.

For $\alpha \neq 0$ however, we find there are no straight invariant lines, and the phenomena occur more or less independently. This results in solutions of Riemann problems that are structurally stable to higher order perturbations of the nonlinearity $F(U)$.

The solution of the Riemann problem involves three classes of admissible elementary waves, namely slow waves, fast waves and undercompressive shocks. Slow and fast waves are combinations of admissible Lax shocks and rarefaction waves. These waves may be represented by wave curves, as described in [5], for example. For instance, consider small $\alpha > 0$, small b, and U_L near the u-axis. There is a slow wave curve $W_1(U_L)$ through U_L representing points $U_- = (u_-, v_-)$ such that U_L may be joined, in the (x,t)-plane to U_- by a slow wave. The curve $W_1(U_L)$ crosses the u-axis transversally, and we may parameterize $W_1(U_L)$ smoothly by v_-.

Since b is fixed, varying U_- along $W_1(U_L)$ varies (b, v_-) on a vertical line in Figure 1, thus selecting a sequence of bifurcation diagrams. These bifurcation diagrams contain information about the possible shock waves with U_- on the left. In particular, we can use the bifurcation diagrams to locate undercompressive shocks and the transitions between undercompressive shocks and fast and slow admissible Lax shocks. It is then straightforward to add admissible fast waves in the (x,t)-plane, to get all possible wave combinations with U_L on the left, and intermediate values of U near the u-axis. Each combination of waves determines a constant value U_R of U on the right. The Riemann problem is solved if the values of U_R cover an appropriate set once for each fixed value of U_L in the domain of interest. Here, U_L lies in a neighborhood of a fixed point $(-1, 0)$, and the set of U_R is a neighborhood of $(u_-, -3u_-) \times \{0\}$.

This procedure, and the resulting covering of the U_R plane for each fixed U_L is represented by a picture of the U_R plane, in which the plane is divided into regions.

Each region is labelled by the specific combination of waves used to construct the solution of the Riemann problem for U_R in that region. For each $\alpha \neq 0$, there are lots of different cases to consider, although most cases differ only in small details. To see why there are so many cases, first note that the curves in Figure 1 represent U_L boundaries in the sense of [5]. That is, for fixed b, the solution of the Riemann problem changes as (b, v_L) crosses a boundary in Figure 1. In addition, the sequence of bifurcation diagrams depends on the location of b in Figure 1. Transitions occur at values of b at which curves in Figure 1 intersect. Given the large number of cases, each with a fairly complex solution of the Riemann problem, we content ourselves here with stating that the solution is well defined, unique, and has continuous dependence in L_1 on the initial data. Going through each of the cases, as (b, v_L) varies within Figure 1, we have proved the following result.

PROPOSITION 4.1. *There is $\delta > 0$ and a neighborhood M of $(-1,3) \times \{0\}$ in $\mathbf{R}^2$ such that for each $U_L = (u_L, v_L), b, \alpha$ satisfying $|u_L + 1| + |v_L| + |b| + |\alpha| < \delta$, and each U_R in M, there is a unique centered solution of the Riemann problem (4.1),(4.2) constructed from admissible elementary waves.*

We show just one example of the U_R plane in Figure 3, choosing (b, v_L) to lie in region 1 of Figure 1, and with the sequence 1,2,7,8,9,10 of bifurcation diagrams, as v_- decreases from v_L. Figure 3 is a picture of the neighborhood M in U_R-space of Proposition 4.1. Note that since M is in reality a narrow horizontal strip, the vertical v_R scale has been exaggerated to show detail. M is divided into regions by various curves. Each region is labelled by the corresponding sequence of waves occuring in the solution of the Riemann problem, for U_R in that region. The construction and notation are by now standard (cf.[5,7,8]) and self-explanatory, with the exception of the curves Σ representing undercompressive shocks, and the curve S_σ.

The undercompressive shock curve Σ has end points $E', \tilde{F}_2'$. Let E, F_2 be the points on $W_1(U_L)$ corresponding to the boundaries E, F_2. That is, if $U = (u_-, v_-)$ at E, F_2 (respectively), then the corresponding point (b, v_-) lies on the boundary E, F_2 of Figure 1. Now there is a slow shock from F_2 to a point $\tilde{F}_2$ as shown, with shock speed s that is characteristic on the left of the shock. Further, by the role of the boundary F_2 in Figure 1, there is an undercompressive shock with F_2 on the left that also has speed s. The endpoint $\tilde{F}_2'$ of Σ is on the right of this shock. Similarly, there is an undercompressive shock from E to the endpoint E'. For each point U_- on $W_1(U_L)$ between F_2 and $\tilde{F}_2$, there is a unique corresponding point U_-' on Σ such that $U_- \longrightarrow U_-'$ is an undercompressive shock.

The curve S_σ cuts off fast Lax shocks where they become inadmissible. The point is that when a saddle-to-node connection $U_- \longrightarrow U_+$ in the phase plane becomes a saddle-to-saddle connection $U_- \longrightarrow U_-'$, say at speed s, then there is also a saddle-to-node connection $U_-' \longrightarrow U_+$ at the same speed s. This second shock is the fast shock from a point on Σ to a point on S_σ.

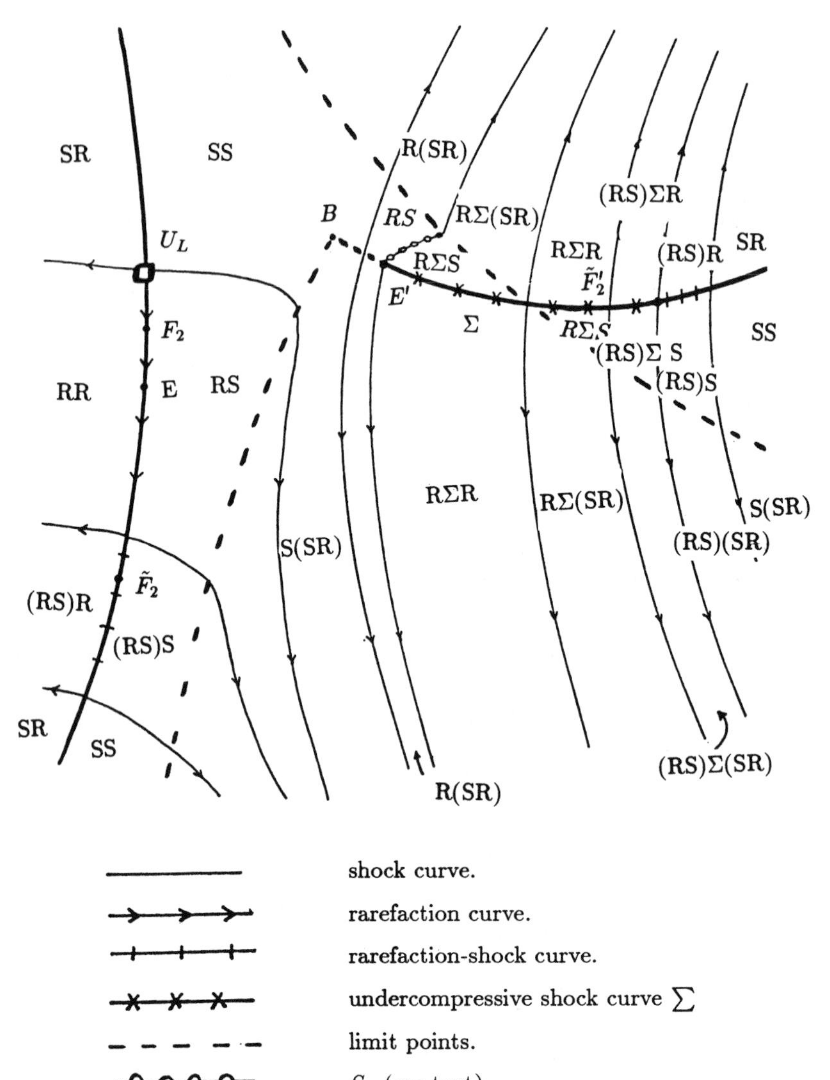

Figure 3. Solution of Riemann Problems: the U_R plane.

REFERENCES

[1] C. C. CHICONE, *Quadratic gradients on the plane are generically Morse-Smale*, J. Differential Equations, 33 (1979), pp. 159–161.

[2] M. GOLUBITSKY AND D.G. SCHAEFFER, *Singularities and Groups in Bifurcation Theory*, Springer, New York, 1985.

[3] E. ISAACSON, D. MARCHESIN, B. PLOHR, *Transitional waves*, preprint.

[4] D.G. SCHAEFFER AND M. SHEARER, *The classification of 2 × 2 systems of nonstrictly hy-*

perbolic conservation laws, with application to oil recovery, Comm. Pure Appl. Math., 40 (1987), pp. 141–178.

[5] D.G. SCHAEFFER AND M. SHEARER, *Riemann problems for nonstrictly hyperbolic* 2×2 *systems of conservation laws*, Trans. Amer. Math. Soc., 304 (1987), pp. 267–306.

[6] S. SCHECTER AND M. SHEARER, *Undercompressive shocks for nonstrictly hyperbolic conservation laws*, IMA preprint.

[7] M. SHEARER, D.G. SCHAEFFER, D. MARCHESIN AND P.J. PAES-LEME, *Solution of the Riemann problem for a prototype* 2×2 *system of nonstrictly hyperbolic conservation laws*, Arch. Rat. Mech. Anal., 97 (1987), pp. 299–320.

[8] M. SHEARER, *The Riemann problem for* 2×2 *systems of hyperbolic conservation laws with Case I quadratic nonlinearities*, J. Differential Equations, 80 (1989), pp. 343–363.

MEASURE VALUED SOLUTIONS TO A BACKWARD-FORWARD HEAT EQUATION: A CONFERENCE REPORT*

M. SLEMROD†

Abstract. We examine the asymptotic behavior of measure valued solutions to the initial value problem for the nonlinear heat conduction equation

$$\frac{\partial u}{\partial t} = \nabla \cdot \mathbf{q}(\nabla u), \qquad x \in \Omega,\ t > 0$$

in a bounded domain $\Omega \subset \mathbf{R}^N$ with boundary condtions of the form

$$u = 0 \text{ on } \partial\Omega \quad \text{or} \quad \mathbf{q}(\nabla u) \cdot \mathbf{n} = 0 \text{ on } \partial\Omega.$$

As major themes of this workshop are ill-posed and mixed initial value problems, it seems appropriate to report on some recent work of mine on a backward-forward heat equation which will appear in Slemrod (1989). To be specific consider the heat conduction equation

$$\frac{\partial u}{\partial t} = \nabla \cdot \mathbf{q}(\nabla u) \qquad x \in \Omega,\ t > 0$$

in a bounded domain $\Omega \subseteq \mathbf{R}^N$. The boundary $\partial\Omega$ of Ω is assumed to be smooth. On $\partial\Omega$ impose either homogeneous Dirichlet boundary values

$$u = 0 \quad \text{on} \quad \partial\Omega,\ t > 0$$

or no flux insulated boundary values

$$\mathbf{q}(\nabla u) \cdot \mathbf{n} = 0 \quad \text{on} \quad \partial\Omega,\ t > 0$$

where $\mathbf{n}(x)$ denotes the exterior unit normal at $x \in \partial\Omega$. Also u will satisfy the initial condition

$$u(x,0) = u_0(x), \quad x \in \Omega.$$

For convenience call the above system P_D and P_N respectively.

The main interest of this paper is that $\mathbf{q}$ will only be assumed to satisfy smoothness and growth conditions and Fourier's inequality

$$\lambda \cdot \mathbf{q}(\lambda) \geq 0.$$

*This research was supported in part by the Air Force Office of Scientific Research, Air Force Systems, USAF, under Contract Grant No. AFOSR-87-0315. The United States Government is authorized to reproduce and distribute reprints for government purposes not withstanding any copyright herein.

†Center for Mathematical Sciences, University of Wisconsin, Madison, WI 53706

No monotonicity requirements such as $(\mathbf{q}(\lambda_1)-\mathbf{q}(\lambda_2))\cdot(\lambda_1-\lambda_2) \geq 0$ for $\lambda_1, \lambda_2 \in \mathbf{R}^N$ will be imposed.

Note the Fourier inequality is consistent with the classical Clausius-Duhem inequality (Truesdell, 1984, p. 116).

Other assumptions on $\mathbf{q}$ are that

(i) $\mathbf{q}$ is (at least) a continuous map : $\mathbf{R}^N \to \mathbf{R}^N$ satisfying growth conditions

(ii)

$$|\mathbf{q}(\lambda)| \leq c_1(1+|\lambda|^\gamma), \quad 1 \leq \gamma < 2$$

(iii)

$$c_2|\lambda|^2 - c_3 \leq |\lambda \cdot \mathbf{q}(\lambda)|$$

for all $\lambda \in \mathbf{R}^N$; here c_1, c_2, c_3 are positive constants.

(iv) $\mathbf{q}$ is the gradient of a $C^1(\mathbf{R}^N;\mathbf{R})$ potential Φ, i.e.

$$\mathbf{q}(\lambda) = \nabla\Phi(\lambda)$$

for all $\lambda \in \mathbf{R}^N$.

(v)

$$\begin{aligned} \ker\{\lambda \cdot \mathbf{q}(\lambda)\} &\stackrel{\text{def}}{=} \{\lambda \in \mathbf{R}^N; \lambda \cdot \mathbf{q}(\lambda) = 0\} \\ &\subset \{\lambda \in \mathbf{R}^N; |\lambda| < \rho_0\} \quad \text{for some} \quad \rho_0 > 0. \end{aligned}$$

The approach taken in Slemrod (1989d) to analyze P_D and P_N is rather straightforward and physically natural. First imbed P_D or P_N respectively in the singularly perturbed systems ($P_{D,\epsilon}$ and $P_{N,\epsilon}$)

$$\frac{\partial u^\epsilon}{\partial t} = \nabla \cdot (\mathbf{q}(\nabla u^\epsilon)) - \epsilon\Delta^2 u, \quad x \in \Omega, t > 0, \ \epsilon > 0,$$

with boundary conditions

$$u^\epsilon = 0, \Delta u^\epsilon = 0 \quad \text{on} \quad \partial\Omega, \ t > 0$$

or

$$\mathbf{q}(\nabla u^\epsilon) \cdot \mathbf{n} = 0, \ \frac{\partial}{\partial \mathbf{n}}(\Delta u^\epsilon) = 0 \quad \text{on} \quad \partial\Omega, \ t > 0,$$

and initial data

$$u^\epsilon(x,0) = u_0(x).$$

The regularization is not ad hoc and is in fact based on a higher order theory of heat conduction due to J. C. Maxwell (1876, eqns. (53), (54)); see also Truesdell and Noll (1965, p. 514).

Problems $P_{D,\epsilon}$ and $P_{N,\epsilon}$ admit two natural "energy" estimates which motivate an attempt to pass to a weak limit as $\epsilon \to 0^+$ and hence obtain a weak solution of P_D and P_N. Unfortunately the presence of the nonlinear terms $\mathbf{q}(\nabla u^\epsilon)$ prevents the success of this venture. However in the spirit of the work of L. Tartar (1979,

1982) and later work of R. DiPerna (1983 a, b, c, 1985), M. Schonbek (1982), and R. DiPerna and A. Majda (1987a, b) the following information on the sequence $\{\mathbf{q}(\nabla u^\epsilon)\}$ is known. Namely if for $0 < T < \infty$,

$$\nabla u^\epsilon \rightharpoonup \nabla \overline{u} \quad \text{in} \quad L^2(Q_T),$$

$Q_T = (0,T)\times\Omega$, where $\rightharpoonup$ denotes weak convergence then for q continuous satisfying growth conditions there is a subsequence $\{\nabla u^{\epsilon_k}\}$ so that

$$q(\nabla u^{\epsilon_k}) \rightharpoonup \langle q(\lambda), \nu_{x,t}(\lambda)\rangle$$

weakly in $L^1(Q_T)$ for a probability measure $\nu_{x,t}(\lambda)$,

$$\langle \mathbf{q}(\lambda), \nu_{x,t}(\lambda)\rangle = \int_{\mathbf{R}^N} \mathbf{q}(\lambda) d\nu_{x,t}(\lambda)$$

and ν is called a Young measure.

The above representation of weak limits of $\mathbf{q}(\nabla u^\epsilon)$ permits a passage to weak limits in $P_{D,\epsilon}$ and $P_{N,\epsilon}$. These limits satisfy a measure theoretic version of P_D and P_N and the associated $\overline{u}$ is called a "measure valued" solution of P_D or P_N (in the sense of DiPerna (1985)). The function $\overline{u}$ lies in $L^\infty((0,\infty); V)$ when $V = H_0^1(\Omega)$ for P_D and $H^1(\Omega)$ for P_N. Moreover it inherits a natural "energy" inequality from the regularized problems $P_{D,\epsilon}$ and $P_{N,\epsilon}$. This inequality can be exploited to establish the trend to equilibrium as $t \to \infty$ of $\overline{u}$ (see also Slemrod, 1989a, b, c).

Notation:

We *endow* $L^2(\Omega)$ with the usual inner product

$$(u,v) = \int_\Omega u(x)v(x)dx \quad \text{for } u,v \in L^2(\Omega),$$
$$\|u\|^2 = (u,u).$$

Q_T *denotes* the cylinder $\Omega \times (0,T)$ and for $u,v \in L^2(Q_T)$

$$(u,v)_{L^2(Q_T)} = \int_{Q_T} u(x,t)v(x,t)dxdt,$$
$$\|u\|^2_{L^2(Q_T)} = (u,u)_{L^2(Q_T)}.$$

For problem P_D we denote

$V = H_0^1(\Omega)$ where $H_0^1(\Omega)$ is endowed with the inner product

$$(u,v)_1 = \int_\Omega \nabla u \cdot \nabla v dx \quad \text{for} \quad u,v \in H_0^1(\Omega),$$
$$\|u\|_1^2 = (u,u)_1.$$

For problem P_N we denote

$V = H^1(\Omega)$ where $H^1(\Omega)$ is endowed with the inner product

$$(u,v)_1 = \int_\Omega \nabla u \cdot \nabla v dx + \left(\int_\Omega u dx\right)\left(\int_\Omega v dx\right) \text{ for } u,v \in H^1(\Omega),$$
$$\|u\|_1^2 = (u,u)_1^2$$

(see Temam (1988)).

For P_D we set

$$(u,v)_{H^1(Q_T)} = \int_{Q_T} \frac{\partial u}{\partial t}\frac{\partial v}{\partial t} + \nabla u \cdot \nabla v dx dt,$$
$$\|u\|^2_{H^1(Q_T)} = (u,u)_{H^1(Q_T)}.$$

For P_N we set

$$(u,v)_{H^1(Q_T)} = \int_{Q_T} \frac{\partial u}{\partial t}\frac{\partial v}{\partial t} + \nabla u \cdot \nabla v dx dt + \int_0^T \left(\int_\Omega u dx\right)\left(\int_\Omega v dx\right) dt,$$
$$\|u\|^2_{H^1(Q_T)} = (u,u)_{H^1(Q_T)}.$$

The subscript b will denote a uniformly bounded subset of an indicated set.

For the problem P_D we denote the space of *test functions* $W = C_0^\infty(\Omega)$.

For the problem P_N we denote the space of *test functions* $W = \{w \in C^\infty(\Omega); \frac{\partial w}{\partial \mathbf{n}} = 0\}$. Let $L^\infty_w(Q_T; M(\mathbf{R}^N))$ *denote the space of weak $*$ measurable mappings* $\mu : Q_T \mapsto M(\mathbf{R}^N)$ that are essentially bounded with norm

$$\|\mu\|_{\infty,M} = \text{ess} \sup_{x,t\in Q_T} \|\mu(x,t)\|_M < \infty.$$

(Recall μ is weak $*$ measurable if $\langle \mu(x,t), f\rangle$ is measurable with respect to $x,t \in Q_T$ for every $f \in C_0(\mathbf{R}^N)$.)

$M(\mathbf{R}^N)$ is the Banach space of bounded Radon measures over $\mathbf{R}^N$. For $\nu \in M(\mathbf{R}^N)$ we write

$$\|\nu\|_M = \int_{\mathbf{R}^N} d|\nu|.$$

Prob $(\mathbf{R}^N)$ is the Banach space of probablity measures over $\mathbf{R}^N$. For $\nu \in$ Prob $(\mathbf{R}^n)$ we write

$$\|\nu\|_M = \int_{\mathbf{R}^N} d\nu.$$

$C_0(\mathbf{R}^N)$ denotes the Banach space of continuous functions $f : \mathbf{R}^N \to \mathbf{R}$ satisfying $\lim_{|\lambda|\to\infty} f(\lambda) = 0$,

$$\|f\|_{C_0(\mathbf{R}^N)} = \sup_{\lambda\in\mathbf{R}^N} |f(\lambda)|.$$

The arrows $\rightarrow, \rightharpoonup, \stackrel{*}{\rightharpoonup}$ denote strong, weak, and weak $*$ convergence respectively.

We now recall the concept of measure valued solution introduced by R. DiPerna (1985).

An element $\overline{u} \in H^1(Q_T) \cap C([0,T]; L^2(\Omega)) \cap L^\infty((0,T); V)$ is a *measure valued solution of* P_D or P_N on Q_T if there exists a measure valued map

$$\nu : x, t \mapsto \nu_{x,t} \in \text{Prob}\,(\mathbf{R}^N)$$

from the physical domain Q_T to Prob $(\mathbf{R}^N)$ the space of probability measures over the state space domain $\mathbf{R}^N$ so that

$$\frac{d}{dt}(\overline{u}, w) + (\langle q(\lambda), \nu_{x,t}(\lambda) \rangle, \nabla w) = 0$$

for all $w \in W$ a.e. in $(0,T)$;

$$\nabla \overline{u} = \langle \lambda, \nu_{x,t}(\lambda) \rangle \quad \text{a.e. in } Q_T;$$
$$\overline{u}(x,0) = u_0(x) \quad x \in \Omega.$$

An element $\overline{u} \in V$ is a *measure valued equilibrium solution of* P_D or P_N if there exists a measure valued map $\nu : x, t \mapsto \nu_{x,t} \in \text{Prob}(\mathbf{R}^N)$ from $\Omega \times [0, \infty)$ to $\text{Prob}(\mathbf{R}^N)$ so that

$$(\langle q(\lambda), \nu_{x,t}(\lambda), \nabla w) = 0$$

for all $w \in W$ a.e. on $(0, \infty)$ and

$$\nabla \overline{u}(x) = \langle \lambda, \nu_{x,t}(\lambda) \rangle \quad \text{a.e. in } \Omega \times [0, \infty).$$

We state the fundamental theorem for Young measures as given by J. M. Ball (1988).

Let $S \subset \mathbf{R}^n$ be Lebesgue measurable. Let $K \subset \mathbf{R}^m$ be closed and let $z^{(j)} : S \mapsto \mathbf{R}^n$, $j = 1, 2, \ldots$ be a sequence of Lebesgue measurable functions satisfying $z^{(j)}(\cdot) \rightarrow K$ in measure as $j \mapsto \infty$, i.e. given any open neighborhood U of K in $\mathbf{R}^m$

$$\lim_{j \rightarrow \infty} \text{ meas } \{y \in S; z^{(j)}(y) \notin U\} = 0.$$

Then there exists a subsequence $z^{(\mu)}$ of $z^{(j)}$ and a family $\{\nu_y\}$, $y \in S$, of positive measures on $\mathbf{R}^m$, depending measurably on y, so that

(i) $\|\nu_y\|_M = \int_{\mathbf{R}^m} d\nu_y \leq 1$ a.e. in $y \in S$;

(ii) supp $\nu_y \subset K$ for almost all $y \in S$;

(iii) $f(z^{(\mu)}) \stackrel{*}{\rightharpoonup} \langle \nu_y, f \rangle = \int_{\mathbf{R}^m} f(\lambda) d\nu_y(\lambda)$ in $L^\infty(S)$ for each continuous function $f \in C_0(\mathbf{R}^m)$.

Suppose further that $\{z^{(\mu)}\}$ satisfies the boundedness condition

$$\lim_{k\to\infty}\ \sup_{\mu}\ \text{meas}\{y\in S\cap B_R : |z^{(\mu)}(y)|\geq k\}=0$$

for every $R>0$ where $B_R=\{y\in \mathbf{R}^m; |y|\leq R\}$. Then $\|\nu_y\|_M=1$ for $y\in S$ (i.e. ν_y is a probability measure) and given any measurable subset A of S

$$f(z^{(\mu)})\rightharpoonup\langle\nu_y,f\rangle \text{ in } L^1(A)$$

for any continuous function $f:\mathbf{R}^m\to\mathbf{R}$ such that $\{f(z^{(\mu)})\}$ is sequentially weakly relatively compact in $L^1(A)$.

As noted in Ball (1988) the boundedness condition is very weak and is equivalent to the following: given any $R>0$ there exists a continuous nondecreasing function $g_R:[0,\infty)\to\mathbf{R}$ with $\lim_{t\to\infty} g_R(t)=\infty$, such that

$$\sup_{\mu}\int_{S\cap B_R} g_R(|z^{(\mu)}(y)|)dy<\infty.$$

Furthermore Ball notes that if A is bounded, the condition that $\{f(z^{(\mu)})\}$ be sequentially weakly relatively compact in $L^1(A)$ is satisfied if and only if

$$\sup_{\mu}\int_A \psi(|f(z^{(\mu)})|)dy<\infty$$

for some continuous function $\psi:[0,\infty)\to\mathbf{R}$ with $\lim_{t\to\infty}\frac{\psi(t)}{t}=\infty$ (de la Vallée Poussin's criterion; cf. MacShane (1947), Dellacherie & Meyer (1975), Natanson (1955)).

To construct measure valued solutions of P_D or P_N we proceed as follows: First let u^ϵ be a classical smooth solution of either $P_{D,\epsilon}$ or $P_{N,\epsilon}$. Then for all $t,\tau\in\mathbf{R}^+$:

$$\begin{aligned}\|u^\epsilon(t+\tau)\|^2-\|u^\epsilon(t)\|^2=&-2\int_0^\tau((\Delta u^\epsilon(s+t),\mathbf{q}(\nabla u^\epsilon(s+t))))ds\\&-2\epsilon\int_0^\tau\|\Delta u^\epsilon\|^2ds\end{aligned}$$

and

$$\begin{aligned}&\epsilon\|\Delta u^\epsilon(t)\|^2+\int_\Omega\Phi(\nabla u^\epsilon(x,t))dx\\&\quad+\int_0^t\|\frac{\partial u^\epsilon(s)}{\partial t}\|^2dx=\epsilon\|\Delta u_0\|^2+\int_\Omega\Phi(\nabla u_0(x))dx.\end{aligned}$$

Furthermore for $P_{N,\epsilon}$

$$\int_\Omega u^\epsilon(x,t)dx=\int_\Omega u_0(x)dx$$

for all $t\in\mathbf{R}^+$.

From the above estimates we observe that for any $T > 0$

$$\{u^\epsilon\} \subset L_b^\infty((0,\infty); V) \cap H_b^1(Q_T),$$
$$\{\frac{\partial u^\epsilon}{\partial t}\} \subset L_b^2((0,\infty); L^2(\Omega)),$$
$$\{\epsilon^{1/2}\Delta u^\epsilon\} \subseteq L_b^\infty((0,\infty); L^2(\Omega)).$$

Furthermore there exists a subsequence of $\{u^\epsilon\}$ also denoted by $\{u^\epsilon\}$ and $\overline{u}$,

$$\overline{u} \in L^\infty((0,\infty); V)) \cap H^1(Q_T) \cap C([0,T]; L^2(\Omega)),$$
$$\frac{\partial \overline{u}}{\partial t} \in L^2((0,\infty); L^2(\Omega)),$$

so that

(a) $u^\epsilon \overset{*}{\rightharpoonup} \overline{u}$ in $L^\infty((0,\infty); V)$;

(b) $\nabla u^\epsilon \overset{*}{\rightharpoonup} \nabla \overline{u}$ in $L^\infty((0,\infty); L^2(\Omega))$;

(c) $\frac{\partial u^\epsilon}{\partial t} \rightharpoonup \frac{\partial \overline{u}}{\partial t}$ in $L^2((0,\infty); L^2(\Omega))$;

(d) $u^\epsilon \rightharpoonup \overline{u}$ in $H^1(Q_T)$;

(e) $u^\epsilon \to \overline{u}$ in $C([0,T]; L^2(\Omega))$;

(f) $\overline{u}(t) \to u_0$ in $L^2(\Omega)$ as $t \to 0^+$.

We can then use (a)-(f) above and the fundamental theorem on Young measures to conclude there exists a further subsequence again denoted by $\{u^\epsilon\}$ and a probability measure $\nu_{x,t}$, $(x,t) \in \Omega \times \mathbf{R}^+$, so that for every bounded subset $A \subset \Omega \times \mathbf{R}^+$ and every $f(\lambda)$ satisfying

$$|f(\lambda)| \leq \text{const.}(1 + |\lambda|^\gamma), \;\; 0 < \gamma < 2, \; \lambda \in \mathbf{R}^N$$

we have

$$f(\nabla u^\epsilon) \rightharpoonup \langle \nu_{x,t}, f \rangle \;\text{ in }\; L^2(A),$$

and

$$\nabla \overline{u} = \langle \nu_{x,t}, \lambda \rangle \;\text{ a.e. in }\; A.$$

We then see that $\overline{u}$ is a measure valued solution of the relevant initial-boundary value problem P_D or P_N. Furthermore if we set

$$g(\lambda) \overset{\text{def.}}{=} \lambda \cdot \mathbf{q}(\lambda) \quad |\lambda| \leq a,$$

$$\frac{\lambda \cdot \mathbf{q}(\lambda) a^3}{|\lambda|^3} \quad |\lambda| > a,$$

then $\overline{u}$ satisfies the "energy" inequality

$$\|\overline{u}(t+T)\|^2 - \|\overline{u}(t)\|^2 \leq -2 \int_0^T \int_\Omega \langle g(\lambda), \nu_{x,s+t} \rangle dt ds.$$

Remark. The idea of minorizing $\lambda \cdot \mathbf{q}(\lambda)$ by $g(\lambda)$ was given to the author by Professor E. Zuazua. It allows the application of the fundamental theorem on Young measure to the energy inequality (3.5) without putting additional growth restrictions on $\mathbf{q}$.

Having established the existence of measure valued solutions we now turn to their asymptotic behavior of measure valued solutions.

In what follows we shall assume u_0 is smooth as is necessary for $P_{D,\epsilon}$ and $P_{N,\epsilon}$ to have classical smooth solutions. $\mathcal{O}^+(u_0) = \bigcup_{t\geq 0} \overline{u}(t; u_0)$ defines the *positive orbit* in V through u_0. We already know $\overline{u} \in L^\infty((0,\infty); V)$ and so $\mathcal{O}^+(u_0) \subseteq B \subset V$, with a metrized weak-$V$ topology with metric d. Define *the w-distance* between two sets $B_1, B_2 \subseteq B$ by

$$\text{w-dist}(B_1, B_2) = \inf_{\substack{b_1 \in B_1 \\ b_2 \in B_2}} d(b_1, b_2).$$

Finally define the *weak ω-limit set* of $\mathcal{O}^+(u_0)$ by $\omega_w(u_0) = \{\chi \in V; \overline{u}(t_n; u_0) \rightharpoonup \chi$ in V as $n \to \infty$ for some sequence $\{t_n\}$, $t_n \to \infty\}$.

It thus follows that $\omega_w(u_0)$ is non-empty and w-dist$(\overline{u}(t; u_0), \omega_w(u_0)) \to 0$ as $t \to \infty$.

Now we are able to state the main result of Slemrod (1989d). Namely we can give a categorization of $\omega_w(u_0)$.

Theorem. Let $\chi \in \omega_w(u_0)$. Then χ is a measure valued equilibrium solution of $P_D(P_N)$ evolution equations, i.e. there is positive probability measure $\overline{\nu}_{x,t}$ with

$$\text{supp}\, \overline{\nu}_{x,t} \subseteq \ker \mathbf{q}$$

hence satisfying

$$\langle \mathbf{q}(\lambda), \overline{\nu}_{x,t} \rangle = 0 \quad \text{a.e. in} \quad Q_T$$

and

$$\nabla \chi(x) = \langle \lambda, \overline{\nu}_{x,t} \rangle \quad \text{a.e. in} \quad Q_T.$$

Moreover if $\ker q_i \subseteq [a_i, b_i]$, $i = 1, \ldots, N$ then

$$a_i \leq (\nabla \chi)_i \leq b_i \quad \text{a.e. in} \quad \Omega, \; i = 1, \ldots, N.$$

From the above theorem we can establish the following corollaries.

Corollary: For problem P_D if for some i, $1 \leq i \leq N$, then $(\nabla \chi)_i = 0$ a.e. in Ω.

Corollary: For Problem P_D: if for each i, $\ker q_i \subseteq R^-$ or $\ker q_i \subseteq R^+$, $1 \leq i \leq N$, then $\omega_w(u_0) = \{0\}$ and for any u_0 $\overline{u}(t, u_0) \rightharpoonup 0$ as $t \to \infty$ in $H_0^1(\Omega)$.

Corollary: For Problem P_N: if $\ker \mathbf{q} = 0$ then $\omega_w(u_0) = c$ (a constant),

$$c = (\text{meas}\, \Omega)^{-1} \int_\Omega u_0(x) dx,$$

and for any u_0 $\overline{u}(t, u_0) \rightharpoonup c$ as $t \to \infty$ in $H^1(\Omega)$.

Examples.

1) Consider the case $N = 1$ and q possessing the graph shown in Figure 1.

For Problem P_D: $\overline{u}(t, u_0) \rightharpoonup 0$ as $t \to \infty$ in $H_0^1(\Omega)$.

For Problem P_N: $\text{dist}(\overline{u}(t, u_0), \omega_w(u_0)) \to 0$ in $H^1(\Omega)$ as $t \to \infty$ where $\omega_w(u_0) \subseteq \{\chi$; measure valued equilibrium solutions of P_N, $0 \le \frac{d\chi}{dx}(x) \le \xi_1$ a.e. in $\Omega\}$.

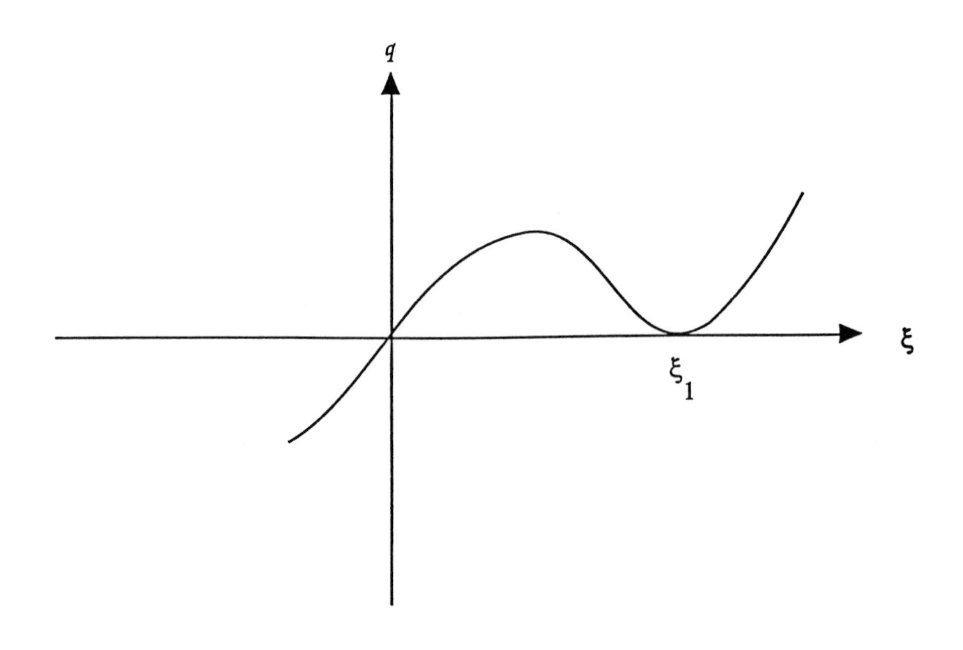

Figure 1

2) Consider the case $N = 1$ and q possessing the graph shown in Figure 3.

For Problems $P_D(P_N)$: weak-dist$(\overline{u}(t, u_0); \omega_w(u_0)) \to 0$ in V as $t \to \infty$ where $\omega_w(u_0) \subseteq \{\chi$; measure valued equilibrium solutions of $P_D(P_N)$, $\xi_0 \le \frac{d\chi}{dx}(n) \le \xi_1$ a.e. in $\Omega\}$.

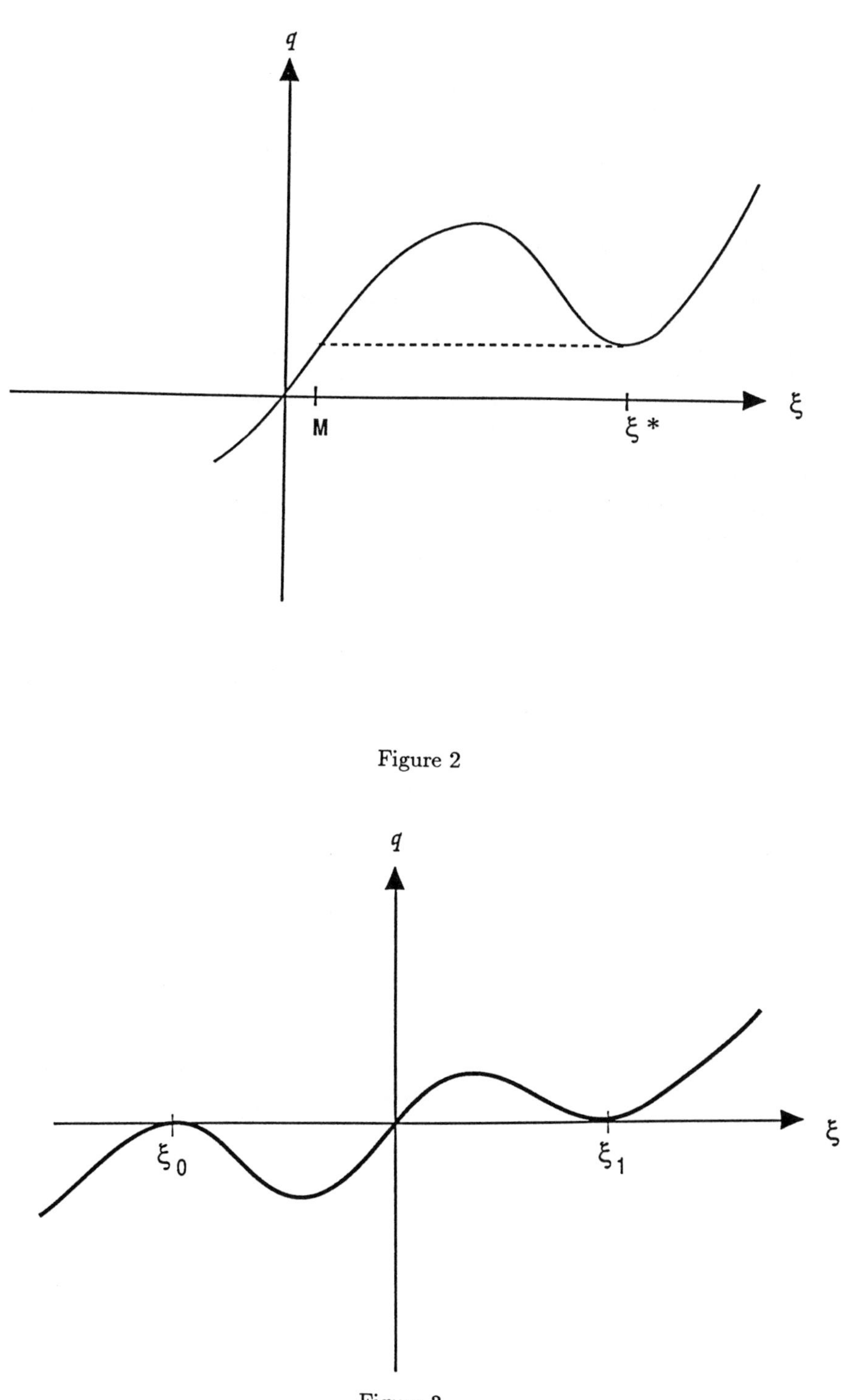

Figure 2

Figure 3

REFERENCES

J. M. Ball (1988), A version of the fundamental theorem for Young measures, to appear Proc. CNRS-NSF Workshop on Continuum Theory of Phase Transitions: Nice, France, January 1988, eds. M. Rascle and D. Serre, Springer Lecture Notes in Mathematics.

C. Dellacherie and P-A. Meyer (1975), Probabilities et Potentiel, Hermann, Paris.

R. J. DiPerna (1983a), Convergence of approximate solutions to conservation laws, Archive for Rational Mechanics and Analysis 82, pp. 27-70.

R. J. DiPerna (1983b), Convergence of the viscosity method for isentropic gas dynamics, Comm. Math. Physics 91, pp. 1-30.

R. J. DiPerna (1983c), Generalized solutions to conservation laws, in Systems of Nonlinear Partial Differential Equations, NATO ASI Series, ed. J. M. Ball, D. Reidel.

R. J. DiPerna (1985), Measure-valued solutions to conservation laws, Archive for Rational Analysis and Mechanics 88, pp. 223-270.

R. J. DiPerna and A. J. Majda (1987a), Concentrations and regularizations in weak solutions of the incompressible fluid equations, Comm. Math. Physics 108, pp. 667-689.

R. J. DiPerna and A. J. Majda (1987b), Concentrations in regularizations for 2-D incompressible flow, Comm. Pure and Applied Math. 40, pp. 301-345.

E. J. MacShane (1947), Integration, Princeton Univ. Press.

J. C. Maxwell (1876), On stresses in rarified gases arising from inequalities of temperature, Phil. Trans. Roy. Soc. London 170, pp. 231-256 = Papers 2, pp. 680-712.

I. P. Natanson (1955), Theory of Functions of a Real Variable, vol. 1, F. Unger Publishing Co., New York.

M. E. Schonbek (1982), Convergence of solutions to nonlinear dispersive equations, Comm. in Partial Differential Equations 7, pp. 959-1000.

M. Slemrod (1989a), Weak asymptotic decay via a "relaxed invariance principle" for a wave equation with nonlinear, nonmonotone clamping, to appear Proc. Royal Soc. Edinburgh.

M. Slemrod (1989b), Trend to equilibrium in the Becker-Döring cluster equations, to appear Nonlinearity.

M. Slemrod (1989c), The relaxed invariance principle and weakly dissipative infinite dimensional dynamical systems, to appear Proc. Conf. on Mixed Problems, ed. K. Kirschgassner, Springer Lecture Notes.

M. Slemrod (1989d), Dynamics of measured valued solutions to a backwards forwards heat equation, submitted to Dynamics and Differential Equations.

L. Tartar (1979), Compensated compactness and applications to partial differential equations in "Nonlinear Analysis and Mechanics", Herior-Watt Symposium IV, Pitman Research Notes in Mathematics pp. 136-192.

R. Temam (1988), Infinite-Dimensional Dynamical Systems in Mechanics and Physics, Springer-Verlag, New York.

C. Truesdell (1984), Rational Thermodynamics, Second Edition, Springer-Verlag, New York.

C. Truesdell and W. Noll (1965), The Non-Linear Field Theories of Mechanics, in Encyclopedia of Physics, ed. S. Flugge, Vol. III/3, Springer-Verlag, Berlin-Heidelberg-New York.

ONE-DIMENSIONAL THERMOMECHANICAL PHASE TRANSITIONS WITH NON-CONVEX POTENTIALS OF GINZBURG-LANDAU TYPE

JÜRGEN SPREKELS*

Abstract. In this paper we study the system of partial differential equations governing the nonlinear thermomechanical processes in non-viscous, heat-conducting, one-dimensional solids. To allow for both stress- and temperature-induced solid-solid phase transitions in the material, possibly accompanied by hysteresis effects, a non-convex free energy of Ginzburg-Landau form is assumed. Results concerning the well-posedness of the problem, as well as the numerical approximation and the optimal control of the solutions, are presented in the paper, in particular in connection with the austenitic-martensitic phase transitions in the so-called "shape memory alloys".

Key words. phase transitions, non-convex potentials, Ginzburg-Landau theory, shape memory alloys, hysteresis, conservation laws.

AMS(MOS) subject classifications. 35L65, 35K60, 73U05, 73B30

1. Introduction.

In this paper we consider thermomechanical processes in non-viscous, one-dimensional heat-conducting solids of constant density ρ (assumed normalized to unity) that are subjected to heating and loading. We think of metallic solids that do not only respond to a change of the strain $\epsilon = u_x$ (u stands for the displacement) by an elastic stress $\sigma = \sigma(\epsilon)$, but also react to changes of the curvature of their metallic lattices by a couple stress $\mu = \mu(\epsilon_x)$. Thus, the corresponding free energy F is assumed in Ginzburg-Landau form, i.e.,

$$F = F(\epsilon, \epsilon_x, \theta), \tag{1.1}$$

where θ is the absolute temperature. In the framework of the Landau theory of phase transitions, the strain ϵ plays the role of an "order parameter", whose actual value determines what phase is prevailing in the material (see [3]).

Since we are interested in solid-solid phase transitions, driven by loading and/or heating, which are accompanied by hysteresis effects, we do not assume that $F(\epsilon, \epsilon_x, \theta)$ is a convex function of the order parameter ϵ for all values of (ϵ_x, θ). A particularly interesting class of materials are the metallic alloys exhibiting the so-called "shape memory effect". Among those there are alloys like $CuZn, CuSn, AuCuZn_2, AgCd$ and, most important, $TiNi$ (so-called Nitinol). In these materials, the metallic lattice is deformed by shear, and the assumption of a constant density is justified. The relation between shear stress and shear strain ($\sigma - \epsilon$-curves) of shape memory alloys exhibit a rather spectacular temperature-dependent hysteretic behavior (see [2] for an account of the properties of shape memory alloys):

---

*Fachbereich Bauwesen, Universität-GHS-Essen, D-4300 Essen 1, West Germany (visiting at IMA). This work was supported by Deutsche Forschungsgemeinschaft (DFG), SPP "Anwendungsbezogene Optimierung und Steuerung".

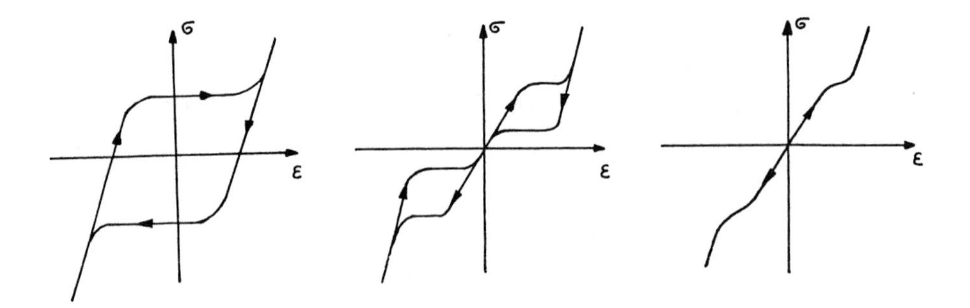

Fig 1. Typical $\sigma - \epsilon$-curves in shape memory alloys, with temperature θ increasing from a) to c).

In addition, for sufficiently small shear stresses σ another hysteresis occurs in the $\epsilon - \theta$-diagrams:

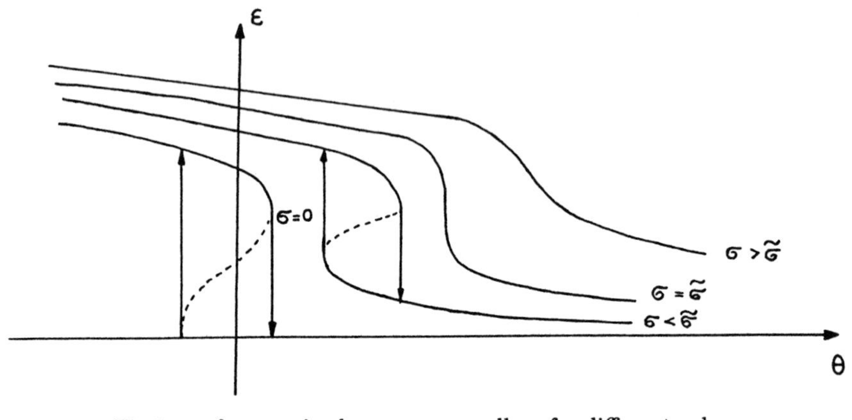

Fig 2. $\epsilon - \theta$ curves in shape memory alloys for different values of σ.

On the microscopic scale, this hysteretic behaviour is ascribed to first-order stress-induced (fig. 1a,b) or temperature-induced (fig. 2) phase transitions between different configurations of the metallic lattice, namely the symmetric high-temperature phase "austenite" (taken as reference configuration) and its two oppositively oriented sheared versions termed "martensitic twins", which prevail at low temperatures (cf., [6], [7]).

The simplest form for the free energy F which matches the experimental evidence given by figs. 1,2 quite well and takes interfacial energies into account is given by (cf., [4], [5])

$$(1.2)\quad F(\epsilon,\epsilon_x,\theta) = -C_V\theta\log(\theta/\theta_2) + C_V\theta + \widetilde{C} + \kappa_1(\theta-\theta_1)\epsilon^2 - \kappa_2\epsilon^4 + \kappa_3\epsilon^6 + \frac{\gamma}{2}\,\epsilon_x^2\,,$$

where C_V denotes the specific heat, $\widetilde{C}$ is some constant, θ_1 and θ_2 are (positive) temperatures and $\kappa_1, \kappa_2, \kappa_3, \gamma$ are positive constants. A complete set of data for the alloy $AuCuZn_2$ is given in [5]. Note that within the range of interesting temperatures, for $\theta \to \theta_1$, F is not convex as function of ϵ.

In the sequel, we assume F in the somewhat more general form (with positive $\kappa_1, \kappa_2, \gamma$)

$$(1.3)\qquad F(\epsilon,\epsilon_x,\theta) = -C_V\theta\log(\theta/\theta_2) + C_V\theta + \widetilde{C} + \kappa_1\theta F_1(\epsilon) + \kappa_2 F_2(\epsilon) + \frac{\gamma}{2}\,\epsilon_x^2\,,$$

where F_1 and F_2 satisfy the hypothesis:

(H1) $F_1, F_2 \in C^4(\mathbf{R})$; $F_2(\epsilon) \geq \overline{c}_1|\epsilon| - \overline{c}_2$, $\forall \epsilon \in \mathbf{R}$, with positive constants $\overline{c}_1, \overline{c}_2$.

The dynamics of thermomechanical processes in a solid are governed by the conservation laws of linear momentum, energy and mass. The latter may be ignored since ρ is constant for the materials under consideration (we assume $\rho \equiv 1$). The two others read

$$(1.4a)\qquad u_{tt} - \sigma_x + \mu_{xx} = f\,,$$

$$(1.4b)\qquad e_t + q_x - \sigma\epsilon_t - \mu\epsilon_{xt} = g\,.$$

Here the involved quantities have their usual meanings, namely: σ-elastic stress, μ-couple stress, u-displacement, f - density of loads, e - density of internal energy, q - heat flux, g -density of heat sources or sinks.

We have the constitutive relations

$$(1.5)\qquad \sigma = \frac{\partial F}{\partial \epsilon}\,,\quad \mu = \frac{\partial F}{\partial \epsilon_x}\,,\quad e = F - \theta\,\frac{\partial F}{\partial \theta}\,,$$

and we assume the heat flux in the Fourier-form

$$(1.6)\qquad q = -\kappa\theta_x\,,\quad \text{where } \kappa > 0 \text{ is the heat conductivity.}$$

Notice that (1.6) implies that the second principle of thermodynamics in form of the Clausius-Duhem inequality is automatically satisfied.

Inserting (1.3), (1.5), (1.6) in the balance laws and assuming a one-dimensional sample of unit length, we obtain the system

$$u_{tt} - \left(\kappa_1 \theta F_1'(\epsilon) + \kappa_2 F_2'(\epsilon)\right)_x + \gamma u_{xxxx} = f \ , \tag{1.7a}$$

$$C_V \theta_t - \kappa_1 \theta F_1'(\epsilon)\epsilon_t - \kappa \theta_{xx} = g, \tag{1.7b}$$

$$\epsilon = u_x \ , \tag{1.7c}$$

to be satisfied in the space-time cylinder Ω_T, where $T > 0$, $\Omega = (0,1)$, and, for $t > 0$, $\Omega_t := \Omega \times (0,t)$.

In addition, we prescribe the initial and boundary conditions

$$u(x,0) = u_0(x), u_t(x,0) = u_1(x), \ \theta(x,0) = \theta_0(x), x \in \overline{\Omega}, \tag{1.7d}$$

$$u(0,t) = u_{xx}(0,t) = u(1,t) = u_{xx}(1,t) = 0, \ t \in [0,T], \tag{1.7e}$$

$$\theta_x(0,t) = 0, \quad -\kappa\theta_x(1,t) = \beta\big(\theta(1,t) - \theta_\Gamma(t)\big), \ t \in [0,T], \tag{1.7f}$$

where $\beta > 0$ is a heat exchange coefficient, and θ_Γ stands for the outside temperature at $x = 1$.

In the following sections we state some results concerning the well-posedness of the system (1.7a–f), including a convergent numerical algorithm for its approximate solution. To abbreviate the exposition, all constants in (1.7a–f) are assumed to equal unity; this will have no bearing on the mathematical analysis.

2. Well-posedness. We consider (1.7a–f). In addition to (H1), we generally assume:

(H2) $u_0 \in \widetilde{H}^4(\Omega) = \{u \in H^4(\Omega) | u(0) = u(1) = 0 = u''(0) = u''(1)\}$;
$u_1 \in \overset{\circ}{H}{}^1(\Omega) \cap H^2(\Omega)$; $\theta_0 \in H^2(\Omega)$, $\theta_0(x) > 0, \quad \forall\, x \in \overline{\Omega}$.

(H3) $\theta_0'(0) = 0, \ \theta_\Gamma(0) = \theta_0(1) + \frac{\kappa}{\beta}\,\theta_0'(1) > 0$ (compatibility).

(H4) $f, g \in H^1(0,T; H^1(\Omega))$, $\quad \theta_\Gamma \in H^1(0,T)$, where $g(x,t) \geq 0$ on Ω_T and $\theta_\Gamma(t) > 0$ on $[0,T]$.

We have the result:

THEOREM 2.1. *Suppose (H1)–(H4) hold. Then (1.7a–f) has a unique solution (u,θ) which satisfies*

$$u \in W^{2,\infty}(0,T; L^2(\Omega)) \cap W^{1,\infty}(0,T; \overset{\circ}{H}{}^1(\Omega) \cap H^2(\Omega)) \cap L^\infty(0,T; \widetilde{H}^4(\Omega)), \tag{2.1a}$$

$$\theta \in H^1(0,T; H^1(\Omega)) \cap L^2(0,T; H^3(\Omega)), \tag{2.1b}$$

$$\theta(x,t) > 0, \quad \text{on} \quad \overline{\Omega}_T. \tag{2.1c}$$

Moreover, the operator $(f, g, \theta_\Gamma) \longmapsto (u, \theta)$ maps bounded subsets of $H^1(0,T; H^1(\Omega)) \times H^1(0,T; H^1(\Omega)) \times \mathcal{M}$ into bounded subsets of $X \times Y$, where $\mathcal{M} := \{z \in H^1(0,T)|$

$z(t) > 0$ on $[0,T]$, $z(0) = \theta_0(1)+\theta_0'(1)\}$, $X := W^{2,\infty}(0,T;L^2(\Omega))\cap W^{1,\infty}(0,T;\overset{\circ}{H}{}^1(\Omega)\cap H^2(\Omega))\cap L^\infty(0,T;\widetilde{H}^4(\Omega))$ and $Y := H^1(0,T;H^1(\Omega))\cap L^2(0,T;H^3(\Omega))$.

Proof. The existence result is easily obtained by combining the Galerkin approximation employed in the proof of Theorem 2.1 in [9] with the a priori estimates derived in the proof of Theorem 2.1 in [12]; the uniqueness is a direct consequence of the subsequent Theorem 2.3. Finally, the boundedness of the mapping $(f,g,\theta_\Gamma) \longmapsto (u,\theta)$ follows from the above-mentioned a priori estimates. □

A sharper existence result, with regards to the smoothness properties of the solution (u,θ), has been established in [12]:

THEOREM 2.2. *Suppose that, in addition to (H1)-(H4), the following assumptions on the data of (1.7a–f) are satisfied:*

$$\text{(2.2)}\qquad \begin{aligned} &u_0 \in H^5(\Omega),\ \ u_1 \in H^3(\Omega),\ \ \theta_0 \in H^4(\Omega),\\ &f_{tt} \in L^2(\Omega_T),\ \ g \in L^2(0,T;H^2(\Omega)),\ \ \theta_\Gamma \in H^2(0,T)\,. \end{aligned}$$

Furthermore, suppose that θ_0 satisfies compatibility conditions of sufficiently high order. Then (1.7a–f) has a unique classical solution (u,θ), and all the partial derivatives appearing in (1.7a–c) belong to the Hölder class $C^{\alpha,\alpha/2}(\overline{\Omega}_T)$, for some $\alpha \in (0,1)$.

Proof. See Theorem 2.1 in [12]. □

We now derive a stability result with respect to the data (f,g,θ_Γ) which guarantees the uniqueness of the solution (u,θ).

THEOREM 2.3. *Suppose the general hypotheses (H1), (H2) and $\theta_0'(0) = 0$ are satisfied. We consider the variational problem*

$$\text{(2.3a)}\qquad \int_\Omega u_t(x,t)\varphi(x,t)\,dx - \int_\Omega u_1(x)\varphi(x,0)\,dx - \int_0^t\int_\Omega u_t\varphi_t dx\,d\tau$$

$$+\int_0^t\int_\Omega \left[\frac{\partial F}{\partial \epsilon}(u_x,\theta)\varphi_x + u_{xx}\varphi_{xx} - f\varphi\right]dx\,d\tau = 0,$$

$$\forall \varphi \in H^1(0,T;\overset{\circ}{H}{}^1(\Omega))\cap L^2(0,T;H^2(\Omega))\,,\quad 0 \le t \le T,$$

$$\text{(2.3b)}\qquad \int_0^t\int_\Omega [\theta_t\eta - g\eta - \theta u_x u_{xt}\eta + \theta_x\eta_x]dx\,d\tau$$

$$+\int_0^t (\theta(1,\tau)-\theta_\Gamma(\tau))\eta(1,\tau)d\tau = 0\,,\quad \forall\,\eta \in L^2(0,T;H^1(\Omega)),\quad 0 \le t \le T\,,$$

$$\text{(2.3c)}\qquad \theta(x,0) = \theta_0(x)\,,\quad u(x,0) = u_0(x),\quad x \in \overline{\Omega}.$$

Suppose the data $(f^{(i)},g^{(i)},\theta_\Gamma^{(i)}), i = 1,2$, satisfy (H3) and (H4), and suppose that $(u^{(i)},\theta^{(i)})$ are solutions to (2.3a–c) corresponding to the data $(f^{(i)},g^{(i)},\theta_\Gamma^{(i)})$, $i =$

1, 2, *such that* $u^{(i)} \in W^{1,\infty}(0,T;\overset{\circ}{H}{}^1(\Omega)) \cap L^\infty(0,T;H^3(\Omega))$, $\theta^{(i)} \in H^1(0,T;L^2(\Omega)) \cap L^\infty(0,T;H^1(\Omega))$, $i = 1,2$. *Then there is some* $C > 0$ *such that*

$$\sup_{t\in(0,T)} \left(\|u_t(t)\|^2 + \|u_{xx}(t)\|^2 + \|\theta(t)\|^2\right) + \int_0^T\int_\Omega \theta_x^2 dx\, dt \tag{2.4}$$

$$+ \int_0^T \theta^2(1,t)\, dt \le C(\|\theta_\Gamma\|^2_{L^2(0,T)} + \|g\|^2_{L^2(\Omega_T)} + \|f\|^2_{L^2(\Omega_T)}),$$

where $\theta = \theta^{(1)} - \theta^{(2)}, u = u^{(1)} - u^{(2)}, \theta_\Gamma = \theta_\Gamma^{(1)} - \theta_\Gamma^{(2)}, f = f^{(1)} - f^{(2)}, g = g^{(1)} - g^{(2)}$.

REMARKS.

1. Here (and throughout) we have omitted the arguments of the involved functions if no confusion may arise.
2. $\| \,.\, \|$ denotes always the $L^2(\Omega)$-norm.
3. Obviously, any solution (u,θ) of (1.7a–f) with (2.1a,b) solves (2.3a–c); consequently, the solution of (1.7a–f) is unique.
4. From the upcoming proof it will become evident that a corresponding stability result holds with respect to the initial data u_0, u_1, θ_0; we restrict ourselves to the data (f, g, θ_Γ) as they are the natural candidates to serve as control variables if the system is to be controlled from the outside.

Proof. Let $\epsilon^{(i)} = u_x^{(i)}$, $i = 1,2$, and $\epsilon = \epsilon^{(1)} - \epsilon^{(2)}$. In terms of the variables (u,θ) introduced in the assertion, (2.3–c) can be rewritten as

$$\int_\Omega u_t(x,t)\varphi(x,t)dx - \int_0^t\int_\Omega u_t\varphi_t\, dx\, d\tau + \int_0^t\int_\Omega (u_{xx}\varphi_{xx} - f\varphi)dxd\tau \tag{2.5a}$$

$$+ \int_0^t\int_\Omega \left(\frac{\partial F}{\partial \epsilon}\,(\epsilon^{(1)},\theta^{(1)}) - \frac{\partial F}{\partial \epsilon}\,(\epsilon^{(2)},\theta^{(2)})\right)\varphi_x\, dx\, d\tau = 0,$$

$$\forall\, \varphi \in H^1(0,T;\overset{\circ}{H}{}^1(\Omega)) \cap L^2(0,T;H^2(\Omega)),\quad 0 \le t \le T,$$

$$\int_0^t\int_\Omega [\theta_t\eta - g\eta + \theta_x\eta_x]\, dx\, dt + \int_0^t (\theta(1,\tau) - \theta_\Gamma(\tau))\eta(1,\tau)d\tau \tag{2.5b}$$

$$+ \int_0^t\int_\Omega [\theta^{(2)}\epsilon^{(2)}\epsilon_t^{(2)} - \theta^{(1)}\epsilon^{(1)}\epsilon_t^{(1)}]\eta\, dxd\tau = 0,\ \ \forall \eta \in L^2(0,T;H^1(\Omega)) 0 \le t \le T,$$

$$\theta(x,0) = u(x,0) = 0\ ,\quad x \in \overline{\Omega}. \tag{2.5c}$$

Next observe that, owing to our assumptions, $u_x^{(i)}, u_{xx}^{(i)}$ and $\theta^{(i)}$ belong to $C(\overline{\Omega}_T), i = 1,2$. Thus, due to (H1), expressions of the form $\frac{\partial^k}{\partial \epsilon^k}\, F_j(\epsilon^{(i)})$, $1 \le k \le 4, j = 1,2, i =$

1, 2, are bounded. Moreover $u_t^{(i)} \in L^\infty(\Omega_T), i = 1, 2$. In the sequel, $C_i, i \in \mathbb{N}$, always denote positive generic constants. We proceed in two steps:

STEP 1: Let $\delta > 0$ be given (to be specified later). We insert $\varphi = u_t$ in (2.5a), integrate by parts and use Young's inequality to arrive at the estimate

$$\frac{1}{2}\|u_t(t)\|^2 + \frac{1}{2}\|u_{xx}(t)\|^2 \le \frac{1}{2}\|f\|^2_{L^2(\Omega_T)} + C_1 \int_0^t \|u_t(\tau)\|^2 d\tau + \delta \int_0^t \int_\Omega \left| \left(\frac{\partial F}{\partial \epsilon}(\epsilon^{(1)}, \theta^{(1)}) - \frac{\partial F}{\partial \epsilon}(\epsilon^{(2)}, \theta^{(2)}) \right)_x \right|^2 dx d\tau. \tag{2.6}$$

Now

$$\begin{aligned} \frac{\partial}{\partial x} \left(\frac{\partial F}{\partial \epsilon}(\epsilon^{(1)}, \theta^{(1)}) - \frac{\partial F}{\partial \epsilon}(\epsilon^{(2)}, \theta^{(2)}) \right) &= \theta_x F_1'(\epsilon^{(1)}) \\ &+ \theta_x^{(2)} (F_1'(\epsilon^{(1)}) - F_1'(\epsilon^{(2)})) + F_2''(\epsilon^{(1)}) u_{xx} \\ &+ u_{xx}^{(2)} (F_2''(\epsilon^{(1)}) - F_2''(\epsilon^{(2)})) + \theta F_1''(\epsilon^{(1)}) u_{xx}^{(1)} \\ &+ u_{xx}^{(1)} \theta^{(2)} (F_1''(\epsilon^{(1)}) - F_1''(\epsilon^{(2)})) + \theta^{(2)} F_1''(\epsilon^{(2)}) u_{xx}. \end{aligned} \tag{2.7}$$

Consequently, invoking the mean value theorem,

$$\int_0^t \int_\Omega \left| \left(\frac{\partial F}{\partial \epsilon}(\epsilon^{(1)}, \theta^{(1)}) - \frac{\partial F}{\partial \epsilon}(\epsilon^{(2)}, \theta^{(2)}) \right)_x \right|^2 dx\, d\tau \le C_2 \int_0^t \int_\Omega (\theta_x^2 + |\theta_x^{(2)}|^2 \epsilon^2 + u_{xx}^2 + \theta^2 + \epsilon^2) dx\, d\tau. \tag{2.8}$$

But

$$\int_0^t \int_\Omega |\theta_x^{(2)}|^2 \epsilon^2 dx dt \le \int_0^t \|\epsilon(\tau)\|^2_{L^\infty(\Omega)} \|\theta_x^{(2)}(\tau)\|^2 d\tau \le C_3 \int_0^t \|\epsilon(\tau)\|^2_{L^\infty(\Omega)}\, d\tau, \tag{2.9}$$

since $\theta^{(2)} \in L^\infty(0, T; H^1(\Omega))$. Now observe that $u(0, t) = 0 = u(1, t)$. Hence, to any $t \in [0, T]$ there is some $x_0(t) \in (0, 1)$ such that $u_x(x_0(t), t) = 0$. Thus,

$$|\epsilon(x, \tau)| \le \|u_{xx}(\tau)\|, \quad 0 \le \tau \le t, \quad x \in \overline{\Omega}. \tag{2.10}$$

Summarizing, we have shown the estimate

$$\|u_t(t)\|^2 + \|u_{xx}(t)\|^2 \le C_4 \|f\|^2_{L^2(\Omega_T)} + C_5 \int_0^t \|u_t(\tau)\|^2 dt + C_6 \cdot \delta \cdot \int_0^t (\|\theta_x(\tau)\|^2 + \|\theta(\tau)\|^2 + \|u_{xx}(\tau)\|^2) d\tau. \tag{2.11}$$

STEP 2: Next we substitute $\eta = \theta$ into (2.5b) to obtain via Young's inequality:

$$\frac{1}{2}\|\theta(t)\|^2 + \int_0^t \|\theta_x(\tau)\|^2 d\tau + \frac{1}{2}\int_0^t \theta^2(1,\tau)\, d\tau \tag{2.12}$$

$$\leq \frac{1}{2}\int_0^t \|\theta(\tau)\|^2 d\tau + \frac{1}{2}\|g\|^2_{L^2(\Omega_T)} + \frac{1}{2}\|\theta_\Gamma\|^2_{L^2(0,T)} + A,$$

where

$$A = \int_0^t\int_\Omega \theta(\theta^{(1)}\epsilon^{(1)}\epsilon_t^{(1)} - \theta^{(2)}\epsilon^{(2)}\epsilon_t^{(2)})dx\, d\tau \tag{2.13}$$

$$= \int_0^t\int_\Omega \frac{d}{dx}\,[\theta(\theta^{(1)}\epsilon^{(1)}u_t^{(1)} - \theta^{(2)}\epsilon^{(2)}u_t^{(2)})]dx d\tau$$

$$- \int_0^t\int_\Omega \theta_x(\theta^{(1)}\epsilon^{(1)}u_t^{(1)} - \theta^{(2)}\epsilon^{(2)}u_t^{(2)})dx d\tau$$

$$- \int_0^t\int_\Omega \theta(\theta_x^{(1)}\epsilon^{(1)}u_t^{(1)} + \theta^{(1)}u_{xx}^{(1)}u_t^{(1)}$$

$$- \theta_x^{(2)}\epsilon^{(2)}u_t^{(2)} - \theta^{(2)}u_{xx}^{(2)}u_t^{(2)})dx d\tau.$$

Since $u_{t|x=0,1} = 0$, the first integral vanishes; the other terms have to be treated individually.

a) We have, since $\theta^{(i)}, \epsilon^{(i)}, u_t^{(i)} \in L^\infty(\Omega_T)$, $i = 1,2$,

$$I_1 := \left|\int_0^t\int_\Omega \theta_x(\theta^{(1)}\epsilon^{(1)}u_t^{(1)} - \theta^{(2)}\epsilon^{(2)}u_t^{(2)})dx d\tau\right| \tag{2.14}$$

$$= \left|\int_0^t\int_\Omega \theta_x[\theta\epsilon^{(1)}u_t^{(1)} + \theta^{(2)}\epsilon u_t^{(1)} + \theta^{(2)}\epsilon^{(2)}u_t]dx\, d\tau\right|$$

$$\leq \delta \int_0^t\int_\Omega \theta_x^2 dx d\tau + C_7 \int_0^t\int_\Omega (\theta^2 + \epsilon^2 + u_t^2)dx\, d\tau.$$

b) Next we estimate

$$|I_2| := \left| \int_0^t \int_\Omega \theta(\theta_x^{(1)}\epsilon^{(1)}u_t^{(1)} - \theta_x^{(2)}\epsilon^{(2)}u_t^{(2)})dxd\tau \right| \tag{2.15}$$

$$\left| \int_0^t \int_\Omega \theta[\theta_x\epsilon^{(1)}u_t^{(1)} + \theta_x^{(2)}\epsilon u_t^{(1)} + \theta_x^{(2)}\epsilon^{(2)}u_t]dxd\tau \right|$$

$$\leq \delta \int_0^t \int_\Omega \theta_x^2 dxd\tau + C_8 \int_0^t \int_\Omega (\theta^2 + \epsilon^2 + u_t^2)dxd\tau$$

$$+ C_9 \int_0^t \int_\Omega |\theta_x^{(2)}|^2\theta^2 dx\ d\tau.$$

Recalling Nirenberg's inequality in one space dimension (cf., [1]), we have with suitable $\alpha_1 > 0, \alpha_2 > 0$:

$$\begin{aligned} \|\theta(\tau)\|^2_{L^\infty(\Omega)} &\leq \alpha_1\|\theta_x(\tau)\|\ \|\theta(\tau)\| + \alpha_2\|\theta(\tau)\|^2 \\ &\leq \delta\|\theta_x(\tau)\|^2 + C_{10}\|\theta(\tau)\|^2. \end{aligned} \tag{2.16}$$

Since $\theta_x^{(2)} \in L^\infty(0,T;L^2(\Omega))$, this implies that

$$\int_0^t \int_\Omega |\theta_x^{(2)}|^2\theta^2 dxd\tau \leq \int_0^t \|\theta(\tau)\|^2_{L^\infty(\Omega)}\|\theta_x^{(2)}(\tau)\|^2 d\tau \tag{2.17}$$

$$\leq C_{10}\delta \int_0^t \int_\Omega \theta_x^2 dxd\tau + C_{11} \int_0^t \int_\Omega \theta^2 dxdt,$$

whence

$$|I_2| \leq C_{11}\delta \int_0^t \|\theta_x(\tau)\|^2 d\tau \tag{2.18}$$

$$+ C_{12} \int_0^t (\|\theta(\tau)\|^2 + \|u_{xx}(\tau)\|^2 + \|u_t(\tau)\|^2)d\tau.$$

c) Finally we have

$$|I_3| := \left| \int_0^t \int_\Omega \theta[\theta^{(1)}u_{xx}^{(1)}u_t^{(1)} - \theta^{(2)}u_{xx}^{(2)}u_t^{(2)}]dxd\tau \right| \tag{2.19}$$

$$= \left| \int_0^t \int_\Omega \theta[\theta u_{xx}^{(1)}u_t^{(1)} + \theta^{(2)}u_{xx}u_t^{(1)} + \theta^{(2)}u_{xx}^{(2)}u_t]dxd\tau \right|$$

$$\leq C_{13} \int_0^t (\|\theta(\tau)\|^2 + \|u_{xx}(\tau)\|^2 + \|u_t(\tau)\|^2)dxd\tau.$$

Summarizing the inequalities (2.12), (2.14), (2.18) and (2.19), we have shown that

$$(2.20)\qquad \frac{1}{2}\,\|\theta(t)\|^2 + \int_0^t \|\theta_x(\tau)\|^2 d\tau + \frac{1}{2}\int_0^t \theta^2(1,\tau)d\tau$$
$$\le C_{14}\cdot\delta\int_0^t \|\theta_x(\tau)\|^2 d\tau + \frac{1}{2}\,\|g\|^2_{L^2(\Omega_T)} + \frac{1}{2}\,\|\theta_\Gamma\|^2_{L^2(0,T)}$$
$$+\,C_{15}\int_0^t (\|\theta(\tau)\|^2 + \|u_{xx}(\tau)\|^2 + \|u_t(t)\|^2)d\tau.$$

Adding (2.11) and (2.20), adjusting $\delta > 0$ sufficiently small, and invoking Gronwall's lemma, we have finally proved the assertion. □

3. Optimal Control. We now turn our interest to optimal control problems associated with the system (1.7a–f). It is of considerable interest in the technological application of shape memory alloys to control the evolution of the austenitic-martensitic phase transitions in the material; in this connection, a typical object is to influence the system via the natural control variables f, g, θ_Γ in such a way, that a desired distribution of the phases in the material is produced. Since the phase transitions are characterized by the order parameter ϵ, it is natural to use ϵ as the main variable in the cost functional. We consider the following control problem:

(CP)

$$\text{Minimize}\ \ J(u,\theta;f,g,\theta_\Gamma) = \int_0^T\int_\Omega L_1\big(x,t,u_x(x,t),\theta(x,t),f(x,t),g(x,t)\big)dxdt$$
$$+\int_0^T L_2\big(t,\theta_\Gamma(t)\big)dt + \int_\Omega L_3\big(x,u_x(x,T),\theta(x,T)\big)dx,$$

subject to (1.7a–f) and the side condition $(f,g,\theta_\Gamma)\in\mathcal{K}$, where $\mathcal{K}$ denotes some nonempty, bounded, closed and convex subset of $H^1(0,T;H^1(\Omega))\times\{g\in H^1(0,T;H^1(\Omega))|g(x,t)\ge 0 \text{ on } \overline{\Omega}_T\}\times\mathcal{M}$.

For $L_1 : \mathbb{R}^6\to\mathbb{R}$, $L_2 : \mathbb{R}^2\to\mathbb{R}$, $L_3 : \mathbb{R}^3\to\mathbb{R}$, we assume:

(H5) (i) L_1, L_2, L_3 are measurable with respect to the variables (x,t), resp. t, resp. x, and continuous with respect to the other variables.

(ii) L_1 is convex with respect to f and g.

(iii) L_2 is convex with respect to θ_Γ.

These assumptions are natural in the framework of optimal control; a typical form for J would be

$$(3.1)\qquad J(u,\theta;f,g,\theta_\Gamma) = \beta_1\|u_x-\overline{u}_x\|^2_{L^2(\Omega_T)} + \beta_2\|\theta-\overline{\theta}\|^2_{L^2(\Omega_T)}$$
$$+\,\beta_3\|u_x(\cdot,T)-\tilde{u}\|^2 + \beta_4\|\theta(\cdot,T)-\widetilde{\theta}\|^2$$
$$+\,\beta_5\|f\|^2_{L^2(\Omega_T)} + \beta_6\|g\|^2_{L^2(\Omega_T)} + \beta_7\|\theta_\Gamma\|^2_{L^2(0,T)},$$

with $\beta_i \geq 0$, but not all zero, and functions $\overline{u}_x, \overline{\theta} \in L^2(\Omega_T), \tilde{u}, \widetilde{\theta} \in L^2(\Omega)$, representing the desired strain and temperature distributions during the evolution and at $t = T$.

There holds

THEOREM 3.1. *Suppose (H1)–(H5) are true, then (CP) has a solution $(\overline{u}, \overline{\theta}; \overline{f}, \overline{g}, \overline{\theta}_\Gamma)$.*

Proof. Let $\{(f_n, g_n, \theta_{\Gamma,n})\} \subset \mathcal{K}$ denote a minimizing sequence, and let (u_n, θ_n) denote the solution of (1.7a–f) associated with $(f_n, g_n, \theta_{\Gamma,n}), n \in \mathbb{N}$. Since $\mathcal{K}$ is bounded, we may assume that

$$
\begin{aligned}
(3.2) \qquad && f_n &\to \overline{f}, && \text{weakly in } H^1(0,T; H^1(\Omega)), \\
&& g_n &\to \overline{g}, && \text{weakly in } H^1(0,T; H^1(\Omega)), \\
&& \theta_{\Gamma,n} &\to \overline{\theta}_\Gamma, && \text{weakly in } H^1(0,T).
\end{aligned}
$$

Due to the weak closedness of the convex and closed set $\mathcal{K}, (\overline{f}, \overline{g}, \overline{\theta}_\Gamma) \in \mathcal{K}$. Let $(\overline{u}, \overline{\theta})$ denote the associated solution of (1.7a–f). Now, owing to the boundedness of $\mathcal{K}$ and Theorem 2.1, $\{(u_n, \theta_n)\}_{n \in \mathbb{N}}$ is a bounded subset of $X \times Y$. Therefore, we may assume that for some $(u, \theta) \in X \times Y$ there holds

$$
(3.3a) \qquad \left.\begin{aligned} u_{n,x} &\to u_x, \\ \theta_n &\to \theta, \end{aligned}\right\} \quad \text{uniformly on } \overline{\Omega}_T,
$$

$$
(3.3b) \qquad \left.\begin{aligned} u_{n,tt} &\to u_{tt}, \\ u_{n,xx} &\to u_{xx}, \\ u_{n,xt} &\to u_{xt}, \\ u_{n,xxxx} &\to u_{xxxx}, \end{aligned}\right\} \quad \text{weakly in } L^2(\Omega_T),
$$

as well as

$$
(3.3c) \qquad \left.\begin{aligned} \theta_{n,x} &\to \theta_x, \\ \theta_{n,t} &\to \theta_t, \\ \theta_{n,xx} &\to \theta_{xx}, \end{aligned}\right\} \quad \text{weakly in } L^2(\Omega_T).
$$

Passing to the limit as $n \to \infty$ in the equations (1.7a–f) shows that (u, θ) solves (1.7a–f) for the data $(\overline{f}, \overline{g}, \overline{\theta}_\Gamma)$, i.e., $u = \overline{u}, \theta = \overline{\theta}$. Hence, $(\overline{u}, \overline{\theta}; \overline{f}, \overline{g}, \overline{\theta}_\Gamma)$ is admissible, and, in view of (H5),

$$
(3.4) \qquad J(\overline{u}, \overline{\theta}; \overline{f}, \overline{g}, \overline{\theta}_\Gamma) \leq \lim_{n \to \infty} \inf J(u_n, \theta_n; f_n, g_n, \theta_{\Gamma,n}).
$$

Thus $(\overline{u}, \overline{\theta}; \overline{f}, \overline{g}, \overline{\theta}_\Gamma)$ is a solution of (CP). □

REMARKS.

5. The above way of arguing follows the lines of [10] where a related result was derived for a much more restricted free energy F.
6. It is natural to look for necessary conditions of optimality for the optimal controls of (CP). A corresponding result has not yet been derived.
7. The problem of the automatic self-regulation of the system via a fixed feedback control regulating the boundary temperature θ_Γ has been considered in [11].

4. Numerical Approximation. In this section we follow the lines of [8]. We assume the free energy in the special form (see (1.2))

$$F(\epsilon, \epsilon_x, \theta) = -\theta \log \theta + \theta + \frac{1}{2}\,\theta\epsilon^2 - \frac{1}{4}\,\epsilon^4 + \frac{1}{6}\,\epsilon^6 + \frac{1}{2}\,\epsilon_x^2. \tag{4.1}$$

Let

$$F_0(\epsilon, \theta) = \frac{1}{2}\,\theta\epsilon^2 - \frac{1}{4}\,\epsilon^4 + \frac{1}{6}\,\epsilon^6. \tag{4.2}$$

Then, for $\epsilon_1 \neq \epsilon_2$,

$$\frac{F_0(\epsilon_1, \theta) - F_0(\epsilon_2, \theta)}{\epsilon_1 - \epsilon_2} = \frac{1}{2}\,\theta(\epsilon_1 + \epsilon_2) + \Psi(\epsilon_1, \epsilon_2), \tag{4.3}$$

where $\Psi(\epsilon_1, \epsilon_2)$ is a polynomial of degree 5 in ϵ_1, ϵ_2.

We are going to construct a numerical scheme for the approximate solution of (1.7a–f). To this end, we assume that (H1)–(H4) and (2.2) hold, so that Theorem 2.2 applies.

Now let $K, N, M \in \mathbb{N}$ be chosen. We put $h = \frac{T}{M}$, $t_m^{(M)} = mh$, $0 \leq m \leq M$, and $x_i^{(N)} = \frac{i}{N}$, $0 \leq i \leq N$.

Define

(4.4)
$$Y_N = \{\text{linear splines on [0,1] corresponding to the partition } \{x_i^{(N)}\}_{i=0}^N \text{ of [0,1]}\},$$

and let

$$Z_K = \mathrm{span}\{z_1, \ldots, z_K\}, \tag{4.5}$$

where z_j denotes the j-th eigenfunction of the eigenvalue problem

$$z'''' = \lambda z \ , \ \text{in } (0,1), z(0) = z''(0) = 0 = z''(1) = z(1). \tag{4.6}$$

We introduce the projection operators

$$\begin{aligned} P_K &= H^4(0,1) - \text{orthogonal projection onto } Z_K, \\ Q_K &= H^2(0,1) - \text{orthogonal projection onto } Z_K, \\ R_N &= H^1(0,1) - \text{orthogonal projection onto } Y_N, \end{aligned} \tag{4.7}$$

and the averages

$$f_M^m(x) = \frac{1}{h} \int_{(m-1)h}^{mh} f(x,t)dt \ , \quad g_M^m(x) = \frac{1}{h} \int_{(m-1)h}^{mh} g(x,t)dt, \tag{4.8}$$

$$\theta_{\Gamma,M}^m = \frac{1}{h} \int_{(m-1)h}^{mh} \theta_\Gamma(t)\, dt.$$

We then consider the discrete problem

$\underline{(D_{M,N,K})}$ Find $u^m = \sum_{k=1}^{K} \alpha_k^m z_k, \theta^m = \sum_{k=0}^{N} \beta_k^m y_k^{(N)}$,

$1 \leq m \leq M$, such that

$$\int_\Omega \Big[\frac{u^m - 2u^{m-1} + u^{m-2}}{h^2} \xi + \frac{1}{2} \theta^{m-1}(u_x^m + u_x^{m-1})\xi_x \tag{4.9a}$$

$$+ \Psi(u_x^m, u_x^{m-1})\xi_x + u_{xx}^m \xi_{xx} - f_M^m \xi \Big] dx = 0, \quad \forall \xi \in Z_K,$$

$$\int_\Omega \Big[\frac{\theta^m - \theta^{m-1}}{h} \eta - \frac{1}{2} \theta^{m-1} \cdot \frac{(u_x^m)^2 - (u_x^{m-1})^2}{h} \eta \tag{4.9b}$$

$$+ \theta_x^m \eta_x - g_M^m \eta \Big] dx + (\theta^m(1) - \theta_{\Gamma,M}^m)\eta(1) = 0, \quad \forall \eta \in Y_N,$$

$$u^0 = P_K(u^0), \frac{u^0 - u^{-1}}{h} = Q_K(u_1), \quad \theta^0 = R_N(\theta_0). \tag{4.9c}$$

The following result has been shown in [8]:

THEOREM 4.1. *Suppose (H1)–(H4) and (2.2) are true, and let N be sufficiently large. Then there exist constants $\widehat{C}_1 > 0, \widehat{C}_2 > 0$, which do not depend on M, N, K, such that for $\frac{1}{6N^2} < \frac{1}{M} \leq \widehat{C}_1$ the discrete problem $(D_{M,N,K})$ has a solution which satisfies*

$$\theta^m(x) \geq 0, \quad \forall x \in \overline{\Omega}, \quad 0 \leq m \leq M, \tag{4.10a}$$

$$\max_{0 \leq m \leq M} \left\{ \left\| \frac{u^m - u^{m-1}}{h} \right\| + \left\| \frac{u_x^m - u_x^{m-1}}{h} \right\|^2 + \left\| u_{xxx}^m \right\|^2 \right\} \leq \widehat{C}_2, \tag{4.10b}$$

$$\max_{0 \leq m \leq M} \{ \|\theta_x^m\|^2 + |\theta^m(1)|^2 \} + \sum_{m=1}^{M} h \left\| \frac{\theta^m - \theta^{m-1}}{h} \right\|^2 \leq \widehat{C}_2. \tag{4.10c}$$

Proof. See Theorem 2.1 in [8]. □

It is now easy to derive convergent approximate solutions. To this end, let $\varphi : \mathbb{N} \to \mathbb{N}$ denote some strictly increasing function. We put $N = \varphi(K)$ and $M = 2N^2$ $\left(\text{which implies } \frac{1}{6N^2} < \frac{1}{M}\right)$ and take $K \in \mathbb{N}$ large enough. Let $\{(u_K^m, \theta_K^m)\}_{m=1}^M$ denote corresponding solutions of $(D_{M,N,K})$ with the above choice of N and M.

We define the linear interpolations

$$\begin{aligned} u_K(x,t) &= (Mt - m + 1)u_K^m(x) + (m - Mt)u_K^{m-1}(x), \\ \theta_K(x,t) &= (Mt - m + 1)\theta_K^m(x) + (m - Mt)\theta_K^{m-1}(x), \\ 0 \leq x &\leq 1, \ \frac{m-1}{M} \leq t \leq \frac{m}{M}, \ m = 1, \ldots, M. \end{aligned} \tag{4.11}$$

Then (4.10b,c) imply that, for any sufficiently large $K \in \mathbb{N}$,

$$\|u_K\|_{W^{1,\infty}\left(0,T;\mathring{H}^1(\Omega)\right) \cap L^\infty\left(0,T;H^3(\Omega)\right)} \leq \widehat{C}_2, \tag{4.12a}$$

$$\|\theta_K\|_{H^1\left(0,T;L^2(\Omega)\right) \cap L^\infty\left(0,T;H^1(\Omega)\right)} \leq \widehat{C}_2. \tag{4.12b}$$

From standard compactness arguments we conclude the existence of some $(\tilde{u},\widetilde{\theta})$ such that, for some subsequence,

$$\text{(4.13a)} \qquad u_{K_n} \to \tilde{u}, \quad \text{weakly} -* \text{ in } W^{1,\infty}(0,T;\overset{\circ}{H}{}^1(\Omega)) \text{ and weakly} -* \text{ in } L^\infty(0,T;H^3(\Omega)),$$

$$\text{(4.13b)} \qquad \theta_{K_n} \to \widetilde{\theta}, \quad \text{weakly in } H^1(0,T;L^2(\Omega)) \text{ and weakly} -* \text{ in } L^\infty(0,T;H^1(\Omega)),$$

$$\text{(4.13c)} \qquad \frac{\partial}{\partial x}\, u_{K_n} \to \frac{\partial}{\partial x}\, \tilde{u}\,, \quad \text{uniformly on } \overline{\Omega}_T, \qquad \theta_{K_n} \to \widetilde{\theta}, \quad \text{uniformly on } \overline{\Omega}_T.$$

It is easy to see that the limit point $(\tilde{u},\widetilde{\theta})$ is a solution of the variational problem (2.3a–c). By virtue of Theorem 2.3, $\tilde{u} = u, \widetilde{\theta} = \theta$, where (u,θ) is the (unique) solution of (1.7a–f). It follows that the whole sequence (u_K,θ_K) converges to (u,θ) in the sense of (4.13a-c).

THEOREM 4.2. *Suppose (H1)–(H4) and (2.2) are true, and assume that to $K \in \mathbb{N}$ we define $N = \varphi(K)$ and $M = 2N^2$, where $\varphi : \mathbb{N} \to \mathbb{N}$ is any strictly increasing function. For sufficiently large $K \in \mathbb{N}$, let (u_K,θ_K) be defined by (4.11). Then (u_K,θ_K) converges in the sense of (4.13a–c) to the solution (u,θ) of (1.7a–f).*

REMARK.

8. Results concerning the order of convergence have not yet been established.

Acknowledgement. The author gratefully acknowledges the financial support and the stimulating atmosphere at the Institute for Mathematics and its Applications of the University of Minnesota.

REFERENCES

[1] R.A. ADAMS, *Sobolev Spaces*, Academic Press, New York (1975).

[2] L. DELAEY AND M. CHANDRASEKARAN (eds.), International Conference on Martensitic Transformations, Proceedings, Les Editions de Physique, Les Ulis (1982).

[3] F. FALK, *Landau theory and martensitic phase transitions*, Journal de Physique C4, 12 (1982), pp. 3–15.

[4] ———, *Ginzburg-Landau theory of static domain walls in shape-memory alloys*, Phys. B-Condensed Matter, 51 (1983), pp. 177–185.

[5] ———, *Ginzburg-Landau theory and solitary waves in shape memory alloys*, Phys. B-Condensed Matter, 54 (1984), pp. 159–167.

[6] I. MÜLLER AND K. WILMAŃSKI, *A model for phase transition in pseudoelastic bodies*, Il Nuovo Cimento, 57B (1980), pp. 283–318.

[7] ———————, *Memory alloys-phenomenology and ersatzmodel*, in *Brulin, O. and Hsieh, R.K.T (eds.)*, Continuum Models of Discrete Systems, Vol. IV, North-Holland, Amsterdam (1981), pp. 495–509.

[8] M. NIEZGÓDKA AND J. SPREKELS, *Convergent numerical approximations of the thermomechanical phase transitions in shape memory alloys*, (to appear).

[9] J. SPREKELS, *Global existence for thermomechanical process with nonconvex free energies of Ginzburg-Landau form*, J. Math. Anal. Appl, (to appear).

[10] ———, *Stability and optimal control of thermomechanical processes with nonconvex free energies of Ginzburg-Landau type*, Math. Meth. in the Appl. Sci., (to appear).

[11] ———, *Automatic control of one-dimensional thermomechanical phase transitions*, to appear in the Proceedings of the Conferences on Free Boundary Problems held at Obidos, Portugal, October (1988).

[12] J. SPREKELS AND S. ZHENG, *Global solutions to the equations of a Ginzburg-Landau theory for structural phase transitions in shape memory alloys*, Physica D, (to appear).

ADMISSIBILITY OF SOLUTIONS TO THE RIEMANN PROBLEM FOR SYSTEMS OF MIXED TYPE
-transonic small disturbance theory-*

GERALD WARNECKE†

Abstract. The transonic small disturbance equation is put into the context of first order systems of conservation laws. The admissibility of shock solutions is studied via Lax inequalities, entropy inequalities, the viscosity method, Oleĭnik's E-condition, Liu's extended entropy condition and a class of generalized entropy inequalities. This includes the case of mixed-type shocks (elliptic-hyperbolic). The admissibility condition of transonic small disturbance theory is shown to arise from an additional conservation law different from the one used in the case of the classical p-system in isentropic gas dynamics.

1 Introduction. 1.1 The transonic small disturbance equation is derived from the equation of steady potential flow via an expansion procedure for flows around slender profiles. This is derived in Cole/Messiter [5], Hayes [20], Cole/Cook [6]. The result is the equation for the small disturbance potential

$$\left[K\phi_x - (\gamma+1)\frac{\phi_x^2}{2}\right]_x + \phi_{yy} = 0, \tag{1.1}$$

with $K = \frac{1-M_\infty^2}{\delta^{\frac{2}{3}}}$ being the transonic similarity parameter, M_∞ the Mach number of the incoming flow at infinity, δ the profile thickness and γ the adiabatic constant (1.4 for dry air). The equation changes type at $\phi_x = \frac{K}{\gamma+1}$. It is elliptic for $\phi_x < \frac{K}{\gamma+1}$ and hyperbolic for $\phi_x > \frac{K}{\gamma+1}$ (cp. Courant/Hilbert [10]). For simplicity of the mathematical discussion we set $K = 1$, $\gamma = 0$. The results are independent of this restriction. Therefore, we study the equation

$$\left[\phi_x - \frac{\phi_x^2}{2}\right]_x + \phi_{yy} = 0. \tag{1.2}$$

Now the change of type occurs at $\phi_x = 1$. We introduce the velocities $u = \phi_x$, $v = \phi_y$ and using $u_y = v_x$ obtain the first order system

$$\begin{aligned} p(u)_x + v_y &= 0 \\ v_x - u_y &= 0. \end{aligned} \tag{1.3}$$

Here $p(u) = u - \frac{u^2}{2}$ was introduced to make the comparison easier with results for the p-system of unsteady one-dimensional isentropic flow (cp. Oleĭnik [39], Wendroff [49], Smoller [48]) or the theory of phase transitions in van der Waals fluids (cp. Slemrod [47], [44], [45]). Note that $-p(\cdot)$ is convex, $p'' \equiv -1$.

---

*This research was supported by a grant of the Stiftung Volkswagenwerk, Germany

†Mathematisches Institut A, Universität Stuttgart, Pfaffenwaldring 57, 7000 Stuttgart 80

1.2. The paper begins with a chapter on hyperbolic systems in order to fix the notations for further use and to introduce the well known concepts that are then applied to the mixed-type problem. The main issues are the Hugoniot curves, the entropy inequalities, the viscosity method, and the connection between the latter two via inequality (2.26).

In the next chapter some simple facts concerning system (1.3) are given. Two additional conservation laws satisfied by smooth solutions of (1.3) are obtained via variational principles. In the following chapter the investigation is put into the context of previous investigations for hyperbolic systems by formally introducing the variable t to replace y. It is then shown that the first additional conservation law leads to the classical admissibility criteria for the p-system in unsteady one-dimensional isentropic gas dynamics. This is nothing new and can be found in the book by Smoller [48]. It is discussed in order to compare it with the admissibility criteria connected with the second additional conservation law. The latter gives the correct admissibility criterion for the transonic small disturbance problem.

For the viscosity method it is sufficient to add the correct second order terms to the first equation in (1.3). It is not necessary to add third order dispersion terms or equivalently take a full regularisation of both equations, as for instance in the theory of van der Waals fluids(cp. Slemrod [47]). This is due to the convexity of $-p(\cdot)$ in our case. The associated dynamical systems problem in the (u, v)-plane can be restricted to a line $\{v = \text{const}\}$ since only two fixed points are involved. There is no need to circumvent a third fixed point by going into the full $(u,\ v)$-plane.

We also give a generalization of Lax inequalities for the kind of mixed type system considered in this paper. They are simply obtained by squaring the relevant inequalites.

In Chapter 5 we give the appropriate analogues of Oleĭnik's E-condition [39] and of Liu's extended entropy condition [33], [34]. Also we include a class of admissibility conditions, including the h-entropy inequalities of Nečas [38], that has been used in numerical computations (cp. Bristeau/Glowinski/Pironneau et. al. [2], [3], [17]).

1.3. It seems that admissibility conditions for this type of problem have previously only been treated by Mock [36] and recently by Keyfitz [26]. Mock used reversed x and t variables (see(6.1)). I will call the corresponding system the inverse p-system. This notation leads to problems with infinite speeds (see the discussion in Section 6.1). Therefore, Mock used reciprocal eigenvalues and reciprocal shock speeds. This makes it difficult to compare the results. Our analysis is applied to the inverse p-system in order to show that no new insight is gained by using it. Note also that Mock [36] used a full elliptic regularization for the viscosity method. In Keyfitz [26] the viscosity criterion for a quite general class of mixed type 2 by 2 systems is discussed.

1.4. It should be mentioned that there have been a number of results obtained for equations of mixed type in recent years (see for example James [24], Hattori [19], Holden [21], Hsiao [22], Hsiao/de Mottoni [23], Keyfitz [25], Pego [42], Shearer [43], Slemrod [45], [47]). In all of these the elliptic region in $(u,\ v)$-space is restricted to a strip, in Holden [21] even to a ball. Holden avoids the discussion of admissibility by studying the whole solution set.

In elastic bar theory (cp. James [24]), phase transitions in van der Waals fluids (cp. Hattori [19], Slemrod [45]) and related examples (Pego [42], Shearer [43]) mixed type systems arise. There the elliptic states are considered unstable and therefore are not a part of the physically relevant solutions. Mixed-type shocks do not occur. As mentioned before, these problems need third order regularizations in the viscosity method since $p(\cdot)$ is non-convex.

1.5. Finally, it is interesting to note that the steady transonic small disturbance problem has a distinguished direction, namely the direction of the flow. This is assumed to be mainly parallel to the x-axis in order to derive the equation. This fact reappears throughout the admissibility discussion. The relevant second additional conservation law is obtained via the x-invariance of the underlying variational principle (see Section 3.4). The Lax conditions for hyperbolic-hyperbolic type shocks are valid if one considers the Riemann problem for $x > 0$ with initial states given at $x = 0$.

2. Hyperbolic systems in two variables.

2.1. Take $(x,\ t) \in \mathbf{R}^2$ and let $\underline{U}(x,\ t) = (u_1(x,\ t),\ \ldots,\ u_n(x,\ t)),\quad n \in \mathbf{N}$. Then for a function $f : \mathbf{R}^n \to \mathbf{R}^n,\ \ f \in C^3(\mathbf{R}^n)$ we define the first order system of equations

$$\underline{U}_t + f(\underline{U})_x = 0. \tag{2.1}$$

We denote by $f'(\underline{U})$ the Jacobian of f. If $f'(\underline{U}_0)$ has only real eigenvalues $\lambda_k(\underline{U}_0),\ \ 1 \leq k \leq n$, system is called **hyperbolic** at $\underline{U}_0 \in \mathbf{R}^n$. In case the real eigenvalues are distinct the system is termed **strictly hyperbolic**. In passing we note that in case $f'(\underline{U}_0)$ has no real eigenvalues the system is **elliptic** (cp. Courant/Hilbert [10] Sec. III.2).

2.2. The eigenvalues $\lambda_k(\underline{U}_0),\ \ \underline{U}_0 \in \mathbf{R}^n$ are commonly called ***k*-th characteristic speeds**. Suppose $f'(\underline{U})$ may be diagonalized locally (e. g. in the strictly hyperbolic case). Then the k-th component of (2.1) is $\frac{\partial u_k}{\partial t} + \lambda_k(\underline{U})\frac{\partial u_k}{\partial x} = 0$. If we take curves $\Gamma = \{(x(t),\ t) \in \mathbf{R}^2 \mid t \in \mathbf{R}\}$, then $\frac{du_k}{dt} = \frac{\partial u_k}{\partial x}\frac{dx}{dt} + \frac{\partial u_k}{\partial t}$ and $u_k = \text{const}$ along Γ iff $\frac{dx}{dt} = \lambda_k(\underline{U})$. The vector field $(\lambda_k,\ 1) \in \mathbf{R}^2$ is called the k-th characteristic field. The solution curves to the ordinary differential equations

$$\frac{dt}{ds} = 1,\ \ \frac{dx}{ds} = \lambda_k(\underline{U}(x,\ t)),\ t(0) = 0,\ x(0) = x_0, \tag{2.2}$$

are called ***k*-th family of characteristics**.

In order to distinguish linear and nonlinear behavior of solutions to (2.1) one says the k-th family of characteristics (or k-th characteristic field) is **genuinely nonlinear** at $\underline{U} \in \mathbf{R}^n$ if

$$\nabla\lambda_k(\underline{U}) \cdot r_k(\underline{U}) \neq 0 \tag{2.3}$$

(cp. Dafermos [12], Lax [31]). Here $r_k(\underline{U})$ denotes the k-th right eigenvector of $f'(\underline{U})$. In case

$$\nabla\lambda_k(\underline{U}) \cdot r_k(\underline{U}) = 0 \tag{2.4}$$

the family (field) is called linearly degenerate at $\underline{U} \in \mathbf{R}^n$.

2.3. Let $\mathbf{R}^+ = \{t \in \mathbf{R} \mid t > 0\}$. A vector function $\underline{U} \in [L^\infty(\mathbf{R} \times \mathbf{R}^+)]^n$ is a **weak solution** of (2.1) if and only if

$$\int\int_{\mathbf{R}\times\mathbf{R}^+} \underline{U}\varphi_t + f(\underline{U})\varphi_x \, dx dt + \int_{\mathbf{R}} \underline{U}(x, 0)\varphi(x, 0)\, dx = 0 \tag{2.5}$$

for all $\varphi \in \left[C_0^\infty(\mathbf{R}^2)\right]^n$. If additionally $\underline{U} \in \left[C^1(\mathbf{R} \times \mathbf{R}^+) \cap C^0(\overline{\mathbf{R} \times \mathbf{R}^+})\right]^n$, then $\underline{U}$ is a solution to (2.1) if and only if it is a solution to (2.5).

2.4. The **Riemann problem** for a system of conservation laws (2.1) consists of the initial value problem with respect to the variable t with initial data of the form

$$\underline{U}(x, 0) = \underline{U}_0(x) = \begin{cases} \underline{U}_L = (u_{1L}, \ldots, u_{nL}) & \text{for } x < 0 \\ \underline{U}_R = (u_{1R}, \ldots, u_{nR}) & \text{for } x > 0 \end{cases} \tag{2.6}$$

with the constant states $\underline{U}_L$, $\underline{U}_R \in \mathbf{R}^n$.

In the context of genuinely nonlinear strictly hyperbolic systems this problem is solved by connecting the states $\underline{U}_L$ and $\underline{U}_R$ at infinity by constant states which are connected via shock and rarefaction waves (cp. Chapter 17 in Smoller [48], Lax [29], [30], [31]).

2.5. A weak solution to (2.1) that is discontinuous along a smooth curve $\Gamma \subset \mathbf{R}^2$ must satisfy the **jump condition**, called **Rankine-Hugoniot condition** in gas dynamics (cp. Smoller [48] Chapter 15.3),

$$s \cdot [\underline{U}] = [\, f(\underline{U})\,] \quad \text{on } \Gamma. \tag{2.7}$$

Here $s = s(\underline{U}_L, \underline{U}_R) = \frac{dx}{dt}$ is the local speed of the discontinuity (shock), i. e. the tangent to Γ with respect to the t-axis. By $[g(\underline{U})]$ we denote the difference $g(\underline{U}_L) - g(\underline{U}_R)$ for any smooth function $g : \mathbf{R}^n \to \mathbf{R}^n$.

We will restrict our investigation of the Riemann problem only to states $\underline{U}_L$, $\underline{U}_R$ that satisfy (2.7) for some $s \in \mathbf{R}$. This means that we only consider solutions taking the constant values $\underline{U}_L$, $\underline{U}_R$ to the left resp. right of the shock curve Γ (right being the direction of the positive x-axis).

We study the solution set of the Rankine-Hugoniot equations

$$s[\underline{U}_L - \underline{U}] + [f(\underline{U}) - f(\underline{U}_L)] = 0 \tag{2.8}$$

for an arbitrary given $\underline{U}_L \in \mathbf{R}^n$. Then (2.8) gives a system of n equations in the $n+1$ unknowns $(s,\ u_1, \dots,\ u_n)$. The Jacobian of (2.8) is

$$J = (\underline{U}_L - \underline{U},\ -s\,\mathrm{Id} + f'(\underline{U}))\,. \tag{2.9}$$

On the line $\mathcal{L} = \{(s,\ \underline{U}_L) \mid s \in \mathbf{R}\}$ the Jacobian has rank n if and only if $s \neq \lambda_1(\underline{U}_L), \dots,\ \lambda_n(\underline{U}_L)$. Suppose the latter to be the case, then the implicit function theorem gives a unique differentiable solution curve locally around $\underline{U}_L$. This curve must be a part of the line $\mathcal{L}$, since all points of $\mathcal{L}$ trivially solve (2.8). These points are not interesting because on $\mathcal{L}$ one has $\underline{U}_L = \underline{U}_R$, i. e. no shock occurs. The interesting solutions are obtained by the study of the bifurcation curves from the line of trivial solutions $\mathcal{L}$. These bifurcations occur at $\tilde{s}_1 = \lambda_1(\underline{U}_L), \dots,\ \tilde{s}_n = \lambda_2(\underline{U}_L)$ since rank $J < n$ there.

For the following discussion we assume that $f'(\underline{U})$ has distinct (simple) real eigenvalues everywhere, i. e. is strictly hyperbolic, and that the system (2.1) is genuinely nonlinear everywhere. It is natural to take s as the bifurcation parameter.

According to the "bifurcation from simple eigenvalues theorem" (cp. Smoller [48] Theorem 13.4, Hale [4] Theorem 5.1) our assumptions imply that there exists a C^2-curve transversal to $\mathcal{L}$ at $(\tilde{s}_i,\ \underline{U}_L)$ for each i, $1 \leq i \leq n$. We denote these curves by $\kappa_i : \mathbf{R} \to \mathbf{R}^{n+1}$ and assume these parametrized by $\varepsilon \in \mathbf{R}$, i. e. $\kappa_i : \varepsilon \to (s_i(\varepsilon), \underline{U}_i(\varepsilon)), (s_i(0), \underline{U}_i(0)) = (\tilde{s}_i, \underline{U}_L)$, $1 \leq i \leq n$. We will call them the **Hugoniot curves**. The implicit function theorem (cp. Hale [4], Theorem 3.5) lets us determine the curve κ_i as long as for $\varepsilon \neq 0$ we have $s_i(\varepsilon) \neq \lambda_j\,(\underline{U}_i(\varepsilon))$ for all j, $1 \leq j \leq n$, i. e. that rank $J = n$.

2.6. The genuine nonlinearity of the system allows for a parametrization of the κ_i to be chosen such that $\dot{\underline{U}}_i\,(0) = r_i(\underline{U}_L)$, $\ddot{\underline{U}}\,(0) = \dot{r}_i\,(\underline{U}_L)$, $s_i(0) = \lambda_i(\underline{U}_L)$ and $\dot{s}_i\,(0) = \frac{1}{2}$ (see Smoller [48] Corollary 17.13, Lax [30]), where $r_i(\underline{U})$ has been normalized to give $\nabla\lambda_i(\underline{U})r_i(\underline{U}) \equiv 1$.

This also implies that for small ε

$$\begin{aligned} \lambda_i\,(\underline{U}_i(\varepsilon)) < s_i(\varepsilon) < \lambda_{i+1}\,(\underline{U}_i(\varepsilon)) \quad &\text{for } \varepsilon < 0 \\ \lambda_{i-1}\,(\underline{U}_i(\varepsilon)) < s_i(\varepsilon) < \lambda_i\,(\underline{U}_i(\varepsilon)) \quad &\text{for } \varepsilon > 0 \end{aligned} \tag{2.10}$$

(cp. proof of Theorem 17.14 in Smoller [48]). These latter inequalities remain valid globally for general systems only under additional assumptions (cp. Liu [34], Mock [35]). The property $s_i(\varepsilon) \neq \lambda_k\,(\underline{U}_i(\varepsilon))$ for all k, $1 \leq k \leq n$, and $\varepsilon \neq 0$ may be used to define a stronger notion of genuine nonlinearity (cp. Gel'fand [15], Mock [35]). This would imply the global existence of the curves κ_i since it implies rank $J = n$ near κ_i for $\varepsilon \neq 0$.

It is common to distinguish **strong shocks** from **weak shocks** by saying that for strong shocks ε is large resp. small for weak shocks. Important for the distinction is the fact that properties like (2.10) may be lost if ε becomes too large.

2.7. The Riemann problem (2.6) for (2.1) is not uniquely solvable unless a further admissibility criterion for the discontinuities is imposed on solutions. A simple example for multiple solutions for the scalar case ($n = 1$) may be found in Smoller [48] Chap. 15.3.

Compatibility considerations involving the characteristic speeds $\lambda_k(\underline{U}_L)$, $\lambda_k(\underline{U}_R)$, $1 \leq k \leq n$, lead to the following inequalities (cp. Lax [29], Smoller [48])

$$\begin{aligned} &\lambda_k(\underline{U}_R) < s(\underline{U}_L, \underline{U}_R) < \lambda_k(\underline{U}_L) \\ &\lambda_{k-1}(\underline{U}_L) < s(\underline{U}_L, \underline{U}_R) < \lambda_{k+1}(\underline{U}_R) \end{aligned} \tag{2.11}$$

for some $k \in \{1, \ldots, n\}$. These are the **Lax shock conditions**. A discontinuity is considered admissible and called a k-shock iff (2.11) is satisfied for some index k.

Comparing this to (2.10) this implies that one considers as admissible only shocks on the $\varepsilon < 0$ branch of the Hugoniot curves κ_k for small ε.

2.8. Another approach to admissibility is the **viscosity method**. This is motivated by the fact that conservation laws arising in inviscid fluid flows may be obtained as limits when in equations for viscous flows the viscosity and heat conduction coefficients tend to zero. The paradigm is that the admissible shock solutions should be approximated by the smooth solutions of the viscous equations in a suitable sense (cp. Becker [1], Gel'fand [15], Gilbarg [16], Weyl [50]).

For the Riemann problem this may be translated into the following. Consider the following parabolic regularization of the system (2.1)

$$\underline{U}_t + f(\underline{U})_x = \varepsilon \mathcal{A}\, \underline{U}_{xx}, \qquad \varepsilon > 0, \quad t \geq 0 \tag{2.12}$$

where $\mathcal{A} \in \mathbf{R}^{n^2}$ is a matrix to be appropriately chosen. Suppose the system (2.12) possesses for each $\varepsilon > 0$ a smooth solution U^ε of the form

$$\underline{U}^\varepsilon = \underline{U}^\varepsilon \left(\frac{x - st}{\varepsilon} \right) \tag{2.13}$$

with

$$\lim_{\tau \to -\infty} \underline{U}^\varepsilon(\tau) = \underline{U}_L \quad \text{and} \quad \lim_{\tau \to \infty} \underline{U}^\varepsilon(\tau) = \underline{U}_R \tag{2.14}$$

where $\underline{U}_L$, $\underline{U}_R \in \mathbf{R}^n$ are connected by an Hugoniot curve. Here $\tau = \frac{x-st}{\varepsilon}$ and $s = s(\underline{U}_L, \underline{U}_R)$ is the shock speed given by (2.7). If the family $\underline{U}^\varepsilon$ converges to a shock solution of (2.6) in the sense of distributions, then the shock solution is admissible.

The special form (2.13), (2.14) seems to be due to Gel'fand [15]. There is considerable freedom as to the suitable choice for the right hand side, i. e. the regularization, in (2.12) (see e. g. Conley/Smoller [7], [8], [9], Keyfitz [26], Mock [35], Slemrod [46], Smoller [48], Wendroff [49]). Conley and Smoller have investigated classes of matrices that give the same admissibility for certain classes of problems. We will say that a shock $(s, \underline{U}_L, \underline{U}_R)$ has a **viscous profile** if the corresponding solution of the Riemann problem (2.6) can be approximated by solutions to (2.12), (2.14) of type (2.13) in the limit $\varepsilon \to 0$.

A simple form of (2.12) is given by choosing A = Id (cp. Gel'fand [15], Conley/Smoller [7], Slemrod [46], Smoller [48]), i. e.

$$\underline{U}_t + f(\underline{U})_x = \varepsilon \underline{U}_{xx}. \tag{2.15}$$

Substitution of (2.13) into (2.15), multiplication by ε and denoting differentiation with respect to τ by $(\bullet)$ we get the second order ordinary differential system

$$-s\,\dot{\underline{U}} + f(\dot{\underline{U}}) = \ddot{\underline{U}}. \tag{2.16}$$

This system may be integrated once to obtain the first order system

$$-s\underline{U} + f(\underline{U}) + C = \dot{\underline{U}} \tag{2.17}$$

with $C \in \mathbf{R}^n$. The conditions (2.11) imply $\lim_{\tau \to \pm\infty} \dot{\underline{U}}(\tau) = 0$ and therefore the left hand side of (2.14) must vanish at $\underline{U}_L$ and $\underline{U}_R$.

Choosing $\underline{U}_L$ we get

$$C = s\underline{U}_L - f(\underline{U}_L). \tag{2.18}$$

Note that the jump conditions (2.7) then give $C = s\underline{U}_R - f(\underline{U}_R)$. One can now study the vector field

$$X = -s(\underline{U} - \underline{U}_L) + f(\underline{U}) - f(\underline{U}_L). \tag{2.19}$$

We recall (cp. Guckenheimer/Holmes [18]) that a curve $\gamma = \gamma(t)$ having in every point p the vector X_p as its tangent is called an orbit of X. The w-limit set of γ is the set

$$w(\gamma) = \{q \in \mathbf{R}^n \mid \exists (t_n)_{n \in \mathbf{N}} \subset \mathbf{R}^n,\ t_n \to \infty \quad \text{such that} \quad \gamma(t_n) \to q\}$$

If $t_n \to -\infty$ is taken one obtains the α-limit set $\alpha(\gamma)$. The vector field X has fixed points (X = 0) at $p = \underline{U}_L$ and $p = \underline{U}_R$. The problem of finding a solution to (2.12), (2.14) is now equivalent to the problem of finding a heteroclinic orbit $\gamma \subset \mathbf{R}^2$ with

$$\alpha(\gamma) = \underline{U}_L \qquad w(\gamma) = \underline{U}_R. \tag{2.20}$$

This dynamical systems approach to admissibility seems to be due to Weyl [50]. A fixed point is called **hyperbolic** if the eigenvalues of the linearization X′ have a non-zero real part. In this case the stability of the vector field is determined by linear field X′ according to the Hartman-Grobman theorem (cp. Guckenheimer/Holmes [18] Theorem 1.4.1, Hale [4] Sect. 3.7).

2.8. In (1.3) the second equation is a compatibility condition, since the system is derived from a second order equation. It seems preferable to avoid taking viscous perturbations of this compatibility condition. As we shall see this will not be necessary, i. e. it will be possible to use a degenerate Matrix $\mathcal{A}$. The dynamical systems approach becomes trivial, i. e. one-dimensional, in the case of a system of two equations. This works because $-p(\cdot)$ is convex. For nonconvex $p(\cdot)$ that appear for instance in the theory of van der Waals fluids this is no longer the case. Slemrod [45] showed that in that case one has to add a third order dispersion term, together with the second order diffusion term, to one equation. In Slemrod [46] this is shown to be essentially equivalent to adding second order terms to both equations. The point in that case is that a dynamical system in the plane is needed to circumvent a third fixed point.

2.9. A further criterion for the admissibility of solutions is provided by **entropy inequalities**. The basis of this approach is the existence of an additional scalar conservation law

$$U(\underline{U})_t + F(\underline{U})_x = 0 \tag{2.21}$$

satisfied by all smooth solutions to (2.1). The functions $U,\ F : \mathbf{R}^n \to \mathbf{R}$ are assumed to be continuously differentiable. Denoting the respective Jacobians by U' and F' the equation (2.21) is satisfied by smooth solutions of (2.1) iff

$$U'f' = F' \tag{2.22}$$

holds. In the case $n = 2$ (2.22) is a system of two partial differential equations of first order for the two unknowns $U,\ F$ which may be used to determine them. In case $n > 2$ the system (2.22) is overdetermined and generically admits no solution. But, for example the Euler equations ($n \geq 3$) admit an additional conservation law for the entropy. This fact leads one to expect that physically relevant systems of conservation laws may generally possess such an additional conservation law. Due to the mentioned entropy law the function U is called an **entropy** for the system (2.1), even if it is not a physical entropy.

Now (2.21) is only valid for smooth solutions. Its validity across a jump discontinuity would lead to another jump relation, i. e. add another equation in (2.7). This will generally eliminate all shock solutions by reducing the Hugoniot curves to $\mathcal{L}$, leaving $\underline{U}_L = \underline{U}_R$ as the only possible solutions.

2.10. A true physical entropy will not be conserved at a shock (cp. Landau/Lifschitz [28], Oswatitsch [40], [41], Zierep [51]). But, due to the 2nd law of thermodynamics it would satisfy an inequality that restricts the physically possible transitions. Using a method due to Kružkov [27] it is possible to derive such an inequality that can be used as a selection principle. Suppose we are given a family $\underline{U}_\varepsilon,\ 0 < \varepsilon \leq \varepsilon_0$, of smooth solutions to (2.12), (2.14) of type (2.13). Further suppose that the family $\underline{U}_\varepsilon$ is uniformly bounded in $L^\infty(\mathbf{R} \times \mathbf{R}^+)$ and converges in $L^1_{\text{loc}}(\mathbf{R} \times \mathbf{R}^+)$ for $\varepsilon \to 0$ to a solution $\underline{U}$ of the corresponding Riemann problem while $U'(\underline{U}^\varepsilon)\mathcal{A}\underline{U}^\varepsilon_x$ remains bounded independently of ε in $L^1_{\text{loc}}(\mathbf{R} \times \mathbf{R}^+)$.

Multiplication of (2.12) with $U'(u^\varepsilon)$ gives

$$(2.23)\qquad \begin{aligned} U(\underline{U}^\varepsilon)_t + F(\underline{U}^\varepsilon)_x &= \varepsilon U'(\underline{U}^\varepsilon)\mathcal{A}\underline{U}^\varepsilon xx \\ &= \varepsilon \left(U'(\underline{U}^\varepsilon)\mathcal{A}\underline{U}^\varepsilon_x\right)_x - \varepsilon(\underline{U}^\varepsilon_x)^\tau U''(\underline{U}^\varepsilon)\mathcal{A}\underline{U}^\varepsilon_x. \end{aligned}$$

If one now assumes

$$(2.24)\qquad U''(\underline{U}_\varepsilon)\mathcal{A} \geq 0,$$

i. e. $(x)^\tau U''(\underline{U}^\varepsilon)\mathcal{A}x \geq 0$ for all $x \in \mathbf{R}^n$, then (2.24) gives

$$(2.25)\qquad U(\underline{U}^\varepsilon)_t + F(\underline{U}^\varepsilon)_x \leq \varepsilon \left(U'(\underline{U}^\varepsilon)\mathcal{A}\underline{U}^\varepsilon_x\right)_x.$$

Choosing an arbitrary $\varphi \in C_0^\infty(\mathbf{R} \times \mathbf{R}^+)$, $\varphi \geq 0$, one obtains the inequality

$$(2.26)\qquad -\int U(\underline{U}^\varepsilon)\varphi_t + F(\underline{U}^\varepsilon)\varphi_x \; dxdt \leq -\varepsilon \int U'(\underline{U}^\varepsilon)\mathcal{A}\underline{U}^\varepsilon_x\,\varphi_x \; dxdt.$$

Due to our boundedness assumptions the right hand side of (2.26) vanishes for $\varepsilon \to 0$ giving

$$-\int U(\underline{U})\,\varphi_t + F(\underline{U})\,\varphi_x \; dxdt \leq 0$$

or the **entropy inequality**

$$(2.27)\qquad U(\underline{U})_t + F(\underline{U})_x \leq 0$$

in the sense of distributions. At a shock the left and right states must therefore satisfy the entropy inequality

$$(2.28)\qquad s[U] - [F] \leq 0$$

which is obtained by the same arguments as those used in deriving the jump conditions (2.7). Note that wherever the solution is locally smooth (2.21) will be satisfied anyway.

THEOREM 2.1. *Suppose that the system (2.1) is strictly hyperbolic and genuinely nonlinear. Also suppose that it posesses a strictly convex entropy function $U : \mathbf{R}^n \to \mathbf{R}$ with associated flux function $F : \mathbf{R}^n \to \mathbf{R}$ such that (2.24) is satisfied. If a weak solution of (2.1) contains a weak shock of speed $s(\underline{U}_L, \underline{U}_R)$ then the Lax shock conditions (2.11) and the entropy inequality (2.27) are equivalent. For $\underline{U}_L$, $\underline{U}_R$ sufficiently close a solution to the Riemann problem (2.6) satisfying (2.11) or (2.28) admits an approximation of the type described in Section 2.7.*

Proof. See Smoller [48] Theorem 20.8, Theorem 24.6.

□

3. The transonic small disturbance system.

3.1 In the introduction we have put the transonic small disturbance equation into the following form with $p(u) = u - \frac{u^2}{2}$

$$
(3.1) \qquad \begin{aligned} p(u)_x + v_y &= 0 \\ v_x - u_y &= 0. \end{aligned}
$$

The system does not have the time variable as a canonical directed variable which appears in the Riemann problem (2.6) or the viscosity equations (2.12). The expansion giving (1.1) is carried out under the assumption that the flow is predominantly directed in the positive x-direction. We will see later that this choice of a distinguished direction for the Riemann problem corresponds directly to the choice of admissibility criteria canonical for the transonic flow problem. It will not matter whether we formally set $t = x$ or $t = y$ in (3.3).

3.2 The small disturbance problem has a canonical admissibility condition to exclude non-compressive shocks, inherited from potential flow theory (cp. Landau/Lifschitz [28], Oswatitsch [40], [41], Zierep [51]). Suppose a shock solution of the Riemann problem (2.6) is given and for some value of y a shock is at $\tilde{x}$. We denote by $u^-(\tilde{x})$ the limit value of u as x approaches $\tilde{x}$ for $x < \tilde{x}$ and $u^+(\tilde{x})$ the value for $x > \tilde{x}$. Then this admissibility condition is

$$
(3.2) \qquad u^- > u^+
$$

(cp. Cole/Cook [6]), i. e. the flow velocity in x direction is decreased through a shock. This corresponds to an increase in density for potential flows. We want to investigate how this condition corresponds to the admissibility conditions given in Chap. 2 for hyperbolic problems. The notation u^-, u^+ will allways refer to the original x, y variables of the transonic flow problem.

We had noted in the introduction that $p(u) = u - \frac{u^2}{2}$ implies that (3.3) changes type. Namely it is

$$
(3.3) \qquad \begin{aligned} \text{strictly hyperbolic} \quad &\text{for} \quad u > 1 \\ \text{elliptic} \quad &\text{for} \quad u < 1 \\ \text{degenerate} \quad &\text{for} \quad u = 1 \end{aligned}
$$

since

$$
(3.4) \qquad \det\left[\begin{pmatrix} p'(u) & 0 \\ 0 & 1 \end{pmatrix} + \lambda \begin{pmatrix} 0 & 1 \\ -1 & 0 \end{pmatrix}\right] = 0
$$

gives the eigenvalues $\lambda_\pm(u) = \pm\sqrt{u-1}$. We divide the $(u,\ v)$-plane into half planes $\mathcal{E} = \{(u,\ v) \in \mathbf{R}^2 | u < 1\}$, $\mathcal{H} = \{(u,\ v) \in \mathbf{R}^2 | u > 1\}$, and the line $\mathcal{Z} = \{(u,\ v) \in \mathbf{R}^2 | u = 1\}$. For convenience we will use the same notation to divide the u-axis. The eigenvalues are distinct in $\mathcal{E}$ and $\mathcal{H}$.

3.3. We would like to again take up the discussion of Hugoniot curves. The Rankine-Hugoniot equations (in the form (2.8)) are

$$\begin{aligned} s[u_L - u] &= -[v_L - v] \\ s[v_L - v] &= [p(u_L) - p(u)]. \end{aligned} \tag{3.5}$$

These imply

$$s^2(u_L - u_R) + p(u_L) - p(u_R) = 0$$

or

$$-s^2 = \frac{p(u_L) - p(u)}{u_L - u} = p'(\zeta) \tag{3.6}$$

with ζ between u_L and u. Since $p' \geq 0$ in $\mathcal{E} \cup \mathcal{Z}$ we can deduce that for $u_L \in \mathcal{E} \cup \mathcal{Z}$ the Hugoniot curves cannot lie in $\mathcal{E} \cup \mathcal{Z}$. This corresponds to the fact $\mathcal{L}$ has no bifurcation points if $u_L \in \mathcal{E}$, since there are no real eigenvalues. For $u_L \in \mathcal{Z}$ $s = 0$ is the only possible bifurcation point of $\mathcal{L}$. The Hugoniot curves cannot enter $\mathcal{E}$ due to the above argument.

Suppose $p(\cdot)$ has the additional property that to each $u \in \mathcal{E}$ there corresponds a $w \in \mathcal{H}$ such that $p(u) = p(w)$. Then for $(u_L,\ v_L) \in \mathcal{E}$ the point $(w,\ v)$ with $p(w) = p(u_L)$ and $v = v_L$ lies on an Hugoniot curve and the Jacobian (2.9) has rank 2 there, since 0 is not an eigenvalue in $\mathcal{H}$. This is the case for our example. So in this case one easily finds a branch of the Hugoniot curves that is detached from $\mathcal{L}$.

For $u_L \in \mathcal{H}$ we have the properties observed in section 2.2, i. e. two curves bifurcating from $\mathcal{L}$ at the eigenvalues (i. e. $s = \lambda_1$ or $s = \lambda_2$).

In our example $p(u) = u - \frac{u^2}{2}$ we may eliminate s in (3.6) in order to give an explict representation of the projections of the Hugoniot curves onto the $(u,\ v)$-plane. One has

$$\begin{aligned} v - v_L &= \pm\sqrt{(u - u_L)\,(p(u_L) - p(u))} \\ &= \pm\sqrt{(u - u_L)\left(u_L - u - \left(\frac{u_L^2}{2} - \frac{u^2}{2}\right)\right)} \\ &= \pm\sqrt{(u - u_L)^2\left(\frac{u_L + u}{2} - 1\right)} \end{aligned}$$

or

$$v = v_L \pm (u - u_L)\sqrt{\frac{u_L + u}{2} - 1}. \tag{3.7}$$

One obtains real-valued algebraic curves if $\frac{u_L+u}{2} \geq 1$ i. e. $u \geq 2 - u_L$. Choosing again $z \in \mathcal{E}$, $w \in \mathcal{H}$ such that $p(z) = p(w)$ one sees that for $u_L = z$ the Hugoniot curves are only one curve lying in the half plane $u \geq w$. For $(u_L = w,\ v_L)$ the two

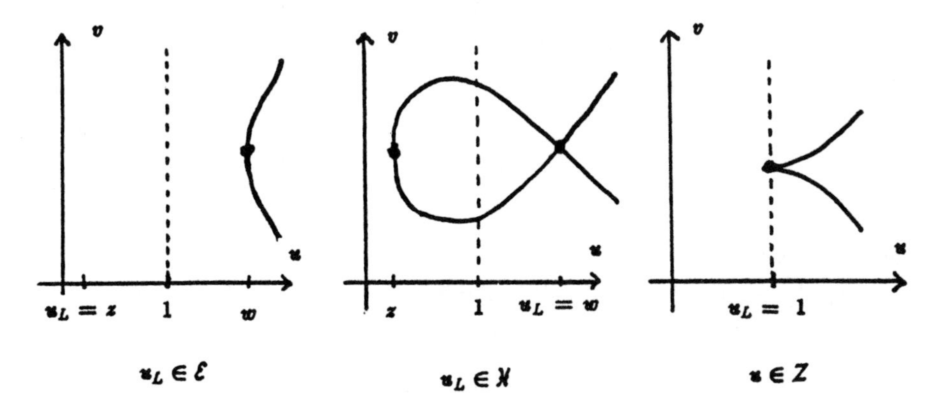

Figure 1

3.4. We set $P(t) = \int_0^t p(s)\,ds$ the primitive of $p(\cdot)$. It is well known that (1.2) is the Euler-Lagrange equation to the functional

$$\int L(\varphi_x,\, \varphi_y)\, dxdy = \int P(\phi_x) + \frac{\varphi_y^2}{2}\, dxdy. \tag{3.8}$$

Since the integrand $L(\varphi_x,\, \varphi_y)$ is not explicitly y or x dependent Noether's theorem gives two additional conservation laws satisfied by smooth solutions to (1.2) resp (3.3) (cp. DiPerna [13], Mock [36]). They are for y-independence

$$\begin{aligned} 0 &= [L - \phi_y L_{\phi_y}]_y + [-\phi_y L_{\phi_x}]_x \\ &= [P(\phi_x) - \frac{\phi_y^2}{2}]_y + [-\phi_y p(\phi_x)]_x \end{aligned} \tag{3.9}$$

and for x-independence

$$\begin{aligned} 0 &= [L - \phi_x L_{\phi_x}]_x + [-\phi_x L_{\phi_y}]_y \\ &= [P(\phi_x) + \frac{\phi_y^2}{2} - \phi_x p(\phi_x)]_x + [-\phi_x \phi_y]_y. \end{aligned} \tag{3.10}$$

We will write these as $(U_i)_x + (F_i)_y = 0,\ i = 1,2$ in the variables $u = \phi_x,\ v = \phi_y$, i. e.

$$\begin{aligned} U_1 &= \frac{v^2}{2} - P(u) \quad & F_1 &= vp(u) \\ U_2 &= -uv \quad & F_2 &= \frac{v^2}{2} + P(u) - up(u). \end{aligned} \tag{3.11}$$

These are not the only additional conservation laws to (3.3) known (cp. Mock [36] Sect. VI).

4. The p-system.

4.1. For better comparison with hyperbolic theory we put our investigations in the context of the well known p**-system** (cp. Oleĭnik [39], Smoller [48], Leibovich [32], Wendroff [49]). We formally introduce the variable t instead of y. This gives the system

$$
\begin{aligned}
u_t - v_x &= 0 \\
v_t + p(u)_x &= 0,
\end{aligned}
\tag{4.1}
$$

where $p: \mathbf{R} \to \mathbf{R}$ is some smooth function, $p'' < 0$ everywhere and $p'(u_{\text{crit}}) = 0$ for one $u_{\text{crit}} \in \mathbf{R}$.[1]

Using the notation $\underline{U} = \binom{u}{v}$ and the flux function $f: \mathbf{R}^2 \to \mathbf{R}^2$, $f(\underline{U}) = \binom{-v}{p(u)}$ the system can be written in the standard conservation law form (cp. (2.1))

$$
\underline{U}_t + f(\underline{U})_x = 0. \tag{4.2}
$$

The Jacobian of the flux function is

$$
f' = \begin{pmatrix} 0 & -1 \\ p'(u) & 0 \end{pmatrix} \tag{4.3}
$$

and has the eigenvalues

$$
\lambda_1 = -\sqrt{-p'(u)}, \quad \lambda_2 = \sqrt{-p'(u)}. \tag{4.4}
$$

The right eigenvectors are

$$
r_1 = \begin{pmatrix} 1 \\ \sqrt{-p'(u)} \end{pmatrix} \qquad r_2 = \begin{pmatrix} 1 \\ -\sqrt{-p'(u)} \end{pmatrix}.
$$

Also

$$
\nabla\lambda_1 \cdot r_1 = \frac{-p''(u)}{2\sqrt{-p'(u)}} \qquad \nabla\lambda_2 \cdot r_2 = \frac{-p''(u)}{2\sqrt{-p'(u)}}
$$

and therefore the system is genuine nonlinear everywhere. For $p'(u) \to 0$ the vectors $\nabla\lambda$ are parallel to the u-axis and the r_i, $i = 1, 2$, approach $\binom{1}{0}$ Therefore genuine nonlinearity, taken as a non-orthogonality condition, holds also for $p'(u) = 0$.

4.2. In case $p'(u) < 0$ the system is strictly hyperbolic since the eigenvalues are real and distinct. This case is extensively discussed in Smoller [48] Chap. 17.A. If $p' > 0$ the eigenvalues are purely imaginary and distinct. In this case the system is elliptic.

Due to the assumption $p'' < 0$ there is at most one $u_{\text{crit}} \in \mathbf{R}$ such that $p'(u_{\text{crit}})=0$. Since we assume the existence of such a u_{crit}, we may divide the (u, v)-plane as in Section 3.3 into the two half planes $\mathcal{E} = \{(u, v) \in \mathbf{R}^2 | \; u < u_{\text{crit}} \text{ i. e. } p'(u) > 0\}$

[1] We interchange u and v w.r.t. Smoller [48] Chapter 17 in order to conform with the standard notation in transonic flows. For the same reason we take $p'' < 0$.

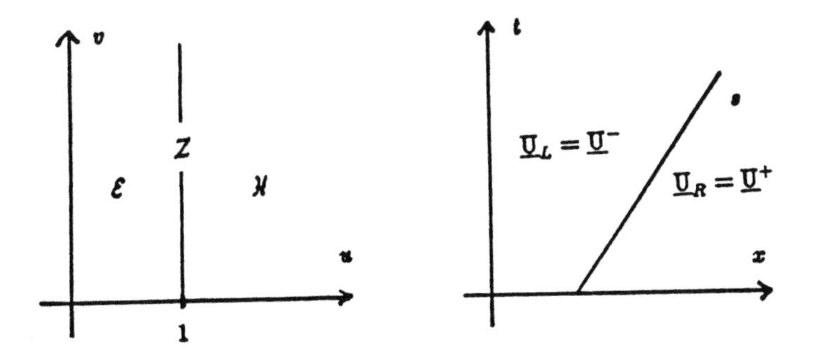

Figure 2

and $\mathcal{H} = \{(u, v) \in \mathbf{R}^2 | u > u_{\text{crit}} \text{ i. e } p'(u) < 0\}$. These planes are separated by the line $\mathcal{Z} = \{(u, v) \in \mathbf{R}^2 | u = u_{\text{crit}} \text{ i. e } p'(u) = 0\}$.

The jump conditions (2.7) for the p-system are (cp. (3.6))

$$
\begin{aligned}
s[u_L - u_R] &= -[v_L - v_R] \\
s[v_L - v_R] &= [p(u_L) - p(u_R)].
\end{aligned} \tag{4.5}
$$

We had seen in Section 3.3 that if both states (u_L, v_L) and (u_R, v_R) lie in the elliptic region $\mathcal{E}$, or one in $\mathcal{E}$ and one in $\mathcal{Z}$, they cannot be connected by a shock since $p' \geq 0$ in $\mathcal{E} \cup \mathcal{Z}$. So in the following we will only consider the cases of hyperbolic-hyperbolic shocks (both states in $\mathcal{H}$) and hyperbolic-elliptic shocks. Shocks between states in $\mathcal{H}$ and $\mathcal{Z}$ will not be discussed explicitly. For the same reason, namely (3.8) and $p'' \neq 0$, the Hugoniot curves κ_i satisfy $s_i(\varepsilon) \neq (\lambda_k(v(\varepsilon))$ for $k = 1, 2$ and $\varepsilon \neq 0$ (cp. Section 2.5). Therefore, they exist for all $\varepsilon \neq 0$. There are no secondary bifurcations, since rank $J = 2$ for $\varepsilon \neq 0$ (cp. (2.9)).

4.3. In the case $p'(u_L), p'(u_R) < 0$, i. e. strict hyperbolicity, the *Lax shock conditions* (2.11) allow for two types of shocks (cp. Lax [29], Smoller [48] Chap. 17.A)

$$
\begin{aligned}
&\text{1-shocks:} \quad s < \lambda_1(u_L), \qquad \lambda_1(u_R) < s < \lambda_2(u_R) \\
&\text{2-shocks:} \quad \lambda_1(u_L) < s < \lambda_2(u_L), \qquad \lambda_2(u_R) < s
\end{aligned} \tag{4.6}
$$

or using (4.4) one commonly makes the distinction

$$
\begin{aligned}
&\text{back shocks:} \quad -\sqrt{-p'(u_R)} < s < -\sqrt{-p'(u_L)}, \quad \text{i. e.} \quad s < 0 \\
&\text{front shocks:} \quad \sqrt{-p'(u_R)} < s < \sqrt{-p'(u_L)}, \qquad \text{i. e.} \quad s > 0.
\end{aligned} \tag{4.7}
$$

If one assumes $p'(u) < 0$ and $p''(u) < 0$ for some subinterval of the real axis containing u_L and u_R then this implies

$$
\begin{aligned}
u_R > u_L \quad &\text{and} \quad s < 0 \quad \text{for back shocks} \\
u_R < u_L \quad &\text{and} \quad 0 < s \quad \text{for front shocks.}
\end{aligned} \tag{4.8}
$$

The inequalities (4.6) resp. (4.7) obviously only make sense for hyperbolic-hyperbolic shocks.

4.4. We will now study the entropy inequalities (2.28) provided by the two additional conservation laws given in (3.11). The first one is usually associatied with the p-system (cp. Smoller [48] Sect. 20.B).

LEMMA 4.1. *The entropy inequality for* U_1, F_1

$$s\left[\frac{v^2}{2} - P(u)\right] - [p(u)v] \leq 0 \tag{4.9}$$

is equivalent to the inequalities

$$\begin{aligned} u_R > u_L \quad &\text{for} \quad s < 0 \\ u_L > u_R \quad &\text{for} \quad s > 0 \end{aligned} \tag{4.10}$$

i. e. (4.8).

Proof. The inequality for (4.9) becomes for $\underline{U}_L = (u_L, v_L), \underline{U}_R = (u_R, v_R)$

$$0 \geq s\left[\frac{v_L^2}{2} - \frac{v_R^2}{2} + P(u_R) - P(u_L)\right] - p(u_L)v_L + p(u_R)v_R.$$

Using the jump conditions (4.5) one gets

$$\begin{aligned} 0 &\geq \frac{v_L + v_R}{2}(p(u_L) - p(u_R)) + s\,[P(u_R) - P(u_L)] - p(u_L)v_L + p(u_R)v_R \\ &= \frac{v_R - v_L}{2}p(u_L) + \frac{v_R - v_L}{2}p(u_R) + s\,[P(u_R) - P(u_L)] \\ &= s\left[(u_L - u_R)\frac{p(u_L) + p(u_R)}{2} + P(u_R) - P(u_L)\right]. \end{aligned} \tag{4.11}$$

Using the mean value theorem one obtains

$$s\left[(u_L - u_R)\frac{p(u_L) + p(u_R)}{2} + P(u_R) - P(u_L)\right] = s\left[(u_L - u_R)^3\,\frac{p''(\xi)}{12}\right] \tag{4.12}$$

for some ξ between u_L and u_R. Since $p'' < 0$, (4.10) gives (4.9) and vice versa.

□

For the transonic flow problem (4.10) would imply $u^- > u^+$ for $s > 0$ and $u^- < u^+$ for $s < 0$. Obviously this does not give the physical shocks (3.2).

4.5. We now proceed to look at the second possibility in (3.13):

LEMMA 4.2. *The entropy inequality for* U_2, F_2

$$s[-uv] - \left[-up(u) + P(u) + \frac{v^2}{2}\right] \leq 0 \tag{4.13}$$

is equivalent to the inequality

$$u_L > u_R. \tag{4.14}$$

Proof. The entropy inequality (4.13) becomes

$$\begin{aligned}0 \geq s\,[u_R v_R - u_L v_L] + u_L p(u_L) - u_R p(u_R) + P(u_R) - P(u_L) + \frac{v_R^2}{2} - \frac{v_L^2}{2} \\ = u_L\,(p(u_L) - s v_L) - u_R\,(p(u_R) - s v_R) \\ + P(u_R) - P(u_L) + \frac{v_R + v_L}{2}\,(v_R - v_L)\,.\end{aligned}$$

Again using the jump relations (4.5) one gets

$$\begin{aligned}0 &\geq (u_L - u_R)\,(p(u_R) - s\,v_R) + P(u_R) - P(u_L) + \frac{v_R + v_L}{2} s\,(u_L - u_R) \\ &= (u_L - u_R)\left(p(u_R) - s\frac{v_R - v_L}{2}\right) + P(u_R) - P(u_L) \\ &= (u_L - u_R)\,\frac{p(u_R) + p(u_L)}{2} + P(u_R) - P(u_L).\end{aligned} \tag{4.15}$$

The mean value theorem again gives the inequality

$$0 \geq (u_L - u_R)^3\,\frac{p''(\xi)}{12} \tag{4.16}$$

for some ξ between u_L and u_R. Since $p'' < 0$ this is equivalent to

$$u_L > u_R.$$

□

The Lax shock conditions (4.6) or (4.7) only make sense if $p'(u_L),\ p'(u_R) < 0$, i. e. only for states that lie in the strictly hyperbolic region $\mathcal{H}$. For the entropy inequalities (4.9) and (4.13) this restriction is not necessary. Note that (4.11) allows u_L to be in $\mathcal{H}$ as well as $\mathcal{E}$, whereas (4.14) requires u_L to be in $\mathcal{H}$ (since not both u_L and u_R may lie in $\mathcal{E}$).

4.6. We had seen in our discussion of the Kružkov argument that the entropy U and the matrix $\mathcal{A}$ should be compatible via the inequality (2.24), i. e. $U''\mathcal{A} \geq 0$. We want to add a diffusion only to the second equation in (4.1). The Hessians of U_1 and U_2 given in (3.11) are

$$U_1'' = \begin{pmatrix} -p'(u) & 0 \\ 0 & 1 \end{pmatrix}, \quad U_2'' = \begin{pmatrix} 0 & -1 \\ -1 & 0 \end{pmatrix}. \tag{4.17}$$

Note that U_2 is not a convex function, whereas U_1 is convex in $\mathcal{H}$, i. e. for $p'(u) < 0$. Compatible matrices $\mathcal{A}$ are given by

$$\mathcal{A}_1 = \begin{pmatrix} 0 & 0 \\ 0 & 1 \end{pmatrix}, \quad \mathcal{A}_2 = \begin{pmatrix} 0 & -1 \\ 0 & 0 \end{pmatrix}.$$

Thereby we obtain the two systems

$$\begin{aligned} u_t - v_x &= 0 \\ v_t + p(u)_x &= \varepsilon v_{xx} \end{aligned} \tag{4.18}$$

and

$$u_t - v_x = 0 \qquad (4.19)$$
$$v_t + p(u)_x = -\varepsilon u_{xx}.$$

Applying the method outlined in Section 2.8 to (4.18) we obtain the system

$$-s\,\dot{u} - \dot{v} = 0$$
$$-s\,\dot{v} + p(\dot{u}) = \ddot{v}$$

or the first order equation

$$-s\,\dot{u} = -s^2 u - p(u) + C \qquad (4.20)$$

with $C = s^2 u_L + p(u_L) = s^2 u_R + p(u_R)$. This one-dimensional dynamical system has only two fixed points u_L, u_R. At u_L there should be a source and at u_R a sink. These requirements imply for $s > 0$

$$-p'(u_L) > s^2 > -p'(u_R) \quad for \quad s > 0 \qquad (4.21)$$
$$-p'(u_L) < s^2 < -p'(u_R) \quad for \quad s < 0.$$

Suppose u_L, $u_R \in \mathcal{H}$. Then we may take roots and obtain with (4.4) the Lax shock conditions (4.7). We have now seen that the Lax shock conditions (4.7), the entropy inequality (4.9) and the viscosity approximation (4.18) give the same admissibility conditions for shock solutions to the Riemann problem if u_L, $u_R \in \mathcal{H}$. This is well known (see Theorem 2.1 in Smoller [48]). The entropy inequality and the viscosity method give equivalent admissibility conditions also for mixed type shocks. These are not the conditions sought for transonic flows.

4.7. We may use (4.21) to introduce **generalized Lax shock conditions** for certain systems of mixed type. Suppose we are given a mixed type 2 by 2 system with two purely imaginary eigenvalues in $\mathcal{E}$, with $\lambda_1 = \bar{\lambda}_2$, and a positive and a negative eigenvalue in $\mathcal{H}$, as in the case of our example. Then we can give shock inequalities that are equivalent to the Lax inequalities for hyperbolic-hyperbolic shocks and are valid even for hyperbolic-elliptic shocks. These are obtained by using (4.21). Suppose Im $\lambda_1 < 0 <$ Im λ_2 for the eigenvalues in $\mathcal{E}$ and $\lambda_1 < 0 < \lambda_2$ in $\mathcal{H}$. We require for a **1-shock** (back shock) that $s < 0$ and

$$[\lambda_1(\underline{U}_R)]^2 > s^2 > [\lambda_1(\underline{U}_L)]^2. \qquad (4.22)$$

For a **2-shock** (front shock) we require $s > 0$ and

$$[\lambda_2(\underline{U}_R)]^2 < s^2 < [\lambda_2(\underline{U}_L)]^2. \qquad (4.23)$$

And for stationary shocks, i. e. $s = 0$, we require either (4.22) or (4.23) to hold.

The above restrictions on the eigenvalues were made to avoid an elaborate discussion of the different cases that may arise otherwise. For the complications that arise in the case of more general complex eigenvalues see Keyfitz [26]. The inequalities (4.22) and (4.23) are equivalent to (4.6) for $\underline{U}_L$, $\underline{U}_R \in \mathcal{H}$ under the above assumptions on the eigenvalues. For the p-system they are also equivalent to (4.21).

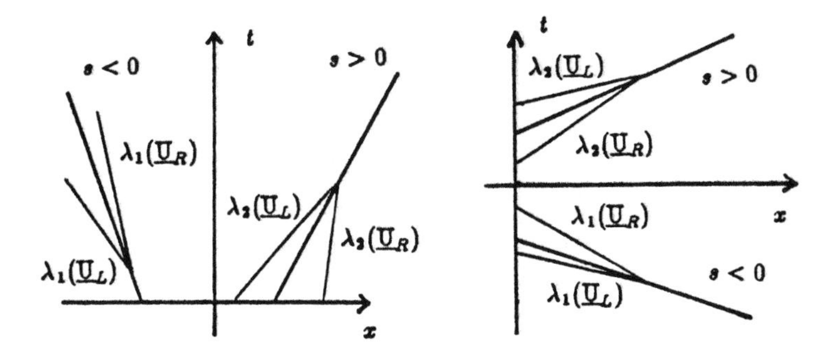

Figure 3

4.8. The entropy inequality (4.13) gave the correct admissibility criterion for transonic flows. Let us look at the system (4.19). We obtain the first order equation

$$\overset{\bullet}{u} = -s^2 u - p(u) + C,$$

with C as before. And we get as analog to (4.21)

$$-p'(u_L) > s^2 > -p'(u_R) \tag{4.24}$$

for all s. We immediately see $u_L \in \mathcal{H}$ and (4.14) follows immediately since $p'' < 0$.

For u_L, $u_R \in \mathcal{H}$ we may take roots to obtain

$$\begin{aligned} \lambda_2(\underline{U}_L) > s > \lambda_2(\underline{U}_R) > 0 \qquad & \text{for } s > 0 \\ 0 > \lambda_1(\underline{U}_R) > s > \lambda_1(\underline{U}_L) \qquad & \text{for } s < 0. \end{aligned} \tag{4.25}$$

Only the first set of inequalities is equivalent to the Lax shock conditions (4.6).

If one thinks of the compatibility considerations between ingoing/outgoing characteristics and the Rankine-Hugoniot conditions that are made to derive the Lax shock conditions (cp. Lax [29] or Smoller [48] Chap. 15.D) there seems to be something wrong here. For $s > 0$ we have a 2-shock with three ingoing and one outgoing characteristic. For $s < 0$ this situation is reversed (see Figure 3).

But, consider the following situation. Suppose we are given the initial data not on the axis $t = 0$, but on the axis $x = 0$. Then one sees (cp. Figure 3) that one has Lax 1-shocks, resp. 2-shocks, as usual.

One clearly sees that the entropy inequality (4.13) and the viscosity method using (4.19) are connected to the flow direction (positive x-direction) as distinguished direction. In the usual hyperbolic theory t is the distinguished direction and we have formally set $y = t$. This direction is only distinguished if a convex entropy like U_1 is used. Use of the nonconvex entropy U_2 makes the x-direction distinguished even though the formal treatment is the same as before. Note also that U_2 was due to the x-invariance of (3.8).

4.9. We again make the assumptions made in Section 4.7 concerning the eigenvalues of a mixed 2 by 2 system. Now we require for **1-shocks** (back shocks) that $s < 0$ and

$$[\lambda_1(\underline{U}_L)]^2 > s^2 > [\lambda_1(\underline{U}_R)]^2. \tag{4.26}$$

For **2-shocks** (front shocks) we require $s > 0$ and

$$[\lambda_2(\underline{U}_L)]^2 > s^2 > [\lambda_2(\underline{U}_R)]^2. \tag{4.27}$$

For stationary shocks $s = 0$ we require either (4.26) or (4.27) to hold. In this case either of the two implies the other. These are equivalent to (4.25) in the case of the p-system.

4.10. We may collect our results in the following theorem:

THEOREM 4.3. *Consider the Riemann problem*

$$\underline{U}(0,\, y) = \underline{U}_0(y) = \begin{cases} \underline{U}_a & \text{for } y < 0 \\ \underline{U}_b & \text{for } y > 0 \end{cases}$$

for the system (4.1). *For* p *we demand* $p'' < 0$. *Assume that* $\underline{U}_a$, $\underline{U}_b \in \mathbf{R}^2$ *lie on a common Hugoniot curve with corresponding shock speed* $s = s(\underline{U}_a, \underline{U}_b)$. *Let* $\underline{U}$ *be the corresponding piecewise constant shock solution, i. e.*

$$\underline{U}(x,\, y) = \begin{cases} \underline{U}_a & \text{for } sy < x \\ \underline{U}_b & \text{for } sy > x. \end{cases}$$

Then the following adimissibility conditions are equivalent:

(a) $u^- > u^+$.

(b) *The entropy inequality* $(U_2)_y + (F_2)_x \le 0$ *holds in the sense of distributions.*

(c) $s[U_2] - [F_2] \le 0$.

(d) $\underline{U}$ *can be obtained as a limit for* $\varepsilon \to 0$ *of solutions* $\underline{U}^\varepsilon$ *to the system (4.18). The limit is achieved in the sense of distributions.*

(e) *The inequalities (4.27) for* $s \le 0$ *or (4.28) for* $s > 0$ *are satisfied.*

Proof. (b) and (c) are obviously equivalent for $\underline{U}$ (see Section 2.10). The equivalence of (a) and (c) was shown in Lemma 4.2.

For the p-system (e) is equivalent to (4.25). Since $p'' < 0$ this implies (a). Now suppose (a) is satisfied. This implies $-p'(u_L) > -p'(u_R)$. Also we have seen that one state must lie in $\mathcal{H}$ (Section 4.2). So at least $u_L \in \mathcal{H}$. In this case for small $\varepsilon \neq 0$ (2.10) is valid. The inequalities (2.10) imply (4.25) and therefore (e) for small ε. Then they must be valid for all $\varepsilon \neq 0$ since $s(\varepsilon)^2$ never becomes equal to $-p'(u(\varepsilon))$ due to (3.7) and the convexity of $-p(\cdot)$.

The equivalence between (4.25) and the solvability of (4.19) for $\varepsilon > 0$ was shown in Section 4.8. It remains to show that $\int_{\mathbf{R}^2_+} \underline{U}^\varepsilon \underline{\varphi}\, dx \to \int_{\mathbf{R}^2_+} \underline{U}\, \underline{\varphi}\, dx$ for every $\underline{\varphi} \in$

$\left[C_0^\infty(\mathbf{R}_+^2)\right]^2$, $\mathbf{R}_+^2 = \{(x, y) \in \mathbf{R}^2 | y > 0\}$. Obviously the sequence $\underline{\mathrm{U}}^\varepsilon$ is uniformly bounded, since $\underline{\mathrm{U}}^\varepsilon(\zeta)$ must lie between u_L and u_R.

Let S_δ be the strip $\{(x, y) \in \mathbf{R}_+^2 \,||x - sy| < \delta\}$. Take $\underline{\varphi} \in C_0^\infty(\Omega)$, then $\operatorname{supp}\varphi$ may be divided into three parts $A_\delta = \operatorname{supp}\varphi \cap \{(x, y) \in \mathbf{R}_+^2 \,| x - sy \geq \delta\}$, $B_\delta = \operatorname{supp}\varphi \cap S_\delta$, $C_\delta = \operatorname{supp}\varphi \cap \{(x, y) \in \mathbf{R}_+^2 \,| x - sy \leq -\delta\}$. Note that $\underline{\mathrm{U}}^\varepsilon(x, y) = \underline{\mathrm{U}}^\varepsilon(\frac{x-sy}{\varepsilon})$. On A_δ we may take, for given $\delta > 0$, ε so small that $|\underline{\mathrm{U}}^\varepsilon(x, y) - \underline{\mathrm{U}}_a|$ is arbitrarily small. The same holds on C_δ for $\underline{\mathrm{U}}_b$.

Since $\underline{\mathrm{U}}^\varepsilon - \underline{\mathrm{U}}$ is bounded on B_δ for any δ the integral

$$\int_{S_\delta} |(\underline{\mathrm{U}}^\varepsilon - \underline{\mathrm{U}})\,\underline{\varphi}|\, dx \tag{4.28}$$

can be made arbitrarily small by choosing δ small. Choose δ to make (4.29) smaller than $\frac{\eta}{3}$ for given $\delta > 0$. Then we choose ε in order to make

$$\int_{A_\delta \cup C_\delta} |\underline{\mathrm{U}}^\varepsilon - \underline{\mathrm{U}}| \cdot |\underline{\varphi}|\, dx < \frac{2}{3}\eta.$$

Therefore $\underline{\mathrm{U}}^\varepsilon$ converges to $\underline{\mathrm{U}}$ in the sense of distributions.

□

5. Further admissibility conditions.

5.1. The convexity of $-p(\cdot)$ can be used to obtain Oleĭnik's E-condition (cp. Oleĭnik [39], Dafermos [11]) for the p-system. The convexity of $-p(\cdot)$ is equivalent to

$$\begin{aligned} p(u) &\geq \left[1 - \frac{u - u_L}{u_R - u_L}\right] p(u_L) + \frac{u - u_L}{u_R - u_L} p(u_r) \\ &= p(u_L) + \frac{u - u_L}{u_R - u_L}\left(p(u_R) - p(u_L)\right) \end{aligned} \tag{5.1}$$

for any u between $u_L, u_R \in \mathbf{R}$.

LEMMA 5.1. *The admissibility condition*

$$u^- > u^+ \tag{5.2}$$

is equivalent to the transonic ***E*-condition**, *namely, the requirement that*

$$\frac{p(u) - p(u_L)}{u - u_L} \leq \frac{p(u_R) - p(u_L)}{u_R - u_L} \tag{5.3}$$

for all u between u_L and u_R.

Proof. The inequality (5.4) is equivalent to $u_L > u_R$. This implies $u_L > u$ for all u between u_L and u_R. Therefore, division of (5.1) by $u - u_L$ for any such u reverses the inequality to give (5.3). Conversely, (5.3) and (5.1) can only be valid simultaneously iff $u_L > u_R$. □

Note that the other set of admissibility conditions, i. e. those connected to Lemma 4.1 are equivalent to

$$(5.4)\qquad \begin{aligned} \frac{p(u)-p(u_L)}{u-u_R} &\geq \frac{p(u_R)-p(u_L)}{u_R-u_L} \quad \text{for } s\leq 0 \\ \frac{p(u)-p(u_L)}{u-u_R} &\leq \frac{p(u_R)-p(u_L)}{u_R-u_L} \quad \text{for } s\geq 0 \end{aligned}$$

(cp. Dafermos [11], Oleĭnik [39], Shearer [43]).

5.2. Another admissibility condition is Liu's **extended entropy condition** (cp. Liu [33], [34], Dafermos [12]). In the context of admissibility conditions compatible with Lemma 4.1 it states that for an admissible shock one must have

$$(5.5)\qquad s(\underline{U}_L,\, \underline{U}_R) \geq s(\underline{U}_L,\, \underline{U})$$

for all $\underline{U}$ that lie on the Hugoniot curve connecting $\underline{U}_L$ and $\underline{U}_R$. Note that we still assume that $-p(\cdot)$ is convex. Hsiao [22] and Hsiao/de Mottoni [23] used a modified version for the case of nonconvex $p(\cdot)$. In order to cover the case $\underline{U}_L \in \mathcal{E}$ the inequality (5.5) is required to hold only for those $\underline{U}$ with u between u_L and u_R for which $s(\underline{U}_L, \underline{U})$ is defined. For the transonic flow problem we have:

LEMMA 5.2. *The admissibility condition*

$$u^- > u^+$$

is equivalent to the requirement that

$$(5.6)\qquad s(\underline{U}_L,\, \underline{U})^2 \geq s(\underline{U}_L,\, \underline{U}_R)^2$$

for all $\underline{U}$ on the Hugoniot curve, that lie between $\underline{U}_L$ and $\underline{U}_R$.

Proof. By (3.6) we have

$$-s\,(\underline{U}_L, \underline{U})^2 = \frac{p(u_L)-p(u)}{u_L-u}.$$

Then Lemma 5.1 states that $u^- > u^+$ is equivalent to

$$s(\underline{U}_L,\, \underline{U})^2 \geq s(\underline{U}_L,\, \underline{U}_R)^2$$

for all $\underline{U}$ on the Hugoniot curve between $\underline{U}_L$ and $\underline{U}_R$ (cp. Section 3.3).

5.3. Another type of admissibility condition can be obtained by generalizing (2.28) in the following manner. We will not require (2.27) to hold for smooth solutions, but instead we replace it by

$$(5.7)\qquad V(\underline{U})_t + G(\underline{U})_x \leq \mathrm{M}$$

for some constant M$\geq$ 0. The idea behind this approach is to choose V, G in a convenient way in order to give (5.2) at a shock. The constant M is then chosen very large so that (5.7) does not become a restriction for smooth parts of the flow. Note that (5.7) still implies an inequality of the type (2.28), i. e.

$$s[V] - [G] \leq 0 \tag{5.8}$$

at a shock.

A very simple example is given by choosing $V(\underline{U}) \equiv 0$ and $G(\underline{U}) = u$. Then (5.7) gives $u_x \leq$M and (5.8) gives $-[u_L - u_R] \leq 0$, i. e. $u_L \geq u_R$. Note that with $u = \phi_x$ the inequality is $\phi_{xx} \leq$ M.

We could also take $V(\underline{U}) = v$ and $G(\underline{U}) = u$. Then we have $v_t + u_x \leq$ M. or $\phi_{yy} + \phi_{xx} \leq$ M. This gives $s[v_L - v_R] - [u_L - u_R] \leq 0$. Using (4.5) we obtain

$$-s^2[u_L - u_R] - [u_L - u_R] = -(s^2+1)[u_L - u_R] \leq 0$$

or $u_L \geq u_R$.

5.4. Inequalities of this type, i. e. $\phi_{xx} \leq$M or $\Delta\phi \leq$M have been uses in variational methods for the calculation of transonic flows (see Bristeau et. al. [2], [3], [17], Feistauer/Nečas [14]). For the calculations one has to ensure that the constant M is chosen so large that it does not interfere with the smooth accelerations present in the transonic flow. Nečas [38] gave a whole class of such inequalities, called **h-entropies** that give the same admissibility condition. Suppose $h : \mathbf{R}_o^+ \to \mathbf{R}$, $\mathbf{R}_o^+ = \{\xi \in \mathbf{R} | \xi \geq 0\}$, is a C^1-function that satisfies $h(\xi) > 0$ for $\xi > 0$ and $h(\xi) + 2\xi h'(\xi) > 0$. Then the **$h$-entropy inequality** is

$$\text{div}\left(h(|\nabla\phi|^2)\nabla\phi\right) \leq \text{M} \tag{5.9}$$

in the sense of distributions (cp. Nečas [37], [38]).

Denoting by $\hat{n}$ a unit normal along a shock curve (5.9) implies

$$\left[h(|\underline{U}|^2)\underline{U}\cdot\hat{n}\right] \leq 0 \tag{5.10}$$

along this curve, i. e. $h(|\underline{U}_L|^2)\underline{U}_L \cdot \hat{n} \leq h(|\underline{U}_R|^2)\underline{U}_R \cdot \hat{n}$. This is the same as (5.8) since $\binom{s}{-1}$ is a normal vector to the shock curve. For (5.11) we may therefore write

$$s\left[h(|\underline{U}|^2)u\right] - \left[h(|\underline{U}|^2)v\right] \leq 0. \tag{5.11}$$

LEMMA 5.3. *The h-entropy inequality (5.9) giving*

$$\left[h(|\underline{U}|^2)\underline{U}\cdot\hat{n}\right] \leq 0 \tag{5.12}$$

where $h(\xi) > 0$ for $\xi > 0$ and $h(\xi) + 2\xi h'(\xi) > 0$, is an admissibility condition equivalent to

(a) $\Delta\phi \leq M$, *resp.* $[\underline{U}\cdot\hat{n}] \leq 0$

(b) $\phi_{xx} \leq M$, *resp.* $u^- > u^+$

when applied to the p-system.

Proof. Choosing $\hat{t} \perp \hat{n}$, both of unit length, we may set $a = \underline{\mathrm{U}} \cdot \hat{n}, \quad b = \underline{\mathrm{U}} \cdot \hat{t}$ and write $h\left(|\underline{\mathrm{U}}|^2\right) \underline{\mathrm{U}} \cdot \hat{n} = h(a^2 + b^2)a$. This function is monotone in a. To see this suppose for given $a, b \in \mathbf{R}$ that $h'(a^2 + b^2) \geq 0$. Then

$$\begin{aligned} \frac{\partial}{\partial a} h(a^2 + b^2)a &= h(a^2 + b^2) + 2h'(a^2 + b^2)a^2 \\ &\geq h(a^2 + b^2) \geq 0. \end{aligned}$$

Now suppose $h'(a^2 + b^2) \leq 0$. Then

$$\begin{aligned} \frac{\partial}{\partial a} h(a^2 + b^2)a &\geq h(a^2 + b^2) + 2h'(a^2 + b^2)(a^2 + b^2) \\ &> 0. \end{aligned}$$

It is strictly monotone for $a^2 + b^2 \neq 0$. This gives the equivalence of (5.12) and (a). We had seen above that (a) implies (b). Since the argument works backwards they are equivalent.

Note that (b) is not included in the class of h-entropy inequalities. Suppose $H(\xi) = \int_0^\xi h(s)\, ds$ is the primitive of $h(\cdot)$. Then the assumptions on h imply that $H(|\underline{\mathrm{U}}|^2)$ is a convex function of $\underline{\mathrm{U}}$. The h-entropy inequality is formally given by the Gâteaux derivative

$$-\langle H'(|\underline{\mathrm{U}}|^2), \underline{\varphi} \rangle = -\langle h(|\underline{\mathrm{U}}|^2)\underline{\mathrm{U}}, \varphi \rangle \leq \langle\, \mathrm{M}, \varphi \rangle$$

for all $\varphi \in C_0^\infty(\mathbf{R}_+^2)$, $\varphi \geq 0$, (cp. Nečas [38]). Choosing $H(u, v)$ a general convex function of (u, v) includes (b) by taking $H(u, v) = \frac{u^2}{2}$.

The other set of admissibility conditions given in Lemma 4.1 imply that u increases through a shock if one passes through the shock in the positive y-direction. This is accomplished by the inequality $u_y = \phi_{xy} \geq \mathrm{M}, \ \mathrm{M} \leq 0$.

6. The inverse p-system.

6.1. For the steady flow problem (1.1) the choice of a t-variable to give (4.1) was arbitrary. Also, we have seen that the x-variable in the transonic problem (1.1) is really the distinguished variable. Therefore, we will take a look at the **inverse p-system**

$$\begin{aligned} p(u)_t + \ v_x &= 0 \\ v_t - \ u_x &= 0 \end{aligned} \tag{6.1}$$

where $p : \mathbf{R} \to \mathbf{R}$ has the same properties as in Section 4.1. The characteristics are now the reciprocals $\lambda_1(\underline{\mathrm{U}}) = -\frac{1}{\sqrt{-p'(u)}}$, $\lambda_2(\underline{\mathrm{U}}) = -\frac{1}{\sqrt{-p'(u)}}$. The type of the system is the same as in Section 4.1. One should not be discouraged by the fact that the eigenvalues become infinite for $p'(u) = 0$. This means that the characteristics become parallel to the x-axis there. In the transonic flow problem one is interested in shocks where the shock speed $\frac{dx}{dy} = 0$, which is possible for mixed type shocks. In the framework of the inverse p-system this amounts to s being infinite. Note also that for mixed type problems involving smooth transition of the solution through the value u_{crit} such that $p'(u_{\text{crit}}) = 0$ the system (5.1) cannot be transformed to (4.1) by inverting

$$\begin{pmatrix} p'(u) & 0 \\ 0 & 1 \end{pmatrix}.$$

6.2. The inverse p-system as a mixed type system was studied by Mock [36]. He avoided the above difficulties by taking reciprocal eigenvalues and reciprocal shock speeds. For our purposes, the discussion of shocks, it will suffice to disregard the state where $p'(u) = 0$ and the infinite shock speeds. Then one may keep the usual notations. The Rankine-Hugoniot jump conditions are

$$\begin{aligned} s[p(u_L) - p(u_R)] &= [v_L - v_R] \\ s[v_L - v_R] &= -[u_L - u_R]. \end{aligned} \tag{6.2}$$

One obtains

$$-\frac{1}{s^2} = \frac{p(u_L) - p(u_R)}{u_L - u_R} = p'(\zeta), \tag{6.3}$$

for a ζ between u_L und u_R. Again this excludes shocks with both states in $\mathcal{E} \cup \mathcal{Z}$.

6.3. We now study the entropy inequality. It is

$$(U_i)_x + (F_i)_t \leq 0 \quad i = 1, 2$$

or

$$s[F_i] - [U_i] \leq 0 \quad i = 1, 2.$$

First we take U_1, F_1. Then we have using (5.2) and the mean value theorem

$$\begin{aligned} 0 \geq\ & s[F_1] - [U_1] \\ &= s[v_L p(u_L) - v_R p(u_R)] + v_R^2 - v_L^2 + P(u_L) - P(u_R) \\ &= \frac{p(u_L) + p(u_R)}{2}(u_R - u_L) + P(u_L) - P(u_R) \\ &= (u_R - u_L)^3 \frac{p''(\xi)}{12}, \end{aligned}$$

ξ between u_L and u_R. For $p'' < 0$ this implies

$$u_L < u_R$$

for shocks. Now u^-, u^+ have to be taken with respect to the t-axis. Therefore one obtains

$$\begin{aligned} u^+ < u^- \quad &\text{for} \quad s > 0 \\ u^- < u^+ \quad &\text{for} \quad s < 0 \end{aligned} \tag{6.4}$$

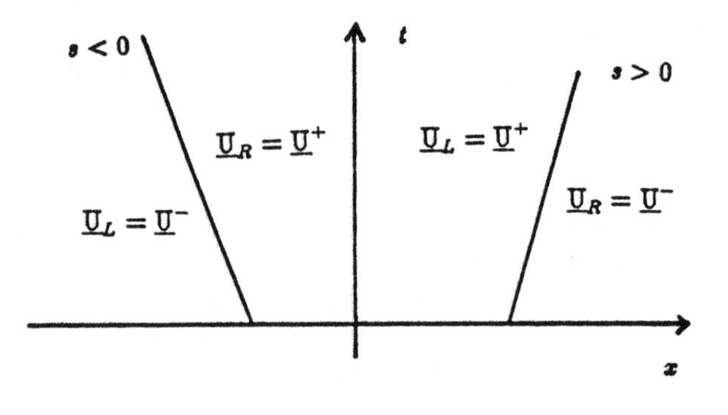

Figure 4

(see Figure 2). This is the same as obtained for the p-system (see Figure 4).

The other entropy inequality gives

$$
\begin{aligned}
0 &\geq s[F_2] - [U_2] \\
&= s\left[-u_L p(u_L) + P(u_L) + \frac{v_L^2}{2} + v_R p(u_R) - P(u_R) - \frac{v_R^2}{2}\right] \\
&\quad + u_L v_L - u_R v_R \\
&= s\left[(u_R - u_L)\frac{p(u_L) + p(u_R)}{2} - P(u_R) + P(u_L)\right] \\
&= (u_R - u_L)^3 \frac{p''(\xi)}{12},
\end{aligned}
$$

ξ between u_L and u_R. Since $p'' < 0$ this implies

$$
\begin{aligned}
u_R &> u_L \quad \text{for} \quad s > 0 \\
u_R &< u_L \quad \text{for} \quad s < 0
\end{aligned}
\tag{6.5}
$$

or

$$
u^- > u^+ \tag{6.6}
$$

since we take u^-, u^+ with respect to the x-axis of the original (x, y) coordinates, i.e. with respect to the current t-axis (see Figure 4).

As a conclusion we see that there is no need to use the inverse p-system, as was done by Mock [36], to study the transonic small disturbance problem.

REFERENCES

[1] R. Becker, *Stoßwelle und Detonation*, Z. f. Phys., 8 (1922), pp. 321–362.
[2] M. O. Bristeau, R. Glowinski, J. Periaux, P. Perrier and O. Pironneau, *On the Numerical Solution of Nonlinear Problems in Fluid Dynamics by Least Squares and Finite*

Element Methods I. Least Squares Formulations and Conjugate Gradient Solution of the Continuous Problem, Comput. Methods Appl. Mech. Engrg., 17/18 (1979), pp. 619–657.

[3] M. O. Bristeau, R. Glowinski, J. Periaux, P. Perrier, O. Pironneau and G. Poirier, *Application of Optimal Control and Finite Element Methods to the Calculation of Transonic Flows and Incompressible Flows*, in *Numerical Methods in Applied Fluid Dynamics*, B. Hunt (Ed.), Academic Press, New York (1980), 203–312.

[4] S.-N. Chow and J. K. Hale, *Methods of Bifurcation Theory*, in Grundlehren der mathematischen Wissenschaften **251**, Springer-Verlag, New York-Heidelberg-Berlin (1982).

[5] J. D. Cole and A. F. Messiter, *Expansion Procedures and Similarity Laws for Transonic Flow*, ZAMP, 8, 1–25 (1957).

[6] J. D. Cole and L. P. Cook, *Transsonic Aerodynamics*, North Holland Series in Applied Mathematics and Mechanics **30**, North-Holland, Amsterdam (1986).

[7] C. C. Conley and J. A. Smoller, *Viscosity Matrices for Two-dimensional Hyperbolic Systems*, Commun. Pure Appl. Math, 23, 867–884 (1970).

[8] C. C. Conley and J. A. Smoller, *Shock Waves as Limits of Progressive Wave Solutions of Higher Order Equations.*, Commun. Pure Appl. Math, 24, 459–472 (1971).

[9] C. C. Conley and J. A. Smoller, *Shock Waves as Limits of Progressive Wave Solutions of Higher Order Equations II*, Commun. Pure Appl. Math, 25, 133–146 (1972).

[10] R. Courant and D. Hilbert, *Methods of Mathematical Physics II*, Interscience Publishers, New York (1962).

[11] C. M. Dafermos, *The Entropy Rate Admissibility Criterion for Solutions of Hyperbolic Conservation Laws*, J. Diff. Eq., 14 (1973), pp. 202–212.

[12] C. M. Dafermos, *Hyperbolic Systems of Conservation Laws*, in *J.M. Ball (Ed.), Reidel-Publ.*, Doordrecht (1983), 25–70.

[13] R. J. DiPerna, *Compensated Compactness and General Systems of Conservation Laws.*, Trans. Am. Math. Soc., 292, 383–420 (1985).

[14] M. Feistauer and J. Nečas, *On the Solvability of Transonic Potential Flow Problems*, Z. Anal. Anw., 4, 305–329 (1985).

[15] I. M. Gel'fand, *Some Problems in the Theory of Quasilinear Equations.*, Usp. Math. Nauk., 14 (1959), pp. 87–158; Engl. transl. in *Amer. Math. Soc. Trans. Ser. 2*, 29, 295-381 (1963).

[16] D. Gilbarg, *The existence and limit behavior of the one-dimensional shock layer*, Amer. J. Math., 7, pp. 256–274 (1951).

[17] R. Glowinski and O. Pironneau, *On the Computation of Transonic Flows*, in *Functional Analysis and Numerical Analysis*, H. Fujita (Ed.), Jap. Soc. Prom. Sci., Tokyo-Kyoto (1978) 143–173.

[18] J. Guckenheimer and P. Holmes, *Nonlinear Oscillations, Dynamical Systems, and Bifurcations of Vector Field*, Springer-Verlag, New York (1983).

[19] H. Hattori, *The Riemann Problem for a van der Waals Fluid with Entropy Rate Admissibility Criterion – Isothermal Case*, Arch. Rat. Mech. Anal., 92, pp. 247–263 (1986).

[20] W. D. Hayes, *La seconde approximation pour les écoulements transsoniques non visqueux*, Journal de Mécanique, 5 (1966), pp. 163–206.

[21] H. Holden, *On the Riemann Problem for a Prototype of a Mixed Type Conservation Laws*, Comm. Pure Appl. Math., 25, pp. 229–264, (1987).

[22] L. Hsiao, *Admissible weak solution for nonlinear system of conservation laws in mixed type*, J. PDE (to appear).

[23] L. Hsiao and P. de Mottoni, *Existence and Uniqueness of Riemann Problem for Nonlinear System of Conservation Laws of Mixed Type*, Trans. AMS (to appear).

[24] R. D. James, *The Propagation of Phase Boundaries in Elastic Bars*, Arch. Rat. Mech. Anal., 73 (1980).

[25] B. L. Keyfitz, *Change of type in three-phase flow: a simple analogue*, J. Diff. Eq. (to appear).

[26] B. L. Keyfitz, *Admissibility conditions for shocks in conservation laws that change type*, in *Problems involving change of type*, K. Kirchgässner (Ed.), Lecture Notes in Mathematics, Springer-Verlag, to appear.

[27] S. N. Kružkov, *First Order Quasilinear Equations in Several Independent Variables*, Math. USSR Sbornik, 10 (1970), pp. 217–243.

[28] L. D. Landau and E. M. Lifschitz, *Lehrbuch der theoretischen Physik VI - Hydrodynamik*, Akademie Verlag, Berlin, 1981.

[29] P. D. Lax, *Hyperbolic Systems of Conservation Laws II*, Comm. Pure Appl. Math., 10, 537–566 (1957).

[30] P. D. LAX, *Shock Waves and Entropy*, in *Contr. to Nonlinear Analysis*, Zarantonello (Ed.), Academic Press, New York-London (1971) 603–634.

[31] P. D. LAX, *Hyperbolic Systems of Conservation Laws and the Mathematical Theory of Shock Waves*, in *Regional Conference Series in Applied Mathematics*, Soc. for Indust. and Appl. Math., Philadelphia (1972).

[32] L. LEIBOVICH, *Solutions of the Riemann Problem for Hyperbolic Systems of Quasilinear Equations without Convexity Conditions*, J. Math. Anal. Appl., 1974, pp. 81–90.

[33] T.-P. LIU, *The Riemann Problem for general* 2×2 *Conservation Laws*, Trans. Amer. Math. Soc., 199, 89–112 (1974).

[34] T.-P. LIU, *The Entropy Condition and the Admissibility of Shocks*, J. Math. Anal. Appl., 53, 78–88 (1976).

[35] M. S. MOCK, *Discrete Shocks and Genuine Nonlinearity*, Michigan Math. J., 25, pp. 131–146 (1978).

[36] M. S. MOCK, *Systems of Conservation Laws of Mixed Type*, J. Diff. Equat., 37 (1980), pp. 70–88.

[37] J. NEČAS, *Entropy Compactification of the Transonic Flow*, Manuscript, Charles University, Praha.

[38] J. NEČAS, *Compacite par Entropie et Ecoulements de Fluides*, Lecture Notes Université de Charles et E.N.S, Paris (1985).

[39] O. A. OLEĬNIK, *Uniqueness and Stability of the Generalized Solution of the Cauchy Problem for a Quasi-Linear Equation*, Am. Math. Soc. Transl. Ser. 2, 33, 285-290 (1963).

[40] K. OSWATITSCH, *Grundlagen der Gasdynamik*, Springer-Verlag, Wien-New York (1976).

[41] K. OSWATITSCH, *Spezialgebiete der Gasdynamik*, Springer-Verlag, Wien-New York, (1977).

[42] R. L. PEGO, *Phase Transitions in One-Dimensional Nonlinear Viscoelasticity: Admissibility and Stability*, Arch. Rat. Mech. Anal., 97 (1987), pp. 353–394.

[43] M. SHEARER, *The Riemann Problem for a Class of Conservation Laws of Mixed Type*, J. Diff. Eq., 46 (1982), pp. 426–443.

[44] M. SLEMROD, *An Admissibility Criterion for Fluids Exhibiting Phase Transitions*, in *Systems of Nonlinear Partial Differential Equations*, J. M. Ball (Ed.), Reidel-Publ., Doordrecht (1983) 423-432.

[45] M. SLEMROD, *Admissibility Criteria for Propagating Phase Boundaries in a van der Waals Fluid*, Arch. Rat. Mech. and Anal., 81 (1983), pp. 301–315.

[46] M. SLEMROD, *Interrelationships among Mechanics, Numerical Analysis, Compensated Compactness, and Oscillation Theory*, in *Oscillation Theory, Computation, and Methods of Compensated Compactness*, C. Dafermos, J. L. Ericksen, D. Kinderlehrer and M. Slemrod (Eds.), Springer-Verlag, New York-Berlin-Heidelberg (1986).

[47] M. SLEMROD, *Admissibility Criteria for Phase Boundaries*, in *Nonlinear Hyperbolic Problems*, C. Carasso, P.-A. Raviart and D. Serre (Eds.), Lecture Notes in Mathematics **1270**, Springer-Verlag, Heidelberg (1987).

[48] J. SMOLLER, *Shock Waves and Reaction-Diffusion Equations*, Grundlehren der mathematischen Wissenschaften **258**, Springer-Verlag, New York-Heidelberg-Berlin (1983).

[49] B. WENDROFF, *The Riemann Problem for Materials with Nonconvex Equations of State I: Isentropic Flow*, J. Math. Anal. Appl., 38 (1972), pp. 445–466.

[50] H. WEYL, *Shock Waves in Arbitrary Fluids*, Commun. Pure Appl. Math., **2** (1949), pp. 103–122.

[51] J. ZIEREP, *Theoretische Gasdynamik*, Braun, Karlsruhe (1976).